PROBABILITY

An Introduction

SAMUEL GOLDBERG

Emeritus Professor of Mathematics,
Oberlin College

Dover Publications, Inc.
New York

Published in Canada by General Publishing Company, Ltd., 30 Lesmill Road, Don Mills, Toronto, Ontario.
Published in the United Kingdom by Constable and Company, Ltd.

This Dover edition, first published in 1986, is an unabridged republication of the work first published in 1960 by Prentice-Hall, Inc., Englewood Cliffs, N.J.

Manufactured in the United States of America
Dover Publications, Inc., 31 East 2nd Street, Mineola, N.Y. 11501

Library of Congress Cataloging-In-Publication Data

Goldberg, Samuel, 1925–
  Probability : an introduction.

  Reprint. Originally published: Englewood Cliffs, N.J. : Prentice-Hall, 1960. (Prentice-Hall mathematics series)
  Includes index.
  1. Probabilities.   I. Title.   II. Series: Prentice-Hall mathematics series.
QA273.G62   1987   519.2   86-24067
ISBN 0-486-65252-1

PREFACE

THIS BOOK is intended for all who require a mathematically sound, but elementary introduction to the theory of probability.

Probability concepts are now of great importance in a wide variety of fields. The theory of probability, as the foundation upon which the methods of statistics are based, should command the attention of those who want to understand as well as apply statistical techniques. Probabilistic theories, making explicit reference to the nature and effects of chance phenomena, are the rule rather than the exception in the physical and biological sciences. Less well known is the fact that probability concepts are finding increased use in the social sciences and business: psychologists develop stochastic models for learning; economists use the techniques of game theory to discuss competition and markets; expected values, variances, and other matters related to random variables turn out to be important in the problem of finding combinations of securities that best meet the needs of the investor; business managers, because their decisions must be made in the face of uncertainty, invoke the theory of probability as an aid in planning inventory, establishing quality control, designing market surveys, etc. We need not go on—it is clear that probability concepts and methods are now widely used and will see even more extensive use in the future.

One noteworthy indication of the importance of our subject is the recent decision of the Commission on Mathematics of the College Entrance Examination Board to recommend that a course in probability and statistical inference be offered in the twelfth grade of the secondary school. Thus, secondary school teachers of mathematics, at some point in their college or in-service training, or in summer institutes (such as those sponsored by the National Science Foundation), should achieve some mastery of the elements of probability theory. Parts of this book were used in courses offered in NSF Institutes held at Oberlin College in 1958 and 1959, and the final manuscript has benefited from the helpful comments of the many teachers who studied preliminary versions.

Although there are a number of excellent textbooks on probability, they are all written for readers who have the mathematical sophistication that comes with a working knowledge of the differential and integral calculus. It seemed to me worthwhile to bring the theory of probability to the attention of those who do not have the calculus prerequisite. It was with this aim in mind that I limited myself to those topics that are accessible to readers with only a good background in high school algebra and a little ability in the reading and manipulation of mathematical symbols. The consequent limitation to finite sample spaces, although severe, facilitates a careful logical treatment of the essentials needed by all who use probability concepts. Furthermore, I have found that an understanding of the basic definitions, theorems, and methods in the finite case makes it much easier for students with the necessary preparation to master the corresponding ideas in the infinite case. I am therefore hopeful that this volume, although written as a basic textbook for courses in probability and statistics for students without calculus, will also prove useful in courses for those who have previous training in calculus.

One further possible use of this book is worthy of mention here. There are many college students who, for one reason or another, can take at most one year of mathematics. These students are often offered a smorgasbord survey course in which they sample one topic after another and learn very little about lots of things. Many teachers, however, prefer to offer a course centering on a few main topics, going into each systematically and deeply enough to give the student a reasonable depth of knowledge in the chosen subjects. Although many topics vie for inclusion in such a program, I believe a strong case can be made for a course that concentrates on sets and probability in the finite case at first, proceeds to an introduction to the calculus, and then applies this calculus to the elements of probability in the infinite case. (In my own course, I also include applications of the differential calculus to simple problems in economics.) Such a course, if properly executed, can give the student a keen sense of the nature and achievements of mathematical thinking, while laying a firm foundation for further study in economics, statistics, operations research, or allied fields. Such a program would therefore be especially valuable for social science and business students, assuming they can devote only a year to mathematics at the college level. I have used this volume in preliminary form in roughly the first third of such a year course at Oberlin College, with students who present less than three years of high school mathematics for entrance. Teachers who share my point of view may also find this book useful in their own introductory mathematics courses.

Since the theory of probability is best formulated using the language and notation of sets, we devote the first chapter to the elementary mathematics of sets. Proofs of laws in the algebra of sets are simplified by the use of so-called membership tables, a device analogous to truth tables in logic. Here we also introduce Cartesian product sets, which are needed at many points throughout the book.

Chapter 2 develops the basic calculus of probability for experiments with only a finite number of possible outcomes (finite sample spaces). A probability measure is first introduced over the events of a sample space and then conditional probability, independent events, and independent trials are carefully defined. Illustrative and problem material is here limited to the simplest experimental situations, and more sophisticated combinatorial techniques are first treated in Chapter 3. The usual order of topics has been reversed because beginning students seem always to have difficulty with the use of permutation and combination formulas, and this difficulty often impairs the learning of the basic probability ideas when both are presented simultaneously. We present the basic ideas first and then, in Chapter 3, offer a set of exercises in which the previously mastered probability theory is applied to a wide variety of situations requiring the use of sophisticated counting techniques. It has been our experience that this procedure makes it considerably easier for the student to learn this basic material.

Chapter 4 is an introduction to the analytic theory of probability in the finite case. Random variables are defined as functions on sample spaces, and probability distributions, means, standard deviations, joint probability functions, covariance, and correlation are discussed. Independence of random variables is defined and, with these ideas extended to the multivariate case, applications to random sampling theory can be included. The sampling distribution of the sample mean is discussed and formulas for its mean and variance are derived for both sampling with and without replacement.

The most important probability function defined on a finite sample space, the binomial distribution, forms the subject matter of the final chapter. The basic properties of a Bernoulli process and a binomially distributed random variable are derived, and the use of tables of cumulative binomial probabilities is discussed. Applications to the testing of statistical hypotheses (significance tests), as well as to a more complex problem of decision-making under uncertainty serve to illustrate how probability methods are applied in statistical investigations.

For some classes, teachers may find it necessary to offer supplementary lessons on the method of mathematical induction and the use of summation signs, as these topics arise in the text. I have also found that it is wise to constantly remind the beginning student of the substitution principle, for example, that from $\text{Var}(X) \geq 0$ for all X it follows that $\text{Var}(2X - 3Y) \geq 0$. Much of the difficulty beginners have with mathematics stems from a lack of understanding of this principle, and it is well worth emphasis.

In all other respects, I have made every effort to have this book self-contained, clear, and readable. Throughout, stress is laid on the explanation of fundamental concepts and patterns of mathematical reasoning, as well as on techniques of problem-solving. Problems at the end of each section are designed to supplement the many worked-out illustrative examples in the text and to enable the reader to check his understanding of new definitions, theorems, and methods. From time to time, problems are included to challenge the better student—the sample variance, maximum likelihood estimation, the hypergeometric distribution, regression functions, and OC-curves for sampling inspection are introduced in problems that are written so as to guide the student toward an understanding of these important topics. Answers (often complete solutions) to half of the 360 problems are collected in a 21-page section at the end of the book. To facilitate computations, tables of ordinary logarithms, logarithms of factorials, and cumulative binomial probabilities are included in the text. A list of books suitable for supplementary reading appears at the end of each chapter. I trust that these features will serve to make the hard job of learning a little less hard.

Comments from readers are always welcome.

SAMUEL GOLDBERG

Cambridge, Mass.

ACKNOWLEDGMENTS

I TAKE this opportunity to gratefully acknowledge my debt to Professor William Feller who, as my teacher, first showed me the beauty and importance of the mathematical theory of probability.

Part of the material on sets and probability in Chapters 1 and 2 was prepared in preliminary form and tested in the classroom under a grant by the Carnegie Corporation of New York to Oberlin College for experimentation in freshman mathematics.

Assistance of various kinds was rendered at Oberlin College during the summers of 1958 and 1959, when I offered a probability course in National Science Foundation Institutes for secondary school teachers of mathematics, by Bruce T. Marcus, David Webster, and especially by Edward T. Wong.

The manuscript was completely rewritten while I held a visiting appointment at the Harvard University Graduate School of Business Administration to teach at the Institute of Basic Mathematics for Application to Business. The Institute, which was sponsored by the Ford Foundation, arranged for the final typing of the manuscript. W. Allen Spivey read part of the manuscript and offered helpful comments. Howard Raiffa made numerous valuable suggestions and the book is much the better for his counsel. Robert Schlaifer kindly gave permission to use the material in the final section of Chapter 5. William A. Ericson read the manuscript and prepared the solutions to problems.

I am grateful to all these friends for their help and to each goes my sincere thanks.

CONTENTS

Chapter 1

SETS

---

1. Examples of sets; basic notation

The concept of a set, whose fundamental role in mathematics was first pointed out in the work of the mathematician Georg Cantor (1845–1918), has significantly affected the structure and language of modern mathematics. In particular, the mathematical theory of probability is now most effectively formulated by using the terminology and notation of sets. For this reason, we devote Chapter 1 to the elementary mathematics of sets. Additional topics in set theory are included throughout the text, as the need for this material becomes apparent.

The notion of a set is sufficiently deep in the foundation of mathematics to defy being defined (at the level of this book) in terms of still more basic concepts. Hence, we can only aim here, by taking advantage of the reader's knowledge of the English language and his experience with the real and conceptual world, to make clear the denotation of the word "set."

A set is merely an aggregate or collection of objects of any sort: people, numbers, books, outcomes of experiments, geometrical figures, etc. Thus, we can speak of the set of all integers, or the set of all oceans, or the set of all possible sums when two dice are rolled and the number of dots on the uppermost faces are added, or the set consisting of the cities of Cambridge and Oberlin and all their residents, or the set of all straight lines (in a given plane) which pass through a given point.

The collection of objects must be *well-defined*, by which we mean that, for any object whatsoever, the question "Does this object belong to the collection?" has an unequivocal "yes" or "no" answer. It is not necessary that we personally have the knowledge required to decide which answer is correct. We must know only that, of the answers "yes" and "no," exactly one is correct.

Let us also agree that no object in a set is counted twice; i.e., the objects are *distinct*. It follows that, when listing the objects in a set, we do not repeat an object after it is once recorded. For example, according to this convention, the set of letters in the word "banana" is a set containing not six letters, but rather the three distinct letters *b*, *a*, and *n*.

The following definition summarizes our discussion to this point and introduces some additional terminology and notation.

Definition 1.1. A *set* is a well-defined collection of distinct objects. The individual objects that collectively make up a given set are called its *elements*, and each element *belongs to* or is a *member of* or is *contained in* the set. If *a* is an object and *A* a set, then we write $a \in A$ as an abbreviation for "*a* is an element of *A*" and $a \notin A$ for "*a* is not an element of *A*." If a set has a finite number of elements, then it is called a *finite* set; otherwise it is called an *infinite* set.

We are relying on the reader's knowledge of the positive integers $1, 2, 3, \cdots$, the so-called counting or natural numbers. This is an infinite set of numbers. To say that a set is finite means that one can enumerate the elements of the set in some order, then count these elements one by one until a *last element is reached*. Let us note that it is possible for a set, like the set of grains of sand on the Coney Island beach, to have a fantastically large number of elements and nevertheless be a finite set.

A set is ordinarily specified either by (i) listing all its elements and enclosing them in braces (the so-called *roster method* of defining the set), or by (ii) enclosing in braces a *defining property* and agreeing that those objects that have the property, and only those objects, are members of the set. We discuss these important ideas further and introduce additional notation in the following examples.

Example 1.1. The set whose elements are the integers 0, 5, and 12 is a finite set with three elements. If we denote this set by *A*, then it is conveniently written using the roster method: $A = \{0, 5, 12\}$. The statements "$5 \in A$" and "$6 \notin A$" are both true.

Example 1.2. If we write $V = \{a, e, i, o, u\}$, then we have defined the set V of vowels in the English alphabet by listing its five elements. To specify V by a defining property we write

$$V = \{x \mid x \text{ is a vowel in the English alphabet}\},$$

which is read "V is the set of those elements x such that x is a vowel in the English alphabet." Braces are always used when specifying a set; the vertical bar $\mid$ is read "such that" or "for which." The symbol x is of course merely a place-holder; any other symbol will do just as well. For example, we can also write

$$V = \{* \mid * \text{ is a vowel in the English alphabet}\}.$$

A slight modification of this notation is often used. Let us first introduce the set A to stand for the set of all letters of the English alphabet. Then we write

$$V = \{* \in A \mid * \text{ is a vowel}\},$$

which is read "V is the set of those elements $*$ of A such that $*$ is a vowel."

Example 1.3. The set $B = \{-2, 2\}$ is the same set as $\{x \in R \mid x^2 = 4\}$, where R is the set of all real numbers. The set $\{x \in R \mid x^2 = -1\}$ has no elements, since the square of any real number is nonnegative. But if C is the set of all complex numbers, then $\{x \in C \mid x^2 = -1\}$ contains the elements $i = \sqrt{-1}$ and $-i$.

Example 1.4. A prime number is a positive integer greater than 1 but divisible only by 1 and itself. A proof of the fact that the set $\{p \mid p \text{ is a prime number}\}$ is an infinite set was given by Euclid (?330–275 B.C.) in the ninth book of his *Elements*. Strictly speaking, the roster method is unavailable for infinite sets, since it is not possible to list all the members and have explicitly before one a totality of elements making up an infinite set. The notation

$$\{2, 3, 5, 7, 11, 13, 17, 19, \cdots\},$$

in which some of the elements of the set are listed followed by three dots which take the place of *et cetera* and stand for obviously understood omissions of one or more elements, is an often used but logi-

cally unsatisfactory way out of this difficulty. (See Problem 1.3.)
To specify an infinite set correctly, one must (as we did when we
introduced the set of prime numbers) cite a defining property of the
set.

Example 1.5. If a rectangular coordinate system (with x-axis and
y-axis) is introduced in a plane, then each point of the plane has an
x-coordinate and a y-coordinate, and can be represented, as in Figure
1(a), by an ordered pair of real numbers. In analytic geometry, one

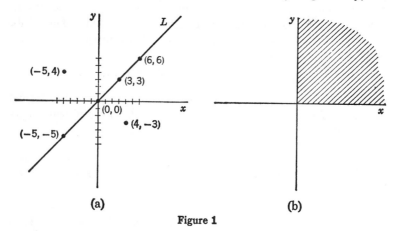

(a)   (b)

Figure 1

is interested in sets of points whose coordinates meet certain requirements. For example, the set $\{(x, y) \mid y = x\}$ is the set of all
points (in a plane) with equal x- and y-coordinates. This infinite set
of points makes up the straight line L, a portion of which is sketched
in Figure 1(a), passing through the origin O and bisecting the first
and third quadrants. We say that the line L is the *graph* of the set
$\{(x, y) \mid y = x\}$. Similarly, the entire x-axis is the graph of the set
$\{(x, y) \mid y = 0\}$, and the *positive* x-axis is the graph of the set
$\{(x, y) \mid x > 0 \text{ and } y = 0\}$. The set $\{(x, y) \mid x > 0 \text{ and } y > 0\}$ is the
set of points whose x- and y-coordinates are both positive. Thus, the
graph of this set is the entire first quadrant (axes excluded), as indicated in Figure 1(b).

We see that a relation (in the form of equalities or inequalities
between x and y) can be considered a *set-selector*, and the graph pic-
tures the set of those points (from among all in the plane) selected by
the requirement that their coordinates satisfy the given relation.

Although it may seem strange at first, it turns out to be convenient to talk about sets that have no members.

Definition 1.2. A set with no members is called an *empty* or *null* set.

The set $\{x \in R \mid x^2 = -1\}$ in Example 1.3 is an empty set. Another example is obtained by considering the set of all paths by which the line drawing of a house in Figure 2 can be traced without lifting one's pencil or retracing any line segment. Whether this set is empty or not is of some interest, since to assert that it is empty is to say that the figure cannot be traced under the prescribed conditions. (Let the reader convince himself that this set is indeed empty.) As our work develops, we shall see many other less frivolous reasons for introducing the notion of an empty set.

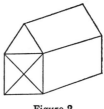

Figure 2

We conclude this ground-breaking section with one more definition.

Definition 1.3. Two sets A and B are said to be *equal* and we write $A = B$ if and only if they have exactly the same elements. If one of the sets has an element not in the other, they are unequal and we write $A \neq B$.

Thus $A = B$ means that every element of A is also an element of B and every element of B is an element of A. Equal sets are identical sets, and this identity is symbolized by the equality sign.

This definition has some interesting consequences. First, it is clear that the order in which we list the elements of a set is immaterial. For example, the set $\{a, b, c\}$ is equal to the set $\{c, a, b\}$, since they do indeed have exactly the same three elements.

Also, when sets are specified by defining properties, they can be equal even though the defining properties themselves are outwardly different. Thus, the set of all even prime numbers and the set of real numbers x such that $x + 3 = 5$ have different defining properties, yet they are equal sets, for each contains the number 2 as its only element.

Up to now, we have been careful to speak of a set having no members as *an* empty set. But it is clear from Definition 1.3 that any two empty sets are equal. For to be unequal it is necessary for one of the sets to contain an element not in the other, and this is impossible since neither set contains any elements. Therefore we are justified in referring to *the* empty set or *the* null set.* We denote the null set by the special symbol ∅.

PROBLEMS

1.1. We list eight sets. For each set, state whether it is finite or infinite. If finite, count the number of elements in the set. Where feasible, write the set using the roster method.

(a) The set of footnotes in Section 1.
(b) The set of letters in the word "probability."
(c) The set of odd positive integers.
(d) The set of prime numbers less than one million.
(e) The set of paths by which the following figure can be traced without lifting one's pencil or retracing any line segment:

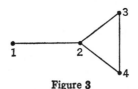

Figure 3

(f) The set of those points (in a given plane) that are exactly five units from the origin O.
(g) The set of real numbers satisfying the equation $x^2 - 3x + 2 = 0$.
(h) The set of possible outcomes of the experiment in which one card is selected from a standard deck of 52 cards.

1.2. The following paragraph was written by a student impressed with the technical vocabulary of set theory. Rewrite in more usual English prose.

Let C be the set of Mr. and Mrs. Smith's children. C was equal to ∅ until March 1, 1958. C contained exactly one element from that date until March 15, 1959 when it increased its membership by two!

---

* The following true story concerns the attempt of a well-known professor of mathematics to teach his five-year-old son the subtle distinction between "a" (or "an") and "the." One day the son answered the telephone, listened a moment and then said, "I'm sorry, but you have the wrong number." (Isn't this what most of us say when someone dials incorrectly?) The father, having overheard, immediately called the boy to him and gently instructed, "What you said would be correct if there were exactly one wrong number. But since there are many possible wrong numbers, it would be more accurate to say, 'I'm sorry, but you have *a* wrong number.' "

1.3. To illustrate the inadequacy of displaying a few elements of a set and indicating the other elements by three dots, consider the set A of all numbers of the form

$$n^2 + (n-1)(n-2)(n-3),$$

where n is any positive integer. Show that the first three elements (i.e., those obtained when $n = 1, 2, 3$) are 1, 4, and 9, so that one is tempted to write $A = \{1, 4, 9, \cdots\}$. If A is written this way on an I. Q. test, we do not hesitate to write the next element as 16. But show that the next element (obtained when $n = 4$) is actually 22 and not 16! Indeed, it is possible to write a defining property for a set so that its fourth element (in order of magnitude) is any number, say 94, although its first three elements are 1, 4, 9. Formulate such a defining property.

1.4. Let $A = \{0, 1, 2, 3, 4\}$. List the elements, if any, of each of the following sets:

(a) $\{x \in A \mid 2x - 4 = 0\}$ (b) $\{x \in A \mid x^2 - 4 = 0\}$
(c) $\{x \in A \mid x^3 - 4x^2 + 3x = 0\}$ (d) $\{x \in A \mid x^2 = 0\}$
(e) $\{x \in A \mid x + 1 > 0\}$ (f) $\{x \in A \mid 2x + 1 \leq 0\}$
(g) $\{x \in A \mid x^2 - 5x + 4 \geq 0\}$ (h) $\{x \in A \mid x^2 - x < 0\}$

1.5. Let x and y be the coordinates of a point in the plane. Identify the following sets and give a geometric interpretation of your results:

(a) $\{(x, y) \mid x + y = 5 \text{ and } 3x - y = 3\}$
(b) $\{(x, y) \mid x + y = 5 \text{ and } 2x + 2y = 3\}$
(c) $\{(x, y) \mid x + y = 5 \text{ and } 2x + 2y = 10\}$

1.6. Show that set equality has the following properties:

(i) Set equality is a *reflexive relation;* i.e., $A = A$ for any set A.
(ii) Set equality is a *symmetric relation;* i.e., for any sets A and B, if $A = B$, then $B = A$.
(iii) Set equality is a *transitive relation;* i.e., for any sets A, B, and C, if $A = B$ and $B = C$, then $A = C$.

(*Note:* A relation that is reflexive, symmetric, and transitive is called an *equivalence relation.*)

1.7. Determine whether $A = B$ or $A \neq B$.

(a) $A = \{2, 4, 6\}$, $B = \{4, 6, 2\}$.
(b) $A = \{1, 2, 3\}$, $B = \{\text{Mars, Venus, Jupiter}\}$.
(c) $A = \{* \mid * \text{ is a plane equilateral triangle}\}$, $B = \{* \mid * \text{ is a plane equiangular triangle}\}$.
(d) $A = \{x \mid x^2 - 2x + 1 = 0\}$, $B = \{x \mid x - 1 = 0\}$.
(e) $A = \{x \mid 2x^2 - 5x + 2 = 0\}$, $B = \{x \mid 2x^3 - 5x^2 + 2x = 0\}$.

1.8. Which of the following are true? Explain.

(a) $2 = \{2\}$, (b) $2 \in \{2\}$, (c) $0 = \emptyset$, (d) $0 \in \emptyset$.

2. Subsets

Each element of the set of vowels in the alphabet is, of course, an element of the set of all letters. Similarly, each number in $\{2, 4, 6\}$ is an element of the set of all even integers, and each real number in $\{x \mid x > 3\}$ is also in $\{x \mid x > 0\}$. In this section, we discuss the simple but important relation between sets illustrated by these examples.

Definition 2.1. A set A is a *subset* of set B, denoted by $A \subseteq B$, if each element of A is also an element of B. We agree to call the null set $\emptyset$ a subset of every set.

For example, we write $\{1, 3\} \subseteq \{1, 2, 3\}$, since each of the two elements in $\{1, 3\}$ belongs to $\{1, 2, 3\}$. Also, $\{1, 3\} \subseteq \{x \mid x \geq 1\}$ and $\{1, 3\} \subseteq \{1, 3\}$. The definition of subset implies that a set is a subset of itself; i.e., $A \subseteq A$ is always true. We can express this fact using the language introduced in Problem 1.6 by saying that set inclusion (i.e., one set being a subset of another set) is a *reflexive* relation. It is also *transitive*, for if $A \subseteq B$ and $B \subseteq C$, it follows that $A \subseteq C$. But set inclusion is not *symmetric*. As a counterexample, let $A = \{a\}$ and $B = \{a, b\}$. Then $A \subseteq B$ is true, but $B \subseteq A$ is false.

It is noteworthy that the definition of set equality in the preceding

TABLE 1

Set A	$n(A)$	Subsets of A	Number of Subsets of A
$\emptyset$	0	$\emptyset$	$1 \, (= 2^0)$
$\{a\}$	1	$\emptyset, \{a\}$	$2 \, (= 2^1)$
$\{a, b\}$	2	$\emptyset, \{a\}, \{b\}, \{a, b\}$	$4 \, (= 2^2)$
$\{a, b, c\}$	3	$\emptyset, \{a\}, \{b\}, \{c\}, \{a, b\}, \{a, c\}, \{b, c\}, \{a, b, c\}$	$8 \, (= 2^3)$
$\{a, b, c, d\}$	4	$\emptyset, \{a\}, \{b\}, \{c\}, \{d\}, \{a, b\}, \{a, c\},$ $\{a, d\}, \{b, c\}, \{b, d\}, \{c, d\}, \{a, b, c\},$ $\{a, b, d\}, \{a, c, d\}, \{b, c, d\}, \{a, b, c, d\}$	$16 \, (= 2^4)$

section was formulated in terms of the subset relation. In fact, it is merely a restatement of Definition 1.3 to say that $A = B$ if and only if $A \subseteq B$ and $B \subseteq A$.

Table 1 illustrates the notion of subset, and also directs our attention to a formula relating the number of subsets of a set to the number of elements in the set. We denote the number of elements in A by $n(A)$.

From the numbers in the last column of this table, we are led to conjecture that if n is any nonnegative integer, then a set with n elements has 2^n subsets. Before proving this result is true, we need to enunciate a principle that is at the heart of most counting procedures, and that is used time and again in computing probabilities.

Fundamental Principle of Counting:

(a) If one task can be completed in N_1 different ways and, following this, another task can be completed in N_2 different ways, then both tasks can be completed in the given order in N_1N_2 different ways.

(b) More generally, suppose a certain job can be done by completing, in some specified order, n smaller units (which we shall call tasks), where n is any positive integer. The first task can be completed in N_1 different ways. Having finished the first task, the second can be completed in N_2 different ways. Having finished the first two tasks, the third can be completed in N_3 different ways. And so on until, having finished all but the last task, this nth task can be completed in N_n different ways. Then the entire job can be done in $N_1N_2N_3 \cdots N_n$ different ways, it being understood that two ways of doing the job are considered different if and only if there is at least one task that is completed differently in the two ways.

The *tree-diagram* in Figure 4 illustrates (a) for the special case $N_1 = 3$ and $N_2 = 2$. Starting from some point, we draw $N_1 = 3$ lines. From each of these lines, we draw $N_2 = 2$ lines. The total number of

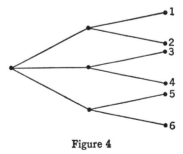

Figure 4

ways of completing task 1 and then task 2 is the same as the total number of branches in the tree.

When there are only two tasks, as in (a), the fundamental principle follows immediately from the definition of multiplication. For each of the N_2 ways of doing task 2, we have N_1 ways of first doing task 1. Hence both tasks can be done in a number of ways equal to $N_1 + N_1 + \cdots + N_1$, where there are N_2 summands. But this number is precisely the product $N_1 N_2$.

The general principle in (b) can be proved by mathematical induction. We leave this for Problem 2.10 and proceed to illustrate how one uses the fundamental principle of counting.

Example 2.1. We roll a green die and then a red die. How many ways can these dice come up? Our job can be thought of as recording the results of the two rolls. This can be done by first recording the number on the green die (task 1), and then recording the number on the red die (task 2). Task 1 can be done in six ways, and then task 2 can also be done in six ways. Hence, there are $6 \cdot 6 = 36$ possible ways that the two dice can come up.

Example 2.2. How many distinct three-letter "words" can be made, using the letters chosen from among those of "number," but with no letter used more than once in a "word"? Our job is to construct a three-letter "word" under the prescribed conditions. This job can be done by selecting the first letter (task 1), then the second letter (task 2), and finally the third letter (task 3). Task 1 can be done in any of six ways, since there are six letters available in "number." Having chosen one letter, there are only five remaining letters, and hence only five ways of completing task 2. Similarly, there are four ways of completing task 3 after the first two letters are chosen. Hence there are altogether $6 \cdot 5 \cdot 4 = 120$ different three-letter "words" that can be formed.

Example 2.3. How many different four-of-a-kind poker hands are there? Our job is to select a hand (subset) of five cards from the ordinary deck (set) of 52 cards in such a way that the hand contains four cards with the same face-value. This job can be done by completing the following tasks in the stated order: (i) Choose one face-value from among the 13 possible face-values; (ii) Select four cards from among those with the face-value chosen in (i), paying no regard to their order; (iii) Choose one card from among the remaining 48 cards. Each time we complete the job this way, we obtain exactly one four-of-a-kind poker hand. Moreover, different ways of completing the job result in different four-of-a-kind poker hands. (It was to make this last assertion true that we described task (ii) as choosing a *set* of four cards, and were not concerned with the *order* in which the cards were selected.) Hence, there are as many different four-of-a-kind poker hands as there are different ways of completing the job. Now, task (i) can be done in 13 ways, task (ii) can be done in only one way (since there is only one set of four cards that can be formed from four given cards), and task (iii) can be done in 48 ways. Hence, there are $13 \cdot 1 \cdot 48 = 624$ different four-of-a-kind poker hands.

We shall return to the fundamental principle of counting in Chapter 3, since it is the basic result from which the formulas of combinatorial analysis are derived. Our main interest here is to use the principle to establish the following theorem.

Theorem 2.1. Let n be any nonnegative integer. If A is a set with n elements, then there are 2^n different subsets of A.

Proof. If $n = 0$, then $A = \emptyset$ and the only subset of $\emptyset$ is $\emptyset$ itself. Since $2^0 = 1$, the theorem is true in this special case. If $n \geq 1$, then let the n elements of A be enumerated in some order. The job of constructing a subset of A can be viewed as made up of the following n tasks. As task 1 we decide whether the first element of A should or should not be an element of the subset. If we decide it should, then let us write down an ϵ; if we decide it should not, then we write ϵ. Then, as task 2, we write ϵ or ϵ, depending upon whether we decide that the second element of A should or should not belong to the subset. Now we move to the third element of A and complete the third task in a similar manner. Since A has n elements, we complete n tasks, and thus obtain a sequence of n decisions, each symbolized by ϵ or ϵ. For example, if $A = \{a, b, c, d\}$, then the sequence $\epsilon\epsilon\epsilon\epsilon$ determines the subset $\{a, b, d\}$, the sequence $\epsilon\epsilon\epsilon\epsilon$ determines the subset A itself, the sequence $\epsilon\epsilon\epsilon\epsilon$ determines the empty subset $\emptyset$. In general, there are as many subsets of A as there are different ways of making the n decisions. Since each decision can be made in two ways (ϵ or ϵ), we conclude by the fundamental principle of counting that there are $2 \cdot 2 \cdot 2 \cdots 2 = 2^n$ ways of making all n decisions, and hence 2^n subsets of A.

Forming subsets of a given set is a method that generates a large number of new sets. In fact, a set with 20 elements has 2^{20} or more than a million different subsets. Ordinarily, however, as the following examples point out, one is interested in studying a small number of subsets from among the many available ones.

Example 2.4. Let a green and a red die be rolled, and let S denote the set of possible outcomes. S has 36 elements which we can enumerate as follows, using the abbreviation (x, y) to stand for "green die showed the number x and red die showed the number y":

$$\begin{array}{cccccc}
(1,1) & (1,2) & (1,3) & (1,4) & (1,5) & (1,6) \\
(2,1) & (2,2) & (2,3) & (2,4) & (2,5) & (2,6) \\
\cdot & \cdot & \cdot & \cdot & \cdot & \cdot \\
\cdot & \cdot & \cdot & \cdot & \cdot & \cdot \\
\cdot & \cdot & \cdot & \cdot & \cdot & \cdot \\
(6,1) & (6,2) & (6,3) & (6,4) & (6,5) & (6,6)
\end{array}$$

By Theorem 2.1, there are 2^{36} subsets of S. But relatively few of these subsets have any special interest, even to players of "craps." Some of these are:

(i) S_1 = the subset made up of those outcomes for which the sum of the numbers on the two dice is 7; i.e.,

$$S_1 = \{(1,6), (2,5), (3,4), (4,3), (5,2), (6,1)\}.$$

(ii) S_2 = the subset containing the outcomes for which the sum of the numbers on the two dice is 11; i.e.,

$$S_2 = \{(5,6), (6,5)\}.$$

(iii) S_3 = the subset containing the outcomes for which the sum of the numbers on the two dice is either 7 or 11; i.e.,

$$S_3 = \{(1,6), (2,5), (3,4), (4,3), (5,2), (6,1), (5,6), (6,5)\}.$$

Whenever any experiment is performed (as in this example), we can think of the set of all possible outcomes of the experiment. We shall see that such sets and their subsets are of great importance in the mathematical theory of probability.

Example 2.5. The annual directory of college X lists the name, hometown, college residence, and telephone number of each of the college's 2000 students. Let A be the set of these 2000 entries, each entry containing the four pieces of information described above. The total number of subsets of A is astronomical, being 2^{2000}. But the housemother in a certain dormitory is mainly concerned with the subset of those entries containing the names of women who are residents of her dormitory, the mathematics department must estimate ahead of time the approximate number in the subset of entries naming students who will elect mathematics courses, a student may be especially interested in the subset of entries naming all freshmen who come from his hometown, etc. If the information for each student is entered by punching holes on certain specially-designed cards, then the cards corresponding to these subsets of A, as well as many others, can be sorted out of the whole set of cards by a machine. In fact, such sorting machines are designed for the purpose of speedily selecting certain subsets from a given set.

We conclude with an example designed to test the reader's grasp of the difference between the notions of set membership (symbolized by ϵ) and set inclusion (symbolized by $\subseteq$).

Example 2.6. Consider the set M of majorities in a committee of four individuals, each having one vote. Let us label the individuals a, b, c, d, and note that a majority is itself a *set* of three or more of these committeemen. Thus, the set M has sets as elements, and we write M using braces within braces:

$$M = \{\{a, b, c\}, \{a, b, d\}, \{a, c, d\}, \{b, c, d\}, \{a, b, c, d\}\}.$$

Thus $\{a, b, c\}$ is an element of M, but although a set, it is *not* a subset of M. (Why?) But $\{\{a, b, c\}\}$ *is* a subset of M since its only element, $\{a, b, c\}$, is indeed also an element of M.

PROBLEMS

2.1. Let S be the set of 36 outcomes of the experiment in which a green and a red die are rolled. (See Example 2.4.) We define certain subsets of S by listing their elements. State a defining property for each subset.

(a) $\{(1, 1), (2, 2), (3, 3), (4, 4), (5, 5), (6, 6)\}$
(b) $\{(1, 1), (1, 2), (1, 3), (1, 4), (1, 5), (1, 6)\}$
(c) $\{(1, 3), (2, 2), (3, 1)\}$
(d) $\{(1, 3), (1, 4), (1, 5), (1, 6), (2, 4), (2, 5), (2, 6), (3, 5), (3, 6), (4, 6)\}$

2.2. Let $A = \{1, 2, 3\}$. Identify the sets B such that $\{1\} \subseteq B, B \subseteq A$, and $B \neq A$.

2.3. You are told that there is only one set A such that $A \subseteq B$. Identify the set B.

2.4. Let A be any set. The set $\{X \mid X \subseteq A\}$ of all subsets of A is called the *power set* of A and is denoted by 2^A. (If A is a set with n elements, then Theorem 2.1 says that the power set 2^A has 2^n elements. This fact accounts for the name "power set" and the symbol 2^A used to denote this set.) Explain the following true statements, assuming $A = \{x, y, z\}$:

(a) $\emptyset \notin A$, but $\emptyset \in 2^A$.
(b) $x \in A$, but $x \notin 2^A$.
(c) $\{x, y\} \notin A$, but $\{x, y\} \in 2^A$.
(d) A is an element, but is not a subset of 2^A.
(e) $\{A\}$ is not an element, but is a subset of 2^A.

2.5. Which of the following are correct and why?

(a) $\{1\} \in \{\{1\}\}$ (b) $\{1\} \subseteq \{\{1\}\}$
(c) $\{1\} \in \{1, \{1\}\}$ (d) $\{1\} \subseteq \{1, \{1\}\}$

2.6. Give an example of two sets A and B such that both $A \in B$ and $A \subseteq B$ are true.

2.7. The graph of the set of points $C = \{(x, y) \mid x^2 + y^2 = 4\}$, where x and y are real numbers, is the circumference of the circle with center at $(0, 0)$ and radius 2 units. Determine the graphs of the following subsets of C:

(a) $\{(x, y) \in C \mid x = 0\}$ (b) $\{(x, y) \in C \mid x = 2\}$
(c) $\{(x, y) \in C \mid x = 3\}$ (d) $\{(x, y) \in C \mid x > 0\}$
(e) $\{(x, y) \in C \mid y \geq 0\}$ (f) $\{(x, y) \in C \mid y = \sqrt{4 - x^2}\}$

2.8. (a) If $z \in A$ and $A \subseteq B$, is it necessarily the case that $z \in B$?
(b) If $z \in A$ and $A \in B$, is it necessarily the case that $z \in B$?

2.9. Draw a tree diagram to illustrate the fundamental principle of counting, assuming $n = 3$ and $N_1 = 4$, $N_2 = 3$, $N_3 = 2$.

2.10. Assume the truth of the fundamental principle of counting for $n = 2$, i.e., for a job made up of only two tasks. Prove the principle for any positive integer n by mathematical induction.

In each of the following problems, state explicitly how the fundamental principle of counting is used in obtaining your answer. Draw a tree diagram where feasible.

2.11. A man has five coins in his pocket. He agrees to give one coin to his son and one to his daughter. In how many ways can this be done?

2.12. In how many different orders can one call out the numbers 1, 2, 3, 4, 5?

2.13. In dialing a telephone number, one has to select seven slots, the first two for the letters of the exchange, and then five digits to identify the telephone in that exchange. The telephone dial contains ten slots, one for each of the digits 0, 1, 2, $\cdots$, 9, but letters appear in only eight of these slots. If the first number cannot be a zero, how many different telephone numbers, distinguishable as dialed, are possible?

2.14. How many three-digit even integers can be formed from the digits 1, 5, 6, and 8, with no digit repeated?

2.15. How many different ways are there of selecting two letters from the set $\{a, b, c\}$? *Let the reader realize that the question as stated is vague and needs to be made precise before it can be answered.* We must know how the letters are selected, and we must decide when results of the selection process will be considered different. We list four possibilities. Answer the question in each case.

  (a) The first letter is chosen and the second is selected from the remaining two letters; i.e., repetitions are not allowed. We count two ways of making the selections different if they result in different *ordered* pairs of letters; i.e., we record not only which two letters were selected, but also the order in which they were selected.

  (b) The first letter is chosen and the second is selected from the entire set of three letters; i.e., repetitions are allowed. We count *ordered* pairs of letters as in (a).

  (c) Repetitions are not allowed, as in (a), and we count two ways of making the selections different only if they result in different *sets* of two letters; i.e., we disregard the order in which the letters were selected.

  (d) Repetitions are allowed, as in (b), and we disregard order as in (c).

2.16. How many different ways are there of selecting two cards from a standard deck of 52 cards? Consider various interpretations of this question, as in the preceding problem.

2.17. How many ways can three coins fall? four coins? n coins, where n is any positive integer?

2.18. Two cards are drawn one after the other from a standard deck of 52 cards. In how many ways can one draw

  (a) first a spade and then a heart?
  (b) first a spade and then a heart or a diamond?
  (c) first a spade and then another spade?

2.19. Repeat the preceding problem, assuming the first card is put back in the deck before the second is drawn.

2.20. Let $A = \{1, 2, 3, \cdots, 365\}$. (a) Two numbers are selected in order, each from the full set A. The result is an *ordered 2-tuple*, or ordered pair of numbers. How many are there? (b) Three numbers are selected in order, each from the full set A. The result is an *ordered 3-tuple*, or ordered triple of numbers. How many are there? (c) r numbers are selected in order, each from the full set A, where r is some positive integer. The result is an *ordered r-tuple*. How many are there? (*Note:* A general definition of an ordered r-tuple is given in Section 5.)

2.21. (a) How many ways are there of placing three distinguishable balls into two numbered cells? Into three numbered cells? Into n numbered cells?

   (b) How many ways are there of placing r distinguishable balls into two numbered cells? Into three numbered cells? Into n numbered cells?

3. Operations on sets

In any particular discussion of sets, it is necessary to define some fixed set of elements (called the *universal set*) to which we limit the discussion. This point has been eloquently made by Langer:

> In ordinary conversation, we assume the limitations of such a universe, as when we say: "Everybody knows that another war is coming," and assume that "everybody" will be properly understood to refer only to adults of normal intelligence and European culture, not to babies in their cribs, or the inhabitants of remote wildernesses. For conversational purposes, the tacit understanding will do; but if the statement is to be challenged, i.e., if someone volunteers to produce a person to whom it is not true, then it becomes important to know just what the limits of its applicability really are. Arguments of this sort have their own technique, by which the opposition marshals contradictory cases—in this example, persons who have no such knowledge—and the asseverator rules them out as "not meant" by his statement. The universe of ordinary discourse is vague enough so that this process can go on as long as the bellicosity of the two adversaries lasts. Logicians and scientists, however, take no pleasure in casuistry. Their universe of discourse must be definite enough to allow no dispute whatever about what does or does not belong to it.*

The fixed universal set we shall denote by $\mathcal{U}$. *Once having decided on the universal set $\mathcal{U}$ for a particular discussion, all other sets in that same discussion must be subsets of $\mathcal{U}$.* But different universal sets can be used for different discussions. $\mathcal{U}$ may be a set of people in one problem, a different set of people in another, a set of numbers in yet another, etc.

---

* Susanne K. Langer, *An Introduction to Symbolic Logic*, 2nd edition, Dover Publications, Inc., 1953, p. 68.

We now define the three basic operations on sets.

Definition 3.1. Let A and B be any subsets of a universal set $\mathfrak{U}$. Then

I. The *complement* of A (with respect to $\mathfrak{U}$) is the set of elements of $\mathfrak{U}$ that do *not* belong to A. The complement of A is denoted by A', the particular universal set being understood from the context. In symbols,

$$A' = \{x \in \mathfrak{U} \mid x \notin A\}.$$

II. The *intersection* of A and B is the set of elements that belong to both A *and* B. The intersection of A and B is denoted by $A \cap B$, which is read "A cap B" or "A intersection B." In symbols,

$$A \cap B = \{x \mid x \in A \text{ and } x \in B\}.$$

III. The *union* of A and B is the set of elements that belong to at least one of the sets A and B, i.e., to A *or* B. The union of A and B is denoted by $A \cup B$, which is read "A cup B" or "A union B." In symbols,

$$A \cup B = \{x \mid x \in A \text{ or } x \in B\}.$$

A comment about the meaning of the word "or" in mathematics is in order here. This logical connective is ambiguous in everyday language, sometimes being used in the inclusive sense (in which "p or q" is taken to mean "p or q, or both p and q") and other times being used in the exclusive sense (in which "p or q" means "p or q, but not both"). As we have explicitly indicated by our wording, the "or" in the definition of union of two sets is to be taken in the inclusive sense, in which it is synonomous with the legal use of "and/or." We adhere to the accepted mathematical usage and shall henceforth always so interpret the word "or." The words "not," "and," and "or" are italicized in Definition 3.1, for they are the key words to remember in the definitions of complement, intersection, and union of sets.

The following examples illustrate how one obtains new sets by applying the operations in Definition 3.1 to given sets.

Example 3.1. Let the universal set $\mathfrak{U}$ be the set of letters in the alphabet, and let A be the subset of vowels, and B the subset containing the first three letters, i.e.,

$$A = \{a, e, i, o, u\}, \qquad B = \{a, b, c\}.$$

Then, by Definition 3.1,

$A' =$ the set of consonants, $\quad B' = \{d, e, f, \cdots, x, y, z\}$,
$A \cup B = \{a, b, c, e, i, o, u\}, \quad A \cap B = \{a\}$.

Example 3.2. Let the universal set $\mathfrak{U}$ be the set of all residents of New York City. Let A denote the set of male New Yorkers, B the set of New Yorkers who live in the borough of Brooklyn, and C the set of baseball fans in New York who are rooting for the Dodgers to win the National League pennant. Then $A' =$ set of female New Yorkers, $B' =$ set of New Yorkers who do not live in Brooklyn, $C' =$ set of New Yorkers who are not baseball fans rooting for the Dodgers, i.e., the set of New Yorkers who either are not baseball fans at all or, if they are baseball fans, are not rooting for the Dodgers to win the pennant, $A \cap B =$ set of male residents of Brooklyn, $A \cup B =$ set of New Yorkers who are male or Brooklynites, and $B \cap C =$ set of Brooklynites who are also baseball fans rooting for the Dodgers to win the National League pennant. (It was erroneously asserted by some bitter elements of set B that, when the Dodgers moved to Los Angeles, it would be true that $B \cap C = \emptyset$.)

Suppose subsets A, B, C of a universal set $\mathfrak{U}$ are given. Since A' and $B \cap C$ are themselves sets, we can form their intersection $A' \cap (B \cap C)$, the set of all elements in $\mathfrak{U}$ that do not belong to A but do belong to both B and C. Similarly, we can take the complement of the intersection $B \cap C$, symbolized by $(B \cap C)'$, and thus obtain the set of objects in $\mathfrak{U}$ that are not in both B and C, i.e., that are not in B or not in C. In general, the three operations we have defined, when applied to sets, produce still other sets to which the operations can again be applied. We illustrate this important point in an example.

Example 3.3. Let $\mathfrak{U} = \{1, 2, 3, 4, 5, 6, 7\}$ be the universal set, and consider the subsets of $\mathfrak{U}$ given by

$$A = \{1, 2, 3\}, \quad B = \{2, 4, 6\}, \quad C = \{1, 3, 5, 7\}.$$

By applying Definition 3.1, we find

$A' = \{4, 5, 6, 7\}, \quad B' = \{1, 3, 5, 7\} = C, \quad C' = \{2, 4, 6\} = B$,
$A \cup B = \{1, 2, 3, 4, 6\}, \quad A \cup C = \{1, 2, 3, 5, 7\}, \quad B \cup C = \mathfrak{U}$,
$A \cap B = \{2\}, \quad A \cap C = \{1, 3\}, \quad B \cap C = \emptyset$.

We can continue forming complements, unions, and intersections of these sets. For example,

$$(A')' = \{4, 5, 6, 7\}' = \{1, 2, 3\} = A,$$
$$(A \cup B)' = \{1, 2, 3, 4, 6\}' = \{5, 7\},$$
$$(B \cap C)' = \emptyset' = \mathfrak{U},$$
$$(A \cap B) \cup C = \{2\} \cup \{1, 3, 5, 7\} = \{1, 2, 3, 5, 7\},$$
$$(A \cup C) \cap (A \cap C) = \{1, 2, 3, 5, 7\} \cap \{1, 3\} = \{1, 3\}, \text{ etc.}$$

When considering sets and operations on sets, it is helpful to represent the sets pictorially. A rectangle is drawn to represent the universal set $\mathfrak{U}$. A subset A of $\mathfrak{U}$ is represented by the region within a circle drawn inside the rectangle. Then A', the complement of A, will be represented by the part of the rectangle outside the circle, as in Figure 5.

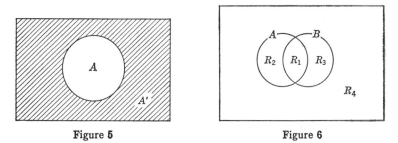

Figure 5   Figure 6

Such diagrams, called *Venn diagrams* (after the English logician John Venn, 1834–1883), can be drawn for a problem involving two subsets, say A and B, of some universal set $\mathfrak{U}$. In Figure 6, we have labeled the four nonoverlapping regions of the rectangle corresponding to the following four possibilities for any element $x \in \mathfrak{U}$:

(1) $x \in A$ and $x \in B$,   i.e., $x \in A \cap B$ (Region R_1)
(2) $x \in A$ and $x \notin B$,   i.e., $x \in A \cap B'$ (Region R_2)
(3) $x \notin A$ and $x \in B$,   i.e., $x \in A' \cap B$ (Region R_3)
(4) $x \notin A$ and $x \notin B$,   i.e., $x \in A' \cap B'$ (Region R_4)

It is important to observe that complements, unions, and intersections of the sets A and B can be represented by combinations of one or more of the regions in Figure 6, as in Table 2. Furthermore, given any set written in terms of operations on A and B, we can easily determine the particular combination of regions representing this set. For example, the set $(A \cap B)'$ contains those elements that are not in $A \cap B$, i.e., not in region R_1. Hence, $(A \cap B)'$ is represented by region R_2 & R_3 & R_4.

Suppose we are told that $A \subseteq B$. Although the Venn diagram is often drawn in this case with the circle representing A entirely within the circle representing B, we prefer to use Figure 6. But since we are told that every element in A is also in B, we conclude that region R_2 represents $\emptyset$, i.e., $A \cap B' = \emptyset$. Furthermore, the regions R_1 & R_2 and R_1 must now represent the same set of points, i.e., $A = A \cap B$. Similarly, R_1 & R_3 and R_1 & R_2 & R_3 also represent equal sets, so that $B = A \cup B$. In

TABLE 2

Set	Region in Figure 6
$\mathfrak{U}$	R_1 & R_2 & R_3 & R_4
A	R_1 & R_2
B	R_1 & R_3
$A \cup B$	R_1 & R_2 & R_3
$A \cap B$	R_1
A'	R_3 & R_4
B'	R_2 & R_4

this way, we see that the following are all equivalent assertions, each giving the information that every element of A is also in B:

(1) $A \subseteq B$,   (2) $A \cap B' = \emptyset$,   (3) $A = A \cap B$,   (4) $B = A \cup B$.

In order to consider another application of Venn diagrams, we need to make an important definition.

Definition 3.2. Two sets A and B are said to be *disjoint* or *mutually exclusive* if they have no elements in common, i.e., if $A \cap B = \emptyset$.

When A and B are disjoint, one customarily draws a Venn diagram with nonoverlapping circles representing A and B. But we can equally well use the diagram in Figure 6, provided we note that region R_1 represents the empty set.

Example 3.4. If S is any set, let us denote by $n(S)$ the number of elements in S. If A and B are disjoint sets, then the number of elements in A or in B is the sum of the number of elements in A and the number of elements in B; i.e.,

(3.1) $n(A \cup B) = n(A) + n(B)$ *if* $A \cap B = \emptyset$.

To find a formula for $n(A \cup B)$ when A and B are not necessarily disjoint, we proceed as follows. $A \cap B'$ and $A \cap B$ are disjoint sets (why?) whose union, as is easily seen from Figure 6, is the set A. Hence by (3.1),

$$n(A \cap B') + n(A \cap B) = n(A).$$

Also, since $A \cap B$ and $A' \cap B$ are disjoint sets whose union is B,

$$n(A \cap B) + n(A' \cap B) = n(B).$$

If we add these equations and subtract $n(A \cap B)$ from both sides of the result, we obtain

$$n(A \cap B') + n(A \cap B) + n(A' \cap B) = n(A) + n(B) - n(A \cap B).$$

But, referring to Figure 6, we recognize the left-hand side of this equation as the number of elements in the set represented by region R_1 & R_2 & R_3. But this region represents the union $A \cup B$. Hence, we obtain the formula

$$(3.2) \qquad n(A \cup B) = n(A) + n(B) - n(A \cap B),$$

which is valid for any sets A and B. Note that (3.2) reduces to (3.1) when A and B are disjoint, for then $n(A \cap B) = n(\emptyset) = 0$.

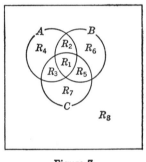

Figure 7

Suppose we pick one card from a standard deck of 52 cards. Let $\mathfrak{U}$ be the set of 52 possible choices, and let A and B denote the set of aces and the set of spades, respectively. Obviously $n(A) = 4$ and $n(B) = 13$. But $n(A \cup B) \neq 17$, since A and B are not disjoint. Indeed, $n(A \cap B) = 1$, since only the ace of spades is common to A and B. By (3.2) we correctly find $n(A \cup B) = 16$. As expected, we obtain an ace or a spade with 16 different cards.

The general Venn diagram in the case of three subsets A, B, and C is found in Figure 7, where we have labeled the eight nonoverlapping regions corresponding to the following eight possibilities for any element $x \in \mathfrak{U}$:

(1) $x \in A$ and $x \in B$ and $x \in C$ (Region R_1)
(2) $x \in A$ and $x \in B$ and $x \notin C$ (Region R_2)
(3) $x \in A$ and $x \notin B$ and $x \in C$ (Region R_3)
(4) $x \in A$ and $x \notin B$ and $x \notin C$ (Region R_4)
(5) $x \notin A$ and $x \in B$ and $x \in C$ (Region R_5)
(6) $x \notin A$ and $x \in B$ and $x \notin C$ (Region R_6)
(7) $x \notin A$ and $x \notin B$ and $x \in C$ (Region R_7)
(8) $x \notin A$ and $x \notin B$ and $x \notin C$ (Region R_8)

It will be helpful in our later work to have clearly in mind the correspondence between various subsets of $\mathcal{U}$ and regions of the Venn diagram in Figure 7. The reader should check the examples in Table 3.

TABLE 3

Set	Region in Figure 7
$\mathcal{U}$	R_1 & R_2 & R_3 & R_4 & R_5 & R_6 & R_7 & R_8
A	R_1 & R_2 & R_3 & R_4
B	R_1 & R_2 & R_5 & R_6
C	R_1 & R_3 & R_5 & R_7
$A \cap B$	R_1 & R_2
$A \cup B$	R_1 & R_2 & R_3 & R_4 & R_5 & R_6
$(A \cup B) \cap C$	R_1 & R_3 & R_5
A'	R_5 & R_6 & R_7 & R_8
$A' \cap (A \cap B)$	None (the set is empty)
$B \cap C$	R_1 & R_5
$(A \cap B') \cap C'$	R_4

We see that by starting with two or more subsets of some universal set and forming their complements, unions, and intersections, many other subsets are obtained. In the next section, we explore certain interesting and important relationships among these subsets. We conclude here with another example in which a Venn diagram proves helpful.

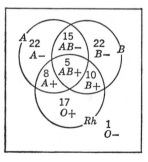

Figure 8

Example 3.5. Persons are classified according to blood type and Rh quality by testing a blood sample for the presence of three antigens: A, B, and Rh. Blood is of type AB if it contains both antigens A and B, of type A if it contains A but not B, of type B if it contains B but not A, and of type O if it contains neither A nor B. In addition, blood is classified as Rh positive ($+$) if the Rh antigen is present, as Rh negative ($-$) otherwise. If we let A, B, and Rh denote the sets of people whose blood contains the A, B, and Rh antigens respectively, then all people are classified into one of the eight categories indicated in the Venn diagram of Figure 8.

Suppose a laboratory technician reports the following statistics after testing blood samples of 100 people:

> 50 contain antigen A
> 52 contain antigen B
> 40 contain antigen Rh
> 20 contain both A and B
> 13 contain both A and Rh
> 15 contain both B and Rh
> 5 contain all three antigens

How many persons of type $A-$ did the technician find? To answer this sort of question, we use the data to fill in the number of people in each of the eight subsets in Figure 8. The trick here is to use the data in reverse order, i.e., work from the bottom of the list to the top. Thus, the last item reported tells us there are five people of type $AB+$. The 15 people reported to have both B and Rh must be of type $AB+$ or $B+$. Since five people are already identified as type $AB+$, we infer that ten people are of type $B+$. In this way we complete the enumeration, and thus obtain the number of people in each of the eight categories. We find there were 22 people of type $A-$.

PROBLEMS

3.1. Let $\mathcal{U} = \{a, b, c\}$, $A = \{a\}$, $B = \{b\}$. List the elements of the following sets: A', B', $A \cup B$, $A \cap B$, $A' \cap B'$, $A' \cap (A \cup B)$.

3.2. Refer to the Venn diagram in Figure 6. Determine the region or combination of regions representing each of the sets (a) $(A \cup B)'$, (b) $A' \cup B'$, (c) $A \cup B'$, (d) $(A')'$, (e) $(A' \cap B)' \cup B$.

3.3. A universal set $\mathcal{U}$ has eight elements corresponding to the eight possible outcomes of the experiment in which a penny, a nickel, and a dime are tossed.

   (a) List the elements of $\mathcal{U}$.
   (b) Suppose subset A contains those elements corresponding to outcomes for which the penny falls heads, subset B those for which all three coins match, and subset C those for which the number of heads exceeds the number of tails. List the elements of the following sets: A', B', $A \cup B$, $A \cup C$, $B \cup C$, $A \cap B$, $A \cap C$, $B \cap C$, $A' \cap C$, $(A \cap B) \cap C$, and $(A \cap B') \cap C$.

3.4. The universal set $\mathfrak{U}$ contains the 52 cards in a standard deck. Let S denote the subset of spades, D the subset of diamonds, and H the subset of honor cards (i.e., ten, jack, queen, king, or ace.)

(a) Identify the following sets and count the number of elements in each: $S \cap H$, S', $D \cap S$, $D \cap S'$, $D \cup S$, $(S \cup D) \cap H$.

(b) Write the following sets in symbolic form as in (a): the set of cards that are not honor cards, the set of cards that are neither spades nor honor cards, the set of clubs or hearts that are not honor cards. (*Note:* It is instructive to try to write this last set in at least three different ways.)

3.5. Table 4 classifies 321 union men with respect to two characteristics: (1) the number of years each has been in the union, and (2) his answer to the question, "Are you willing to picket to help some other shop get organized or get a raise in pay?".

TABLE 4

Response to Question	Number of Years in the Union				Total
	Less than 1	1–3	4–10	Over 10	
Yes	27	54	137	28	246
No	14	18	34	3	69
Don't know	3	2	1	0	6
Total:	44	74	172	31	321

[Source: Arnold M. Rose, *Union Solidarity*, University of Minnesota Press, 1952, p. 77.]

Let the 321 men in this survey be the elements of our universal set $\mathfrak{U}$, and define the following subsets of $\mathfrak{U}$:

Y = set of men who answer "yes,"
N = set of men who answer "no,"
A = set of men who are in the union less than one year,
B = set of men who are in the union 1–3 years,
C = set of men who are in the union 4–10 years.

(a) Find the number of men in each of the following sets: (i) $Y \cap B$, (ii) $Y \cup B$, (iii) $(Y \cup N)' \cap A$, (iv) $(N \cap C)'$.

(b) Write each of the following sets, using only the symbols A, B, C, Y, N, $'$, $\cup$, and $\cap$. [Example: the set of men who answer "yes" and are in the union less than four years is the set $Y \cap (A \cup B)$.]

(1) the set of men who answer "yes" and are in the union 4–10 years.

(2) the set of men who answer "yes" and are in the union at least four years.

(3) the set of men who answer "don't know."

(4) the set of men who answer "don't know" and who are in the union over ten years. (What does the survey tell you about this set?)

3.6. Let $\mathfrak{U}$ be the set of all points in the plane. Relative to some fixed rectangular coordinate system, we can write

$$\mathfrak{U} = \{(x, y) \mid x \in R \text{ and } y \in R\},$$

where R is the set of all real numbers. Let subsets of $\mathfrak{U}$ be defined as follows:

$$A = \{(x, y) \mid x \geq 0\}, \qquad B = \{(x, y) \mid y \geq 0\},$$
$$C = \{(x, y) \mid x + 2y \leq 6\}, \qquad D = \{(x, y) \mid y - x \geq 0\}.$$

Sketch graphs of the sets (a) A, (b) B, (c) C, (d) D, (e) $A \cap B$, (f) $(A \cap B) \cap C$, (g) $[(A \cap B) \cap C] \cap D$.

3.7. (a) Express $n(\mathfrak{U})$, the number of elements in the universal set $\mathfrak{U}$, in terms of $n(A)$, the number of elements in subset A, and $n(A')$, the number of elements in the complement of A.

(b) The formula you wrote in (a) can be deduced from Formula (3.2) of the text. Show how to do this.

3.8. A psychologist ran 50 mice through a maze experiment and reported the following data: 25 mice were male, 25 were previously trained, 20 turned left (at the first choice-point), 10 were previously trained males, 4 males turned left, 15 previously trained mice turned left, and 3 previously trained males turned left. Draw an appropriate Venn diagram and determine the number of female mice who were not previously trained and who did not turn left.

3.9. Of 63 member colleges of the Council For The Advancement of Small Colleges, Inc., 24 were founded before 1931, were coed, and reported annual student costs of less than $1,000; 41 were founded before 1931 and were coed; 27 were founded before 1931 and had student costs of less than $1,000; 45 were founded before 1931; 52 were coed; 34 had student costs of less than $1,000; 4 were not founded before 1931, were not coed, and had student costs of at least $1,000. (Data reported in Supplement Section 11, *The New York Times*, October 11, 1959.)

A high school senior wants to attend a coed college that is relatively new, say founded 1931 or later. His annual student costs must be less than $1,000. How many of the 63 small colleges meet his requirements?

3.10. Show that each of the following pairs of sets are represented by the same region in the Venn diagram of Figure 7:

(a) $(A \cup B)'$ and $A' \cap B'$,

(b) $(A \cap B)'$ and $A' \cup B'$,

(c) $(A \cup B) \cup C$ and $A \cup (B \cup C)$,

(d) $A \cap (B \cup C)$ and $(A \cap B) \cup (A \cap C)$.

3.11. A, B, and C are subsets of a universal set $\mathfrak{U}$. Arrange the following sets in sequential order so that each set in the sequence is a subset of the next set: $A \cup B$, $\mathfrak{U}$, $A \cap B$, $\emptyset$, B, $A \cup (B \cup C)$, $(A \cap B) \cap C$, $(A \cup B) \cup C$, $\emptyset'$, $B \cap A$.

3.12. (a) Show that $\mathfrak{U}' = \emptyset$ and $\emptyset' = \mathfrak{U}$.

(b) Show that if $A \subseteq B$, then $B' \subseteq A'$.

(c) Suppose $A \cup B = \emptyset$. What conclusion can you draw about the sets A and B?

(d) Suppose $A \cap B = \emptyset$. Does it follow that $A = \emptyset$ or $B = \emptyset$?

3.13. Let $\mathfrak{U}$ be the set of all people and

$$M = \text{the set of all males},$$
$$C = \text{the set of all college students},$$
$$I = \text{the set of all intelligent people},$$
$$S = \text{the set of sorority members},$$
$$B = \text{the set of beer drinkers},$$
$$P = \text{the set of professors},$$
$$W = \text{the set of well-dressed people}.$$

Translate each of the following sentences into an equation or an inequality using only the letters standing for sets and the symbols $=$, $\neq$, $\emptyset$, $'$, $\cap$, $\cup$. (For example, the sentence "All college students are intelligent" means that the set of college students is a subset of the set of intelligent people, i.e., $C \subseteq I$. But we are not permitted use of the set inclusion symbol. Hence we refer to the discussion on p. 20 and rewrite the sentence in any of the equivalent forms $C \cap I = C$, $C \cup I = I$, or $C \cap I' = \emptyset$. Similarly, the sentence "Some college students are intelligent" means there is at least one member of the intersection $C \cap I$. Hence this sentence is translated into $C \cap I \neq \emptyset$.)

(a) All professors are beer drinkers.

(b) No males are sorority members.

(c) No male college student is well dressed.

(d) Sorority members are neither intelligent nor male.

(e) Some professors are beer drinking males.

(f) Some professors who drink beer are not males.

(g) Some professors who drink beer are neither intelligent nor well dressed.

(h) College students and professors are beer-drinkers.

(i) If a person is a beer-drinker, then that person is intelligent.

(j) If a person is intelligent, then that person is a beer drinker.

(k) A person is a beer drinker if and only if he is intelligent.

3.14. Read the following discussion carefully. Then to test your understanding, do the exercises.

A finite set $\mathcal{U}$ of n people is given. $\mathcal{U}$ will be called a *decision-making body*. Let $\mathcal{V} = 2^{\mathcal{U}}$ be the power set of $\mathcal{U}$, as defined in Problem 2.4. Each element of $\mathcal{V}$ is a subset of $\mathcal{U}$ and will be called a *coalition*. (In particular, the empty set $\emptyset$ and the set $\mathcal{U}$ itself are coalitions. There are 2^n coalitions altogether.)

We select a subset W of $\mathcal{V}$ and write $\mathcal{V} = W \cup W'$, where W' is the complement of W *with respect to* $\mathcal{V}$. Since W and W' are disjoint, each coalition is in exactly one of W (the set of *winning coalitions*) or W' (the set of *nonwinning coalitions*).

Now consider the set W'. An element of W' will be called a *losing coalition* if its complement (*with respect to* $\mathcal{U}$) is a winning coalition. Thus L, the set of losing coalitions, is defined by

$$L = \{A \mid A \in W' \text{ and } A' \in W\}.$$

[Note that W' means the complement of W with respect to $\mathcal{V}$, whereas A' means the complement of A with respect to $\mathcal{U}$. This confusion arises from the fact that $\mathcal{U}$ serves as universal set for all sets whose elements are people, whereas $\mathcal{V}$ serves as universal set for all sets whose elements are coalitions (sets of people).]

Finally, a coalition that is nonwinning itself and whose complement is also nonwinning is called a *blocking coalition*. Thus B, the set of blocking coalitions, is defined by

$$B = \{A \mid A \in W' \text{ and } A' \in W'\}.$$

Some decision-making bodies contain important persons who get special names: A person $x \in \mathcal{U}$ is said to be a *dictator* if $\{x\} \in W$, i.e., if x is the sole member of a winning coalition. A person $y \in \mathcal{U}$ is said to *have veto power* if $\{y\} \in B$, i.e. if y is the sole member of a blocking coalition.

In each of the following exercises, a particular decision-making body is described and its voting rules specified. *Interpret a winning coalition to mean a set of persons who control enough votes to carry a proposal.* Find all winning coalitions, losing coalitions, blocking coalitions. Determine if any members are dictators or have veto power.

Exercise 1. A committee consists of four people, each with one vote. Majority rule applies; i.e., three votes are needed to carry a proposal.

Exercise 2. A small corporation with 100 shares of stock outstanding has three shareholders. Individual a owns 50 shares, b owns 30 shares, and c owns 20 shares. Each share has one vote, and simple majority rule applies.

Exercise 3. Same as Exercise 2, except that b has sold one of his shares to a.

Exercise 4. A student-faculty committee consists of five students and four faculty members. For a proposal to be passed on to the entire faculty for its consideration, at least three students and three faculty members must vote for the proposal. Each member has one vote.

4. The algebra of sets

We have studied a number of ways of obtaining other sets, once a universal set $\mathfrak{u}$ is given. There are the many sets that can be constructed by performing the operations of complement, union, and intersection on subsets of $\mathfrak{u}$. The reader must suspect by now (especially in view of the results in Problems 3.10–3.12) that there are many relationships among the sets obtained in this way. These relationships form the subject matter of the present section.

We begin by listing a number of important laws obeyed by sets. All follow from our definitions of the empty set $\emptyset$, the universal set $\mathfrak{u}$, the operations denoted by $'$, $\cap$, and $\cup$, together with the definition of set equality.

Theorem 4.1. Let A, B, and C be any subsets of a universal set $\mathfrak{u}$. Then the following laws hold:

Identity laws:

1a. $A \cup \emptyset = A$ 1b. $A \cap \mathfrak{u} = A$
2a. $A \cup \mathfrak{u} = \mathfrak{u}$ 2b. $A \cap \emptyset = \emptyset$

Idempotent laws:

3a. $A \cup A = A$ 3b. $A \cap A = A$

Complement laws:

4a. $A \cup A' = \mathfrak{u}$ 4b. $A \cap A' = \emptyset$
5a. $(A')' = A$

Commutative laws:

6a. $A \cup B = B \cup A$ 6b. $A \cap B = B \cap A$

De Morgan's laws:

7a. $(A \cup B)' = A' \cap B'$ 7b. $(A \cap B)' = A' \cup B'$

Associative laws:

8a. $A \cup (B \cup C) = (A \cup B) \cup C$
8b. $A \cap (B \cap C) = (A \cap B) \cap C$

Distributive laws:

9a. $A \cup (B \cap C) = (A \cup B) \cap (A \cup C)$
9b. $A \cap (B \cup C) = (A \cap B) \cup (A \cap C)$

Before proving these laws, let us note that we are familiar with many of their names from the ordinary algebra of numbers. Thus, addition and multiplication of numbers is commutative, i.e.,

$$a + b = b + a \quad \text{and} \quad a \times b = b \times a,$$

for any numbers a and b. Analogously, Laws 6a and 6b assert that the order in which two sets are written does not affect their union or intersection. For any numbers a, b, and c, we recall the associative laws

$$a + (b + c) = (a + b) + c \quad \text{and} \quad a \times (b \times c) = (a \times b) \times c.$$

The analogy with 8a and 8b is clear. The associative law 8a asserts that the same set is obtained if we form the union of A with the union of B and C or if we form instead the union of the union of A and B with C. In ordinary algebra, we have only one distributive law, namely, $a \times (b + c) = a \times b + a \times c$. This is analogous to 9b, one of the two distributive laws for sets.

Since adding zero to any number yields that same number as sum, 0 is called an identity number with respect to addition. Similarly, since $a \times 1 = a$ for any number a, we say that 1 is an identity number with respect to multiplication. As Laws 1a and 1b show, the empty set $\emptyset$ is an identity set with respect to union and the universal set $\mathcal{U}$ is an identity set with respect to intersection.

Because of these analogies, $A \cup B$ is sometimes called the logical sum and $A \cap B$ the logical product of the sets A and B. But the analogy with ordinary algebra is not perfect, as a glance at the idempotent laws shows. If a is a number, then $a + a = 2a$; if A is a set, then $A \cup A = A$.

It is instructive to try to translate these laws into prose form. For example, 7a asserts that the complement of the union of any two sets is equal to the intersection of their complements, and 7b asserts that the complement of the intersection of any two sets is equal to the union of their complements.

Finally, let us note that (except for 5a) the laws in Theorem 4.1 are listed in pairs, 1a and 1b, 2a and 2b, etc. We shall comment on the significance of this fact after we discuss the proof of these laws.

Our method of proof involves the use of *membership tables*. The basic membership tables for complement, intersection, and union appear in Tables 5–7. In the first column of Table 5, we symbolize the two possibilities for any element x of the universal set $\mathfrak{U}$: either

TABLE 6   TABLE 7

TABLE 5

A	A'
$\in$	$\notin$
$\notin$	$\in$

A	B	$A \cap B$
$\in$	$\in$	$\in$
$\in$	$\notin$	$\notin$
$\notin$	$\in$	$\notin$
$\notin$	$\notin$	$\notin$

A	B	$A \cup B$
$\in$	$\in$	$\in$
$\in$	$\notin$	$\in$
$\notin$	$\in$	$\in$
$\notin$	$\notin$	$\notin$

$x \in A$ or $x \notin A$. If $x \in A$, then $x \notin A'$ and if $x \notin A$, then $x \in A'$. These facts follow from the definition of complement and are summarized in the two rows of Table 5, the membership table for A'.

With respect to the sets A and B, each element x of the universal set $\mathfrak{U}$ falls into exactly one of the following categories: (1) $x \in A$ and $x \in B$, (2) $x \in A$ and $x \notin B$, (3) $x \notin A$ and $x \in B$, and (4) $x \notin A$ and $x \notin B$. These are the four possibilities symbolized in the four rows to the left of the double vertical line in Table 6, the membership table for $A \cap B$. To the right, in the column headed $A \cap B$, is summarized the membership status of x with respect to the intersection of A and B. That is, by the definition of intersection, $x \in A \cap B$ in the case (row 1) when $x \in A$ and $x \in B$; $x \notin A \cap B$ in all other cases (rows 2–4).

Table 7, the membership table for $A \cup B$, is similarly interpreted. We know that $x \in A \cup B$ if and only if x belongs to at least one of the sets A and B. Hence, an $\in$ appears under $A \cup B$ in rows 1–3 of Table 7, but an $\notin$ appears in row 4.

Using these basic tables, we can construct membership tables for other sets. The details of this construction, as well as the rationale behind the use of membership tables for proving equality between sets, are best explained in the context of some examples.

Example 4.1. To prove De Morgan's law,

$$(A \cup B)' = A' \cap B',$$

we proceed as follows. Since this law involves two arbitrary sets A and B, we start by listing in columns (1) and (2) of Table 8 the four

TABLE 8

(1)	(2)	(3)	(4)	(5)	(6)	(7)
A	B	$A \cup B$	$(A \cup B)'$	A'	B'	$A' \cap B'$
$\in$	$\in$	$\in$	$\notin$	$\notin$	$\notin$	$\notin$
$\in$	$\notin$	$\in$	$\notin$	$\notin$	$\in$	$\notin$
$\notin$	$\in$	$\in$	$\notin$	$\in$	$\notin$	$\notin$
$\notin$	$\notin$	$\notin$	$\in$	$\in$	$\in$	$\in$

possibilities for an element $x \in \mathcal{U}$. Since the set $(A \cup B)'$ is obtained by first forming $A \cup B$ and then taking its complement, we include a column for $A \cup B$ and another for $(A \cup B)'$. The entries in column (3) are obtained from (1) and (2) by use of the basic membership table for $A \cup B$. The entry in each row of column (4) is obtained from the entry in the corresponding row of column (3) by using Table 5.

The set $A' \cap B'$ appearing in the law we are trying to prove is obtained by forming A' and B', and then taking their intersection. Hence we have columns (5)–(7) in Table 8. The entry in each row of column (5) is obtained from the entry in the corresponding row of column (1) by use of the basic membership table for A'. Column (6) is similarly obtained from column (2). Finally, from (5) and (6) we get column (7) by using the basic membership table for intersection.

The crucial observation is that columns (4) *and* (7) *are identical:* whenever a row contains an $\in$ in column (4) it also contains an $\in$ in column (7) and likewise for the occurrences of $\notin$. We conclude that whenever an element of $\mathcal{U}$ belongs to $(A \cup B)'$, it also belongs to $A' \cap B'$, i.e.,

(4.1) $$(A \cup B)' \subseteq A' \cap B'.$$

Moreover, whenever an element does not belong to $(A \cup B)'$, then it does not belong to $A' \cap B'$. It follows (why?) that every element that does belong to $A' \cap B'$ must also belong to $(A \cup B)'$, i.e.,

(4.2) $A' \cap B' \subseteq (A \cup B)'.$

From (4.1) and (4.2) we conclude that

$$(A \cup B)' = A' \cap B'$$

and this completes the proof of De Morgan's law 7a.

Before considering another example, we note the striking similarity between the method of proof just illustrated and the method of verifying relations between sets by means of Venn diagrams. In Figure 9, we apply the latter method to the De Morgan law we have

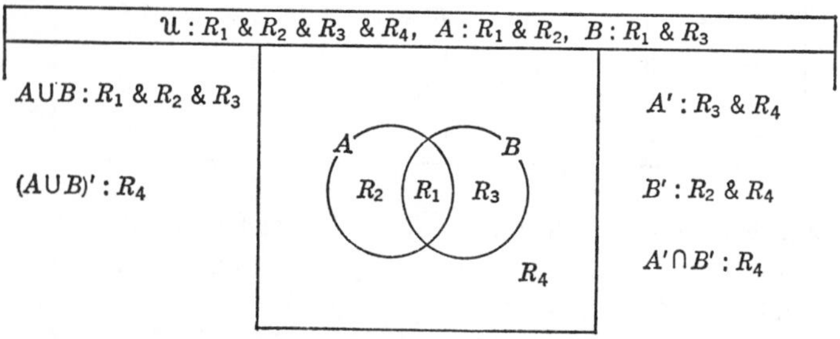

Figure 9

just proved. In the space above the usual rectangle, we list our data: the universal set $\mathcal{U}$ is represented by the entire rectangle, the set A by the region R_1 & R_2, the set B by the region R_1 & R_3. (We use the colon as shorthand for "is represented by" when it separates a set and a region in the Venn diagram.) To the left of the rectangle, we list the steps required to find the region represented by the left-hand side of De Morgan's law; to the right of the rectangle, we find the region represented by the right-hand side of the law. We observe that $(A \cup B)'$ and $A' \cap B'$ are both represented by the same region R_4. This fact constitutes the verification of De Morgan's law 7a by means of a Venn diagram.

Since the four regions in Figure 9 are numbered to correspond to the four rows of Table 8, we can follow in the Venn diagram each step in the construction of Table 8. Thus, the fact that column (1) contains ϵ's in rows 1 and 2 is expressed in Figure 9 by the fact that

A is represented by R_1 & R_2. That columns (4) and (7) are identical and contain ϵ's only in row 4 is expressed in Figure 9 by the fact that both $(A \cup B)'$ and $A' \cap B'$ are represented by region R_4. Although the method of membership table suffices for proving all the laws we shall encounter, the method of Venn diagrams is often a helpful aid in understanding these laws. We give one more example in which both methods are used.

Example 4.2. To prove the distributive law 9b in Theorem 4.1, we construct the membership table with eight rows (since three arbitrary sets are involved) in Table 9. The law is proved by noting that the columns headed $A \cap (B \cup C)$ and $(A \cap B) \cup (A \cap C)$ are identical.

TABLE 9

A	B	C	$B \cup C$	$A \cap (B \cup C)$	$A \cap B$	$A \cap C$	$(A \cap B) \cup (A \cap C)$
ϵ	ϵ	ϵ	ϵ	ϵ	ϵ	ϵ	ϵ
ϵ	ϵ	$\notin$	ϵ	ϵ	ϵ	$\notin$	ϵ
ϵ	$\notin$	ϵ	ϵ	ϵ	$\notin$	ϵ	ϵ
ϵ	$\notin$	$\notin$	$\notin$	$\notin$	$\notin$	$\notin$	$\notin$
$\notin$	ϵ	ϵ	ϵ	$\notin$	$\notin$	$\notin$	$\notin$
$\notin$	ϵ	$\notin$	ϵ	$\notin$	$\notin$	$\notin$	$\notin$
$\notin$	$\notin$	ϵ	ϵ	$\notin$	$\notin$	$\notin$	$\notin$
$\notin$	$\notin$	$\notin$	$\notin$	$\notin$	$\notin$	$\notin$	$\notin$

In Figure 10, this same distributive law is verified by using an appropriate Venn diagram. We have numbered the eight regions in

$\mathcal{U}: R_1$ & R_2 & R_3 & R_4 & R_5 & R_6 & R_7 & R_8, $A: R_1$ & R_2 & R_3 & R_4, $B: R_1$ & R_2 & R_5 & R_6, $C: R_1$ & R_3 & R_5 & R_7

$B \cup C: R_1$ & R_2 & R_3 & R_5 & R_6 & R_7

$A \cap (B \cup C): R_1$ & R_2 & R_3

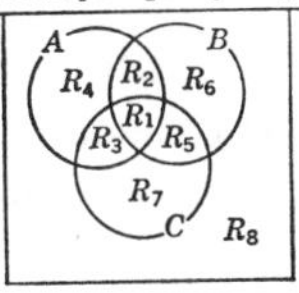

$A \cap B: R_1$ & R_2

$A \cap C: R_1$ & R_3

$(A \cap B) \cup (A \cap C): R_1$ & R_2 & R_3

Figure 10

Figure 10 to correspond to the eight rows of Table 9 in order to bring out here, as in Example 4.1, the similarities in the membership table and Venn diagram methods.

We leave the proofs of the other laws in Theorem 4.1 as problems for the reader. In the next examples, we illustrate how to prove still other laws directly from those already known to be true.

Example 4.3. If A and B are any subsets of a universal set $\mathfrak{U}$, then

(4.3) $A = (A \cap B) \cup (A \cap B')$.

Proof.

$$
\begin{aligned}
(A \cap B) \cup (A \cap B') &= A \cap (B \cup B') && \text{[by 9b.]} \\
&= A \cap \mathfrak{U} && \text{[by 4a.]} \\
&= A && \text{[by 1b.]}
\end{aligned}
$$

Example 4.4. If A and B are any subsets of a universal set $\mathfrak{U}$, then

(4.4) $A = (A \cup B) \cap (A \cup B')$.

Proof.

$$
\begin{aligned}
(A \cup B) \cap (A \cup B') &= A \cup (B \cap B') && \text{[by 9a.]} \\
&= A \cup \emptyset && \text{[by 4b.]} \\
&= A && \text{[by 1a.]}
\end{aligned}
$$

These examples enable us to make a point concerning the pairing of the laws in Theorem 4.1. This was done in order that we may note the so-called *duality principle:* If in any law we replace $\emptyset$ by $\mathfrak{U}$, $\mathfrak{U}$ by $\emptyset$, $\cup$ by $\cap$, and $\cap$ by $\cup$ wherever they occur, then the result is again a law. The new law is said to be the *dual* of the original law. Thus in Theorem 4.1, law 1b is the dual of law 1a, 1a is the dual of 1b, and so on for all a and b laws in our list. The dual of 5a is itself; law 5a is therefore said to be self-dual.

Note that (4.3) and (4.4) are dual laws, and that the proof of (4.4) can be obtained from the proof of (4.3) by replacing each statement by its dual. Since Theorem 4.1 contains the dual of every one of its laws, we can justify each step in proving (4.4) by appealing to the dual of the law justifying the corresponding step in the proof of (4.3). In this way, we could prove the dual of *any* law whose proof followed from Theorem 4.1. Indeed, this is the essence of the duality principle.

The importance of the duality principle cannot be fully appreciated until the algebra of sets we have been discussing is treated formally as a mathematical system. In this more abstract study, known as *Boolean algebra* (after the English logician George Boole, 1815–1864), the algebra of sets becomes just one concrete interpretation of an

abstract system which also has other important interpretations.*
The interested student can consult the references listed at the end of
this chapter for readings on Boolean algebra.

In our later work, we shall need to consider the union and intersection of more than two sets. Because of the associative laws 8a and
8b in Theorem 4.1, it is not necessary to use parentheses to show how
more than two sets separated by $\cup$ or by $\cap$ are paired. For example,
$A \cap B \cap C \cap D$ can be interpreted as $(A \cap B) \cap (C \cap D)$ or as
$(A \cap (B \cap C)) \cap D$ or as $((A \cap B) \cap C) \cap D$, since all of these
sets are equal. (See Problem 4.9.) Similar considerations apply to
the union of more than two sets, so we make the following general
agreement.

Definition 4.1. Let n be any positive integer and suppose B_1,
$B_2, \cdots, B_n$ are given sets. Then the set of elements belonging to
all the given sets is denoted by

$$B_1 \cap B_2 \cap \cdots \cap B_n$$

and the set of elements belonging to *at least one* of the given sets is
denoted by

$$B_1 \cup B_2 \cup \cdots \cup B_n.$$

Many of the laws in Theorem 4.1 can now be generalized to hold
for unions and intersections of more than two sets. We collect some
of these formulas in the following theorem.

Theorem 4.2. Let n be any positive integer and suppose $A, B_1, B_2,$
$\cdots, B_n$ are subsets of a universal set $\mathfrak{U}$. Then

(4.5) $(B_1 \cup B_2 \cup \cdots \cup B_n)' = B_1' \cap B_2' \cap \cdots \cap B_n'.$

(4.6) $(B_1 \cap B_2 \cap \cdots \cap B_n)' = B_1' \cup B_2' \cup \cdots \cup B_n'.$

(4.7) $A \cup (B_1 \cap B_2 \cap \cdots \cap B_n)$
$$= (A \cup B_1) \cap (A \cup B_2) \cap \cdots \cap (A \cup B_n).$$

(4.8) $A \cap (B_1 \cup B_2 \cup \cdots \cup B_n)$
$$= (A \cap B_1) \cup (A \cap B_2) \cup \cdots \cup (A \cap B_n).$$

Proof. We prove (4.5) and leave the others as problems. Our

---

* The so-called statement calculus in logic is another interpretation of Boolean
algebra, and we can therefore expect that the logical analysis of statements and
the study of sets will have many common features. The similarities between the
use of truth tables in logic and our use of membership tables is but one of many
examples.

method of proof is by mathematical induction. When $n = 1$, both sides of (4.5) reduce to B_1', and so (4.5) is certainly true. If $n = 2$, then (4.5) reduces to De Morgan's law 7a in Theorem 4.1 (applied to the sets B_1 and B_2), and hence is again true.

Now suppose (4.5) is true when $n = k$, where k is any positive integer. We complete our proof by mathematical induction if we show that (4.5) must also be true when $n = k + 1$. By grouping the sets as indicated, we obtain

$$(B_1 \cup B_2 \cup \cdots \cup B_{k+1})' = [(B_1 \cup B_2 \cup \cdots \cup B_k) \cup B_{k+1}]'.$$

Applying De Morgan's law 7a to the sets $(B_1 \cup B_2 \cup \cdots \cup B_k)$ and B_{k+1}, we get

$$(B_1 \cup B_2 \cup \cdots \cup B_{k+1})' = (B_1 \cup B_2 \cup \cdots \cup B_k)' \cap B_{k+1}'$$
$$= (B_1' \cap B_2' \cap \cdots \cap B_k') \cap B_{k+1}',$$

the last equality following by our induction hypothesis that (4.5) is true when $n = k$. But the parentheses are not necessary in this last expression, so we can write

$$(B_1 \cup B_2 \cup \cdots \cup B_{k+1})' = B_1' \cap B_2' \cap \cdots \cap B_{k+1}',$$

which is precisely (4.5) when $n = k + 1$.

We have now shown (4.5) is true when $n = 1$ and $n = 2$ and, furthermore, that its truth when $n = k + 1$ follows from its truth when $n = k$ for any positive integer k. We conclude by the principle of mathematical induction that (4.5) is true for all positive integers n. This completes the proof.

Note that (4.5) and (4.6) are generalizations of De Morgan's laws, whereas (4.7) and (4.8) are generalized distributive laws.

PROBLEMS

4.1. Only laws 7a and 9b of Theorem 4.1 are proved in the text. Construct membership tables for the other laws, and thus complete the proof of Theorem 4.1. Also verify each law by means of an appropriate Venn diagram whose regions are numbered to correspond to the rows of the membership table for that law.

4.2. How many rows are required in a membership table for a law involving four arbitrary sets? five sets? n sets, where n is any positive integer?

4.3. Construct membership tables and thus show that the following laws hold for any subsets A, B, C of a universal set $\mathfrak{U}$.

(a) $(A' \cap B')' = A \cup B$
(b) $[A' \cap (A \cup B)]' = A \cup B'$
(c) $(A \cap B) \cap (A \cap B') = \emptyset$
(d) $A \cap (A \cup B) = A \cup (A \cap B) = A$
(e) $(A' \cap (B \cap C))' = A \cup B' \cup C'$

4.4. Prove each of the laws in the preceding problem by using only the laws in Theorem 4.1. Indicate the law in Theorem 4.1 which justifies each step in your proof.

4.5. Verify each of the laws in Problem 4.3 by the method of Venn diagrams.

4.6. (a) It is clear from a Venn diagram that if A and B are disjoint sets, then $A \cap C$ and $B \cap C$ are also disjoint. But prove this result by showing that if $A \cap B = \emptyset$, then $(A \cap C) \cap (B \cap C) = \emptyset$. Justify each step in your proof by appealing to either the hypothesis or to one of the laws in Theorem 4.1.

(b) Show by examples that $A \cap B \cap C = \emptyset$ does not imply $A \cap B = \emptyset$ or $A \cap C = \emptyset$ or $B \cap C = \emptyset$.

4.7. (a) Consider the following valid argument.
Hypotheses. (1) All college students are beer drinkers.
(2) All beer drinkers are well dressed.
Conclusion. Therefore all college students are well dressed.
Write the hypotheses and the conclusion using symbols of set theory (see Problem 3.13), and then prove the argument is valid; i.e., the conclusion is true whenever the hypotheses are both true. Each step in your proof should be justified by appealing to one of the hypotheses or to one of the laws in Theorem 4.1.

(b) Following the procedure outlined in part (a), prove that the following argument is valid.
Hypotheses. (1) All college students are beer drinkers.
(2) No beer drinkers are well-dressed.
Conclusion. Therefore no college students are well-dressed.

4.8. If A and B are subsets of a universal set $\mathfrak{U}$, the *symmetric difference* of A and B, denoted by $A \Delta B$, is defined as the set

$$A \Delta B = (A \cap B') \cup (A' \cap B).$$

(a) Construct a membership table for $A \Delta B$.
(b) In an appropriate Venn diagram, identify the region representing the set $A \Delta B$.
(c) By means of membership tables, prove each of the following laws:
(i) $A \Delta \emptyset = A$
(ii) $A \Delta \mathfrak{U} = A'$
(iii) $A \Delta A = \emptyset$
(iv) $A \Delta A' = \mathfrak{U}$

(v) $A \Delta B = B \Delta A$

(vi) $A \Delta (B \Delta C) = (A \Delta B) \Delta C$

(vii) $A \cap (B \Delta C) = (A \cap B) \Delta (A \cap C)$

(d) Prove each of the laws in part (c) by showing that they follow from the laws in Theorem 4.1.

(e) Use Venn diagrams to verify each of the laws in part (c).

(f) Show that

$$A \Delta (B \cap C) = (A \Delta B) \cap (A \Delta C)$$

is *not* a law, i.e., there are sets A, B, and C for which the equality does not hold. By means of a membership table, or otherwise, determine what additional information about the sets A, B, and C suffices to guarantee that the equality does hold.

4.9. Prove that $(A \cap B) \cap (C \cap D)$ and $(A \cap (B \cap C)) \cap D$ are equal sets, supporting each step in your proof by citing a law in Theorem 4.1.

4.10. Prove Formulas (4.6)–(4.8) by mathematical induction.

4.11. You are given at least one subset of a specified universal set $\mathcal{U}$ and are instructed to form all possible sets from the given sets by using the operations denoted by $\cap$, $\cup$, and $'$. Any new sets you obtain this way are also to be used (with other new sets or with the given sets) to form still other sets by using these same operations. Identify all the different sets you end up with by continuing this process indefinitely if you are originally given the following set(s): (a) $\mathcal{U}$, (b) $\emptyset$, (c) A (= a subset of $\mathcal{U}$), (d) A and A'.

4.12. Let $\mathcal{C}$ be a set of subsets of some fixed universal set $\mathcal{U}$, i.e., the elements of $\mathcal{C}$ are subsets of $\mathcal{U}$. The set $\mathcal{C}$ is called an *algebra of sets* if it satisfies the following conditions:

(1) $\mathcal{C}$ is not empty.

(2) If $A \in \mathcal{C}$, then $A' \in \mathcal{C}$.

(3) If $A \in \mathcal{C}$ and $B \in \mathcal{C}$, then $A \cup B \in \mathcal{C}$.

Prove the following theorems, assuming $\mathcal{C}$ is an algebra of sets.

(a) $\mathcal{U} \in \mathcal{C}$.

(b) $\emptyset \in \mathcal{C}$.

(c) If $A \in \mathcal{C}$ and $B \in \mathcal{C}$, then $A \cap B \in \mathcal{C}$.

4.13. Show that each of the following is an algebra of sets. (Cf. Problems 4.11 and 4.12.)

(a) $\mathcal{C} = \{\mathcal{U}, \emptyset\}$.

(b) $\mathcal{C} = \{\mathcal{U}, \emptyset, A, A'\}$, where A is a subset of $\mathcal{U}$.

(c) $\mathcal{C} = 2^{\mathcal{U}}$, the set of all subsets of $\mathcal{U}$.

5. Cartesian product sets

A pair of objects in which we distinguish one of the objects as the first and the other (which need not be different) as the second is called an *ordered pair*. If the first object is called a and the second b, then the ordered pair is written (a, b). This ordered pair is quite different from the set $\{a, b\}$ containing the two objects a and b. There is no first element in $\{a, b\}$, since order is immaterial when listing elements of a set. Although $\{a, b\} = \{b, a\}$, we want to distinguish between (a, b) and (b, a). We shall define two ordered pairs to be equal if and only if their first objects are the same and their second objects are the same, i.e.,

(5.1) $(a, b) = (c, d)$ if and only if $a = c$ and $b = d$.

We have already used ordered pairs, and they are needed even more in our later work. In Example 1.5 and Problems 1.5 and 2.7, we considered ordered pairs of real numbers. Relative to some rectangular coordinate system, we interpreted (x, y) as representing a point in the plane determined by the coordinate axes. According to (5.1), two such ordered pairs of real numbers are equal if and only if they represent the same point.

The objects in an ordered pair need not be numbers. For example, when we toss a coin twice, we can represent the outcome as one of the ordered pairs (H, H), (H, T), (T, H), (T, T), where we write H for "heads" and T for "tails." We agree that the first object in the ordered pair denotes the result of the first toss and the second object denotes the result of the second toss. Since we want to distinguish between the outcomes (H, T) and (T, H), the use of ordered pairs is essential.

It is of some interest that the concept of an ordered pair can be *defined* in terms of sets and that (5.1) can then be *proved*. To characterize an ordered pair, it is sufficient to state what two objects make up the pair and also which is to be considered as the first object. Thus the ordered pair (a, b) is determined if we know the set $\{a, b\}$ of objects in the ordered pair and the set $\{a\}$ identifying the first object. We are thereby led to the following definition.

Definition 5.1. Let a and b be any objects. The *ordered pair* (a, b) is defined by

(5.2) $(a, b) = \{\{a, b\}, \{a\}\}.$

Theorem 5.1. Two ordered pairs (a, b) and (c, d) are equal if and only if $a = c$ and $b = d$.

Proof. From the definition, we have

$$(c, d) = \{\{c, d\}, \{c\}\}$$

and this set is clearly identical to the set defining (a, b) if $a = c$ and $b = d$. This observation constitutes the proof of the "if" part of the theorem. To prove the "only if" part, we assume $(a, b) = (c, d)$, i.e.,

$$(5.3) \qquad \{\{a, b\}, \{a\}\} = \{\{c, d\}, \{c\}\},$$

and proceed to prove that $a = c$ and $b = d$. Now two sets are equal only when they have the same elements. Therefore it follows from (5.3) that either (1) $\{a, b\} = \{c, d\}$ and $\{a\} = \{c\}$, or (2) $\{a, b\} = \{c\}$ and $\{a\} = \{c, d\}$.

If case (1) holds, then from $\{a\} = \{c\}$ we conclude $a = c$, and $\{a, b\} = \{c, d\}$ then implies $b = d$. Thus the theorem is true in case (1).

If case (2) holds, we start from $\{a, b\} = \{c\}$ and, recalling that $\{a, a\} = \{a\}$, conclude that $a = b = c$. Then $\{a\} = \{c, d\}$ becomes $\{a\} = \{a, d\}$, from which $d = a$ follows. Hence in case (2) we have $a = b = c = d$, and the theorem is certainly true. This completes the proof.

Whenever we have two sets, we can always form ordered pairs by taking the first object of the pair from one of the sets and the second object from the second set. This simple observation turns out to be quite important and, as usual, involves a special notation and terminology.

Definition 5.2. If A and B are sets, then the set of all ordered pairs (a, b) such that a belongs to A and b belongs to B is called the *Cartesian product* of A and B, and is denoted by $A \times B$. In symbols,

$$A \times B = \{(a, b) \mid a \,\epsilon\, A \text{ and } b \,\epsilon\, B\}.$$

We now have still another way of obtaining new sets from given sets: form Cartesian product sets.

Example 5.1. If $A = \{H, T\}$ and $B = \{1, 2, 3\}$, then

$A \times B = \{(H, 1), (H, 2), (H, 3), (T, 1), (T, 2), (T, 3)\}$,
$B \times A = \{(1, H), (1, T), (2, H), (2, T), (3, H), (3, T)\}$,
$A \times A = \{(H, H), (H, T), (T, H), (T, T)\}$,
$B \times B = \{(1, 1), (1, 2), (1, 3), (2, 1), (2, 2), (2, 3), (3, 1), (3, 2), (3, 3)\}$.

This example enables us to indicate the reason for the importance of Cartesian product sets in probability. Consider the following experiments: (1) toss a coin, and (2) choose a number from among the first three positive integers. Each element of A represents an outcome of the coin-tossing experiment, each element of B an outcome of the second experiment. Now think of the composite experiment in which we first toss a coin and then choose a number. Outcomes can be represented by ordered pairs, like (H, 2), indicating the result of each part of the composite experiment. Thus the outcomes of the composite experiment are given by the Cartesian product set $A \times B$.

Note that $B \times A$ is not the same set as $A \times B$. The set $B \times A$ yields outcomes of the different composite experiment in which we first choose a number and then toss a coin. If we toss the coin twice, we obtain an outcome represented by an element of $A \times A$. Finally, if we choose a number and then choose another number, each from the set B, then outcomes of this composite experiment correspond to elements of $B \times B$. We hasten to add that not all composite experiments lead to Cartesian product sets. For example, if we choose a number from the set B and then choose another number from the *remaining* numbers, the set of possible outcomes is

$$\{(1, 2), (1, 3), (2, 1), (2, 3), (3, 1), (3, 2)\},$$

which is *not* a Cartesian product set, although its elements are ordered pairs. These matters are taken up more fully in the next chapter.

Example 5.2. We have previously mentioned our interpretation of ordered pairs of real numbers as points in a plane. If R is the set of all real numbers, then there is one and only one point corresponding to each ordered pair in $R \times R$, and one and only one ordered pair in $R \times R$ corresponding to each point. Indeed, this one-to-one correspondence between points and ordered pairs of real numbers is the fundamental idea of plane analytic geometry. A plane with axes is called a Cartesian plane, after René Descartes (1596–1650), one of the inventors of analytic geometry. The graph of any subset of $R \times R$ is defined as the set of points corresponding to ordered pairs of the subset. For example, the set

$$\{(x, y) \mid x^2 + y^2 = 9\}$$

is a subset of $R \times R$ whose graph is the circle with center at the origin and radius 3 units. The graph of $R \times R$ itself is the entire plane.

We can define ordered triples, and in general, ordered r-tuples, in terms of ordered pairs. An ordered triple (or 3-tuple), for example, is an ordered pair whose first member is an ordered pair, i.e.,

(5.4) $$(a, b, c) = ((a, b), c).$$

Similarly, we define ordered quadruples (4-tuples) as an ordered pair whose first member is an ordered triple, i.e.,

$$(a, b, c, d) = ((a, b, c), d).$$

In general, an ordered r-tuple is defined as follows:

(5.5) $$(a_1, a_2, \cdots, a_r) = ((a_1, a_2, \cdots, a_{r-1}), a_r).$$

From these definitions it can be proved that two ordered r-tuples are equal if and only if their corresponding objects are equal, i.e.,

$$(a_1, a_2, \cdots, a_r) = (b_1, b_2, \cdots, b_r)$$

if and only if

$$a_1 = b_1, \quad a_2 = b_2, \quad \cdots, \quad a_r = b_r.$$

We leave the proof for the problems.

Since the Cartesian product of two sets A and B is a set of ordered pairs (2-tuples) it comes as no surprise that we can define the Cartesian product of r sets as a set of r-tuples.

Definition 5.3. We suppose that r is a positive integer greater than 1 and that $A_1, A_2, \cdots, A_r$ are sets. The set of all ordered r-tuples $(a_1, a_2, \cdots, a_r)$ such that a_1 belongs to A_1, a_2 belongs to A_2, $\cdots$, a_r belongs to A_r is called the Cartesian product of the sets $A_1, A_2, \cdots, A_r$, and is denoted by $A_1 \times A_2 \times \cdots \times A_r$. In symbols,

$$A_1 \times A_2 \times \cdots \times A_r = \{(a_1, a_2, \cdots, a_r) \mid a_j \in A_j \text{ for } j = 1, 2, \cdots, r\}.$$

Example 5.3. Suppose $A = \{H, T\}$. Then the Cartesian product set $A \times A \times A$ is the set of ordered 3-tuples in which each of the three objects is either an H or a T. One such 3-tuple is (H, T, H); there are eight altogether. As in the discussion following Example 5.1, each such 3-tuple can be thought of as representing an outcome of the composite experiment in which a coin is tossed three times.

If the number of elements in each of the sets $A_1, A_2, \cdots, A_r$ is given, then we ought to be able to determine the number of ordered

r-tuples in the Cartesian product of these sets. We denote by $n(A)$ the number of elements in the set A.

Theorem 5.2. If r is any positive integer and $A_1, A_2, \cdots, A_r$ any sets, then

$$(5.6) \qquad n(A_1 \times A_2 \times \cdots \times A_r) = n(A_1)n(A_2) \cdots n(A_r).$$

Proof. There are as many elements in $A_1 \times A_2 \times \cdots \times A_r$ as there are r-tuples $(a_1, a_2, \cdots, a_r)$ where $a_1 \in A_1, a_2 \in A_2, \cdots, a_r \in A_r$. The object a_1 can be chosen in $n(A_1)$ different ways. We can then choose the object a_2 in $n(A_2)$ different ways, and we continue in this way until we come to the last object a_r which can be chosen in $n(A_r)$ different ways. Hence, by the fundamental principle of counting (p. 9), the number of r-tuples is given by the product $n(A_1)n(A_2) \cdots n(A_r)$ and the theorem is proved.

Example 5.4. Let

$$A = \{H, T\} \quad \text{and} \quad B = \{1, 2, 3, 4, 5, 6\}.$$

Then the Cartesian product set $A \times A \times B$ has

$$n(A)n(A)n(B) = 2 \cdot 2 \cdot 6 = 24 \text{ elements.}$$

Each element is a 3-tuple which can be interpreted as representing one of the 24 possible outcomes of the experiment in which we throw a coin twice and then roll a die. For example, $(H, H, 6)$ would denote that both tosses resulted in heads and the die showed the number 6 on its uppermost face.

PROBLEMS

5.1. Let $A = \{1, 2\}$, $B = \{2, 3\}$, and $C = \{3\}$ be subsets of the universal set $\mathfrak{U} = \{1, 2, 3\}$. List the elements of the following sets:

(a) $A \times A$ (b) $C \times C$
(c) $A \times B$ (d) $B \times A$
(e) $(A \times B) \cap (B \times C)$ (f) $(A \times B) \cup (B \times C)$
(g) $(A \times \mathfrak{U}) \cap (\mathfrak{U} \times B)$ (h) $A \times B \times C$

5.2. Sketch the graph of the sets in (a)–(d) of the preceding problem. How would you sketch the graph of the set in part (h)?

5.3. (a) Show that $A \times B = B \times A$ if $A = B$. Is the converse true?
(b) Show that $A \times B = \emptyset$ if and only if $A = \emptyset$ or $B = \emptyset$.
(c) Show that $A \times B \subseteq C \times D$ if $A \subseteq C$ and $B \subseteq D$. Is the converse true?

5.4. (a) Prove that $A \times (B \cap C) = (A \times B) \cap (A \times C)$.
  (b) Prove that $A \times (B \cup C) = (A \times B) \cup (A \times C)$.

5.5. Let A and B be subsets of some universal set $\mathfrak{U}$. Prove that

$$(A \times \mathfrak{U}) \cap (\mathfrak{U} \times B) = A \times B.$$

5.6. The ordered pair (a, b) is defined as a certain set by Formula (5.2). How many different elements does the set (a, b) contain? (Do not fail to consider the case when $a = b$.)

5.7. (a) With ordered 3-tuples defined by Formula (5.4), show that

$$(a, b, c) = (d, e, f)$$

  if and only if $a = d$, $b = e$, and $c = f$.
  (b) More generally, if r is any positive integer greater than 1, prove by mathematical induction that two ordered r-tuples are equal if and only if their corresponding objects are equal.

5.8. Let A be a set with n elements. How many elements are in each of the following sets?

  (a) $A \times A$
  (b) $\{(x, y) \mid x \in A, y \in A, \text{ and } x \neq y\}$
  (c) $A \times A \times A$
  (d) $\{(x, y, z) \mid x \in A, y \in A, z \in A, x \neq y, x \neq z, y \neq z\}$
  (e) For each set in (a)–(d), describe an experiment whose outcomes can be represented by elements of the set.

SUPPLEMENTARY READING

1. Breuer, J., *Introduction to the Theory of Sets*, translated by H. F. Fehr, Prentice-Hall, Inc., 1958.

2. Kemeny, J. G., J. L. Snell, and G. L. Thompson, *Introduction to Finite Mathematics*, Prentice-Hall, Inc., 1957.

3. Mathematical Association of America, Committee on the Undergraduate Program, *Elementary Mathematics of Sets*, 1958.

4. May, K. O., *Elements of Modern Mathematics*, Addison-Wesley Publishing Company, Inc., 1959.

5. Suppes, P., *Introduction to Logic*, D. Van Nostrand Company, Inc., 1957.

Chapter 2

PROBABILITY IN FINITE SAMPLE SPACES

---

1. Sample spaces

Probability questions arise when we think of real or conceptual experiments and their outcomes. Therefore, our first task in the precise formulation of probability theory must be to discover a suitable mathematical way by which an experiment can be specified.

Think of tossing a coin. We ordinarily agree to regard "head" and "tail" as the only possible outcomes. If we denote these outcomes by H and T respectively, then each outcome of the experiment would correspond to exactly one of the elements of the set {H, T}. This *set* is called a *sample space* for the experiment.

Now let us toss a penny and a nickel. How shall we record the outcome of this experiment? Each time we toss the coins, we can write down the number of heads obtained. Accordingly, each outcome of the coin-tossing experiment corresponds to exactly one of the elements of the set $S_1 = \{0, 1, 2\}$. S_1 is a sample space for the experiment. We say *a* rather than *the* sample space, since we can think of other ways of describing the outcomes of this same experiment. Indeed, were we to toss the coins and record, let us say, only that we obtained one head, we are then embarrassed by the question, "Did the penny fall heads?" Our method of classifying outcomes was too coarse; we lost information by merely recording the number of heads obtained.

We get a finer classification by recording whether both coins fall heads (HH), the penny falls heads and the nickel tails (HT), the penny falls tails and the nickel heads (TH), or both coins fall tails (TT). Now each outcome of the experiment corresponds to exactly one element in the set

$$S_2 = \{\text{HH, HT, TH, TT}\}.$$

S_2 is another sample space for this experiment. We recognize S_2 as the Cartesian product $A \times A$, where $A = \{\text{H, T}\}$ and where we have introduced simplified notation for ordered pairs, writing HH for (H, H), HT for (H, T), etc. When, as in this example, there is no possibility of misinterpretation, we shall often use this less cumbersome notation for ordered r-tuples.

This situation is typical of most examples. Whether to classify outcomes one way or another is not a question our theory answers. Let us therefore agree at the outset that there is no one correct sample space for a given experiment. Different people or even the same person at different times may describe the outcomes differently. We insist only that any sample space meet the requirements in the following definition.

Definition 1.1. A *sample space* S associated with a real or conceptual experiment is a set such that (1) each element of S denotes an outcome of the experiment, and (2) any performance of the experiment results in an outcome that corresponds to one and only one element of S.

Although many sample spaces may meet these requirements, and hence serve to describe the same experiment, we have seen that one may be more suitable than another. In general, it is a safe guide to include as much detail as possible in the description of the outcomes of the experiment. Imagine that you are recording the outcome in a notebook and insist that what you write enables you to answer all pertinent questions concerning the result of the experiment.

Example 1.1. Let a green die and a red die be rolled. The set

$$S_1 = \{0 \text{ sixes, exactly 1 six, 2 sixes}\}$$

is a set that meets the requirements of Definition 1.1, and hence can serve as a sample space for this experiment. So can the set

$$S_2 = \{2, 3, 4, 5, 6, 7, 8, 9, 10, 11, 12\},$$

if we understand that an element of S_2 stands for the sum of the numbers on the dice. But neither S_1 nor S_2 involves a fine enough classification of outcomes to answer the question, "Is the number on the red die greater than the number on the green die?" To take care of all relevant questions, we should record the numbers on each of the dice. We are thus led to take as sample space the set S (defined on p. 12) containing 36 ordered pairs, it being understood that (x, y) denotes the outcome in which the green die shows the number x and the red die the number y. Since x and y are themselves integers in the set

$$D = \{1, 2, 3, 4, 5, 6\},$$

the sample space S can be written as a Cartesian product:

(1.1) $$S = D \times D = \{(x, y) \mid x \in D \text{ and } y \in D\}.$$

Note that D itself can serve as a sample space for the experiment in which *one* die is rolled. Finally, let us observe that

$$S_3 = \{0 \text{ sixes}, 2 \text{ sixes}\}$$

and

$$S_4 = \{0 \text{ sixes}, (1, 6), \text{ exactly 1 six}, (6, 6)\}$$

are examples of sets that cannot serve as sample spaces for the two-dice experiment. Both sets violate condition (2) in Definition 1.1: the outcome $(1, 6)$, for example, corresponds to no element of S_3 and to two elements of S_4.

A sample space can be an infinite set. For example, toss a coin until it falls heads for the first time. It is logically conceivable that we get an unending sequence of tails and that a head is never obtained. Call this outcome ω. If a head is obtained, we specify the outcome by recording the number of the toss that produced the first head. Our sample space is

$$S = \{\omega, 1, 2, 3, \cdots\},$$

which is clearly an infinite set. As another example, let an experiment consist of selecting one point from among the points on some line of unit length. (This conceptual experiment can be carried out, at least with our mind's eye, by imagining an exceptionally pointed dart thrown at a line segment.) Since we can associate a unique real number with each point on the line, we can take as sample space the infinite set

$$S = \{x \,\epsilon\, R \mid 0 \le x \le 1\},$$

where R denotes the set of all real numbers.

As our discussion indicates, one way of formulating in a precise way the notion of an experiment is to write down an associated sample space. Since we shall always speak of the probability of an event in connection with some real or conceptual experiment, *our mathematical theory begins with the specification of a sample space by which we define the experiment.* Although the general theory of probability deals with both finite and infinite sample spaces, *in this book we restrict our attention to finite sample spaces only.*

Example 1.2. From a large group of people, r are selected and their birthdays (but not birth years) are recorded. We want to specify a sample space associated with this experiment. Let us number the days of the year 1, 2, 3, $\cdots$, 365 and omit people born in leap years on February 29. A typical outcome of the experiment might be the ordered r-tuple (17, 3, 131, $\cdots$, 78), the first number of which is the birthday of the first person selected, the second number the birthday of the second person selected, etc. Each birthday is an element of the set

$$A = \{1, 2, 3, \cdots, 365\},$$

but an outcome of the experiment is recorded only when we write down an r-tuple of numbers, each selected from A. (Cf. Problem I.2.20.)* Hence we define the sample space given by the Cartesian product set

$$(1.2) \quad S = A \times A \times \cdots \times A = \{(x_1, x_2, \cdots, x_r) \mid x_i \,\epsilon\, A$$
$$\text{for } i = 1, 2, \cdots, r\}.$$

The sample space S contains 365^r elements. If $r \ge 4$, then S contains more than a billion r-tuples. Nevertheless, S is a finite sample space for all values of r. Therefore, probability questions about this

---

* To refer in any chapter to a theorem, definition, example, problem, or formula in that *same* chapter, we use only the number by which it is identified in the text. But to refer to one of these items appearing in *another* chapter, we prefix its identifying number with a roman numeral identifying the chapter. For example, we write Problem I.2.20 to denote the twentieth problem in Section 2 of Chapter 1. Were we to write Problem 2.20 here, we would mean to refer to the twentieth problem in Section 2 of the present chapter, namely Chapter 2.

experiment will be answered by our theory. (We continue this problem in Example 2.1.)

Example 1.3. What sample space should we use for the experiment in which a bridge hand is dealt from an ordinary deck of cards? Since we care only about which 13 cards make up the hand, and not about the order in which they are dealt, we can consider a bridge hand as a subset of 13 cards from the set of 52 cards in the deck. Let us write A_s, K_s, $\cdots$, 2_s to denote the ace, king, $\cdots$, deuce of spades, reserving subscripts h, d, and c to indicate cards that are hearts, diamonds, and clubs, respectively. Then

$$(1.3) \quad D = \{A_s, \cdots, 2_s, A_h, \cdots, 2_h, A_d, \cdots, 2_d, A_c, \cdots, 2_c\}$$

is a set of 52 elements representing the full deck. For our experiment, we take as sample space the set S of all 13-element subsets of D. In symbols,

$$(1.4) \quad S = \{B \mid B \subseteq D \text{ and } n(B) = 13\}.$$

Probability questions concerning bridge hands will be answered by our theory, since S is a finite set. The problem of determining $n(S)$, i.e., of counting the number of possible bridge hands, is taken up in Chapter 3.

PROBLEMS

1.1. We describe certain experiments. In each case specify an appropriate sample space for this experiment.

(a) A card is selected from a standard deck of cards.
(b) Three coins are tossed.
(c) A boy has a penny, a nickel, a dime, and a quarter in his pocket. He takes two coins out of his pocket, one after the other.
(d) Two distinguishable objects are distributed in two numbered cells.
(e) Two indistinguishable objects are distributed in two numbered cells.
(f) A survey of families with two children is made and the sexes of the children (the older child first) are recorded.
(g) A survey of families with three children is made and the sexes of the children (in order of age, oldest child first) are recorded.
(h) A survey of families with r children is made and the sexes of the children (in order of age, oldest child first) are recorded.
(i) r coins are tossed.
(j) A poker hand (five cards) is dealt from an ordinary deck of cards.

1.2. Six boys in a club select a committee of three. The boys are A, B, C, D, E, and F.

   (a) List the 20 elements of the appropriate sample space S for this experiment.

   (b) Find the subset of S containing those outcomes in which A is selected. How many elements does this subset contain?

   (c) Find the subset of S containing those elements in which both A *and* B are selected. How many elements does this subset contain?

   (d) Find the subset of S containing those outcomes in which A *or* B is selected. How many elements in this subset?

   (e) Find the subset of S containing those outcomes in which A is *not* selected. How many elements in this subset?

1.3. Refer to part (c) of Problem 1.1. For how many outcomes of the sample space is it the case that the boy takes less than 15 cents out of his pocket?

1.4. Refer to the sample space S of 36 elements in Example 1.1 of the text. Let E denote the subset of S whose elements denote outcomes for which the sum of the numbers on the dice is greater than 9, and F the subset whose elements denote outcomes for which the numbers on the dice are equal. Determine the elements in the following sets:

   (a) $E \cap F$ (d) $E \cap F'$
   (b) $E \cup F$ (e) E'
   (c) $E' \cap F$ (f) F'

1.5. An experiment consists of selecting one chip from a hat containing six chips numbered 1, 2, 3, 4, 5, and 6. Of the following sets, state which are suitable sample spaces for this experiment and which are unsuitable.

   (a) $S = \{1, 2, 3, 4, 5, 6\}$
   (b) $S = \{1, 2, 3, 4, 5\}$
   (c) $S = \{\text{odd number, even number}\}$
   (d) $S = \{1, 3, 5, \text{even number}\}$
   (e) $S = \{1, 2, \text{number less than 6}, 6\}$
   (f) $S = \{\text{number less than 3}, 3, \text{number greater than 3}\}$

1.6. An experiment consists of selecting r light bulbs from the lot produced by a machine and testing them. A bulb can be good (G) or bad (B). Define a sample space for this experiment and compare your sample space with those in Problems 1.1 (h) and 1.1 (i). What observation do you make and what lesson is learned thereby?

1.7. Urn 1 contains one black and two white balls. Urn 2 contains two black and one white ball. An experiment consists of first selecting an urn and then drawing a ball from this urn. Define a suitable sample space for this experiment.

2. Events

The theory of probability begins when a sample space S, the mathematical counterpart of an experiment, is specified. *The sample space serves as the universal set for all questions concerned with the experiment.*

We may be interested in the occurrence of a variety of events when an experiment is under consideration. For example, think of the experiment of tossing a coin three successive times and let

(2.1) $S = \{$HHH, HHT, HTH, THH, HTT, THT, TTH, TTT$\}$

be the associated sample space. We may be interested in the event "the number of heads exceeds the number of tails." For any outcome of the experiment we can determine whether this event does or does not occur. We find that HHH, HHT, HTH, and THH are the only elements of S corresponding to outcomes for which this event does occur; if the experimental outcome corresponds to one of the other elements of S, then the event in question does not occur. Thus, to say that the event "the number of heads exceeds the number of tails" occurs is the same as saying the experiment results in an outcome corresponding to an element of the set

$$A = \{\text{HHH, HHT, HTH, THH}\}.$$

We recognize A as a *subset* of the sample space S. The subset A can be taken as the mathematical counterpart of the event "the number of heads exceeds the number of tails." Similarly, we find the following correspondence between various *events* and *subsets* of S:

Verbal Description of Event	*Corresponding Subset of S*
Number of heads exceeds number of tails	$A = \{$HHH, HHT, HTH, THH$\}$
Number of heads is exactly 2	$B = \{$HHT, HTH, THH$\}$
Number of heads is at least 2	$C = \{$HHH, HHT, HTH, THH$\} = A$
Second toss is heads	$D = \{$HHH, HHT, THH, THT$\}$
All tosses show the same face	$E = \{$HHH, TTT$\}$
Number of heads is less than 2	$C' = \{$HTT, THT, TTH, TTT$\}$
Second toss is not heads	$D' = \{$HTH, HTT, TTH, TTT$\}$
Second toss is heads and the number of heads is exactly 2	$D \cap B = \{$HHT, THH$\}$

Second toss is heads or the
    number of heads is exactly 2 $D \cup B$ = {HHH, HHT, HTH, THH, THT}

In the light of this example, we introduce the following general terminology.

Definition 2.1. Let a sample space S be given. An *event* is a subset of S. We say the *event E occurs* if the outcome of the experiment corresponds to an element of the subset E.

Because of this definition, the language and notation of set theory can be expected to find extensive use in the theory of probability. To illustrate this point, suppose an experiment specified by the sample space S results in an outcome denoted by the element $o \in S$. The reader must be certain that he understands the correspondence between the everyday language on the left and the set language and symbolism on the right in the following glossary:

Event E	Subset E of the sample space S
Event F	Subset F of the sample space S
Event E occurs	$o \in E$
Complementary event of E (not–E)	E' (the complement of set E)
Event E does not occur	$o \in E'$
Event E or event F	$E \cup F$ (the union of sets E and F)
Either event E or event F occurs (at least one of E and F occurs)	$o \in (E \cup F)$
Event E and event F	$E \cap F$ (the intersection of sets E and F)
Both event E and event F occur	$o \in (E \cap F)$
Event E is impossible	$E = \emptyset$
Event E is certain	$E = S$
E and F are mutually exclusive events	$E \cap F = \emptyset$
If event E occurs, then event F occurs (E implies F)	$E \subseteq F$ (E is a subset of F)

Because of its intuitive appeal, we shall continue to use the everyday language listed in the left column. But let us recognize that each such phrase is *defined* by the corresponding set-theoretic equivalent in the right column and is thereby given precise meaning within our mathematical theory.

Example 2.1. We return to the experiment in which r people are selected and their birthdays recorded. In Example 1.2, we defined a sample space S for this experiment and found that it had 365^r elements. Now let E be the event that at least two among the r people selected have the same birthday. We want to determine $n(E)$, the number of elements in the subset E. It turns out to be easier to calculate $n(E')$, the number of elements in the complementary event of E. We then use the formula (Cf. Problem I.3.7)

$$n(E) + n(E') = n(S) = 365^r$$

to determine $n(E)$.

Now E' is the event that no two among the r people selected have the same birthday. Hence, $n(E')$ is equal to the number of ways of selecting r *different* numbers (birthdays), each being chosen from the full set of 365 possible birthdays. The first man's birthday can be chosen in 365 ways, the second man's in 364 ways, the third man's in 363 ways, and so on until we select the rth man. His birthday, in order for it to be different from all the others, can be chosen in only $365 - (r - 1)$ or $365 - r + 1$ ways. Invoking the fundamental principle of counting, we conclude that

$$n(E') = 365 \cdot 364 \cdot 363 \cdots (365 - r + 1).$$

Finally, we find

(2.2) $n(E) = 365^r - 365 \cdot 364 \cdot 363 \cdots (365 - r + 1)$

for the number of different ways that the r selected people include at least two having the same birthday. (Continued in Example 3.6.)

PROBLEMS

2.1. Refer to the sample space of Problem 1.1, part (a), and determine the subsets defining the following events.

(a) The card selected is a spade.
(b) The card selected is a jack, queen, or king.
(c) The card selected is the ace of spades.

2.2. A green and a red die are thrown. (Cf. Example 1.1.) Let A be the event that the sum of the numbers on the faces is even, and B the event that the number on the green die is odd.

(a) List the elements of subsets A and B.
(b) Give a concise verbal description of the event $A \cap B$.
(c) How many elements of the sample space S are in the event $A \cup B$?

2.3. Refer to the sample space S in Example 1.2. Let E be the event that the first person selected is born on January 3 and F that the second person selected is born on January 28. (a) Write E, F, and $E \cap F$ as Cartesian product sets. (b) Count the number of elements in E, F, $E \cap F$, and $E \cup F$.

2.4. Let A, B, and C be any events of a sample space S. Using only the symbols $\cap$, $\cup$, $'$, A, B, C, write expressions for the events that of A, B, and C:

(a) At least one occurs.   (b) Only A occurs.
(c) A and B occur, but not C.   (d) All three occur.
(e) None occurs.   (f) Exactly one occurs.
(g) Exactly two occur.   (h) At most two occur.

(*Hint:* Refer to Figure 7 on p. 21 and determine the region representing each event.)

2.5. Refer to the sample space of Problem 1.1(e) and let E be the event that the first cell is empty, F the event that the second cell is empty, and G the event that the second cell contains both objects. Show that the following relations among these events are true.

(a) $E \cap F = \emptyset$ (c) $F \subseteq G'$
(b) $E \subseteq G$ (d) $S = E' \cup F'$

2.6. Let S be the sample space defined in Problem 1.7. Suppose E is the event that the first urn is selected, and F that a white ball is drawn. Describe the following events in words, and list their elements: $E \cap F$, E', $E' \cap F$, F', $E \cup F$.

3. The probability of an event

If a (real or conceptual) experiment is under consideration, there are many events in whose occurrence we may have some interest. Using our mathematical language, this amounts to saying that if a sample space S is given, then we can form many subsets of S. In fact, if

$$(3.1) \qquad\qquad S = \{o_1, o_2, \cdots, o_n\}$$

is a finite sample space containing the n elements $o_1, o_2, \cdots, o_n$, then there are 2^n different subsets of S and since each subset is an event, there are 2^n different events. In this section, we are finally in a position to define what is meant by "the probability of an event."

Our first step is to distinguish certain special events that form building blocks from which other events can be constructed.

Definition 3.1. Let a sample space S be given. We shall mean by a *simple event* a *unit subset* of S, i.e., a subset containing only one element of S.

With S defined in (3.1), there are exactly n simple events, *viz.*,

(3.2) $\{o_1\},\ \{o_2\},\ \cdots,\ \{o_n\}.$

The event $\{o_1, o_2\}$ is not a simple event, but it is the union of two simple events:

$$\{o_1, o_2\} = \{o_1\} \cup \{o_2\}.$$

Similarly, the event

$$\{o_2, o_4, o_6\} = \{o_2\} \cup \{o_4\} \cup \{o_6\}.$$

We see that all nonempty events are either simple events or unions of two or more different simple events. In addition to the nonempty events, there is also the null event $\emptyset$. The union of all n simple events enumerated in (3.2) is the entire sample space, i.e.,

$$S = \{o_1\} \cup \{o_2\} \cup \cdots \cup \{o_n\}.$$

Probabilities are assigned to the simple events first.

Definition 3.2. Let the sample space S be defined as in (3.1). To each simple event $\{o_j\}$ we assign a *number* denoted by $P(\{o_j\})$ and called the *probability* of the event $\{o_j\}$. These numbers (probabilities) can be assigned arbitrarily, except that they must satisfy the following two conditions:

(i) The probability of each simple event is a nonnegative number, i.e.,

(3.3) $P(\{o_j\}) \geq 0 \quad (j = 1, 2, \cdots, n).$

(ii) The sum of the probabilities assigned to all simple events of the sample space is 1. In symbols,

(3.4) $\displaystyle\sum_{j=1}^{n} P(\{o_j\}) = P(\{o_1\}) + P(\{o_2\}) + \cdots + P(\{o_n\}) = 1.$

We shall say that an assignment of probabilities to the simple events of S is *acceptable* if it satisfies (i) and (ii).

In view of condition (ii), it is clear that the probability of each simple event is not only at least 0, as required by condition (i), but also at most 1; i.e.,

(3.5) $0 \leq P(\{o_j\}) \leq 1 \quad (j = 1, 2, \cdots, n).$

It is most important to recognize that in spite of these restricting conditions, there are many possible assignments of probabilities to the simple events. We give some examples.

Example 3.1. Let a coin be tossed. We define the sample space $S = \{H, T\}$ and so have two simple events: $\{H\}$ and $\{T\}$. Each of the following assignments of probabilities to these simple events is acceptable:

(1) $P(\{H\}) = P(\{T\}) = \frac{1}{2}$,
(2) $P(\{H\}) = \frac{1}{3}$ and $P(\{T\}) = \frac{2}{3}$,
(3) $P(\{H\}) = 1$ and $P(\{T\}) = 0$.

In fact, if p is *any* real number between 0 and 1 inclusive, then

$$P(\{H\}) = p \quad \text{and} \quad P(\{T\}) = 1 - p$$

is an acceptable assignment of probabilities to the two simple events $\{H\}$ and $\{T\}$. Therefore, we see that there are infinitely many possible acceptable assignments, one for each choice of the number p.

Most people would find the choice $P(\{H\}) = P(\{T\}) = \frac{1}{2}$ the "natural" choice. We do not go into the psychological reasons behind this feeling, but merely make three points:

(1) This choice is neither more nor less acceptable than any other acceptable choice. An assignment of probabilities to simple events is either acceptable or not; there are no degrees of acceptability. Definition 3.2 requires only that we meet the two conditions for assigning probabilities to simple events.

(2) This "natural" choice is *not* dictated by experience with real coins. Experience with real coins shows that they usually fall short of being "ideal" coins: they are not perfectly circular and symmetrically weighted; heads and tails are not equally likely.

(3) Nevertheless, we often *do* choose to develop the theory for such "ideal" or "fair" coins since, as we shall see, the theory then is both logically and psychologically appealing. But more important is the fact that the theory for "ideal" coins supplies a basis for testing real coins and deciding, depending upon the extent to which the predictions of the theory are borne out by the empirical results with the real coin, whether it is reasonable to assume that the real coin is "fair." This last point leads us to expect correctly that probability theory will prove useful in problems of statistical inference.

Example 3.2. A green and red die are rolled, and we use as sample space the set S defined in Example 1.1. There are 36 simple events. The "natural" assignment of probabilities to these simple events is the one in which each simple event is assigned probability $\frac{1}{36}$. This is an acceptable assignment, since both required conditions are fulfilled: the probability of each simple event is nonnegative and the sum of the probabilities of all simple events is 1. Of course, here too there are infinitely many other acceptable assignments of probabilities to the 36 simple events.

Example 3.3. A person is selected from the population of a certain country and asked the question, "Do you think there will be another world war?" We classify each answer into one of the three categories "Yes," "No," "Don't Know." Our sample space S contains three elements, one for each possible answer:

$$S = \{Y, N, DK\}.$$

There are three simple events. This example illustrates the fact that we often do not have any basis for the assignment of probabilities to the simple events of an experiment. In the absence of information about the opinions of people in the country, we can do no more than guess appropriate values for these probabilities. However, we can make the assignment

$$P(\{Y\}) = p, \qquad P(\{N\}) = q, \qquad P(\{DK\}) = r,$$

where we know only that p, q and r are nonnegative real numbers whose sum is 1,

$$p \geq 0, \qquad q \geq 0, \qquad r \geq 0, \qquad p + q + r = 1.$$

If we are told that 60 percent of the population of the country expect another world war, 30 percent do not expect another world war, and 10 percent are uncertain, then it seems natural to choose $p = 0.6$, $q = 0.3$, and $r = 0.1$.

It is now an easy step to define the probability of *any* event. Let a finite sample space S be given, and suppose an acceptable assignment of probabilities has been made for the simple events of S. Let E be *any* event. Then E is either (i) the empty set $\emptyset$, (ii) a simple event, or (iii) the union of two or more different simple events. In case (ii) the probability of E has already been assigned. The following definitions take care of cases (i) and (iii).

Definition 3.3. The probability of the empty set $\emptyset$, denoted by $P(\emptyset)$, is defined to be zero; i.e., $P(\emptyset) = 0$.

Definition 3.4. If E is the union of two or more different simple events, then the probability of E, denoted by $P(E)$, is the sum of the probabilities of those simple events whose union is E. (It is understood that each simple event is counted exactly once.)

Example 3.4. A green and a red die are rolled, and we choose the sample space S containing the by now familiar 36 ordered pairs. Let us assign the probability $\frac{1}{36}$ to each of the 36 simple events of S. If A is the event "sum of numbers on dice is 7," then

$$A = \{(1, 6)\} \cup \{(2, 5)\} \cup \{(3, 4)\} \cup \{(4, 3)\} \cup \{(5, 2)\} \cup \{(6, 1)\}.$$

Hence, by Definition 3.4, we find

$$P(A) = \tfrac{1}{36} + \tfrac{1}{36} + \tfrac{1}{36} + \tfrac{1}{36} + \tfrac{1}{36} + \tfrac{1}{36} = \tfrac{1}{6}.$$

Similarly, if B is the event "sum of numbers on dice is 11," then

$$B = \{(5, 6)\} \cup \{(6, 5)\}$$

and, again by Definition 3.4,

$$P(B) = \tfrac{1}{36} + \tfrac{1}{36} = \tfrac{1}{18}.$$

Example 3.5. A card is selected from a standard deck of 52 cards. We take the set D in (1.3) as sample space, and seek the probability that the card selected is a spade. Let us assign equal probabilities to each of the 52 simple events of D. Then each simple event has probability $\frac{1}{52}$. The event that the card is a spade is the union of the following 13 simple events:

$$\{A_s\}, \{K_s\}, \cdots, \{2_s\}.$$

Hence

$$P(\text{card selected is a spade}) = \underbrace{\tfrac{1}{52} + \tfrac{1}{52} + \cdots + \tfrac{1}{52}}_{13 \text{ times}} = \tfrac{13}{52} = \tfrac{1}{4}.$$

Similarly, the event that the card selected is an ace *or* a spade is the union of the following 16 simple events: $\{A_s\}, \{A_h\}, \{A_d\}, \{A_c\}, \{K_s\}, \{Q_s\}, \cdots, \{2_s\}$. Hence

$$P(\text{card selected is an ace or a spade}) = \underbrace{\tfrac{1}{52} + \tfrac{1}{52} + \cdots + \tfrac{1}{52}}_{16 \text{ times}} = \tfrac{16}{52} = \tfrac{4}{13}.$$

Example 3.6. We continue the birthday problem of Example 2.1, and now compute the probability of the event E that at least two among the r people selected have the same birthday. We assume that each ordered r-tuple of the sample space S is as likely as any other to represent the outcome of the experiment. (This assumption would of course be false if the r people were selected at a convention of twins. But even if the selection is made from the entire population of the United States, the fact that the proportion of all births occurring in a given month varies from month to month still makes our mathematical model only a first approximation to the actual state of affairs.) In any case, we assign to each of the 365^r simple events of S the same probability $1/(365)^r$. Since the number of simple events in E is equal to $n(E)$, the number of elements in E (why?), it follows from Definition 3.4 that $P(E)$ is the sum of $n(E)$ probabilities, each equal to $1/(365)^r$. Hence

$$P(E) = n(E) \frac{1}{(365)^r},$$

and if we substitute the expression for $n(E)$ found in Formula (2.2), we obtain

$$(3.6) \qquad P(E) = 1 - \frac{365 \cdot 364 \cdots (365 - r + 1)}{(365)^r}.$$

The probability $P(E)$ is given (to two decimal-place accuracy) for various values of r in Table 10. Note the rather surprising fact that

TABLE 10

r	10	20	22	23	24	30	40	50
$P(E)$	.12	.41	.48	.51	.54	.71	.89	.97

in as small a group as 23 people, the probability of finding at least two people with the same birthday is greater than $\frac{1}{2}$.

Example 3.7. Two coins are tossed. We define the sample space S given by

$$S = \{HH, HT, TH, TT\}$$

and ask for the probability of the event E that at least one head occurs.

Solution 1. We assign probability $\frac{1}{4}$ to each of the four simple events of S. The event E is the union of three simple events,

$$E = \{HH\} \cup \{HT\} \cup \{TH\}.$$

Hence $P(E) = \frac{1}{4} + \frac{1}{4} + \frac{1}{4} = \frac{3}{4}.$

Solution 2. We assign probabilities to the simple events as follows:

$$P(\{HH\}) = P(\{HT\}) = \frac{1}{2}, \quad P(\{TH\}) = P(\{TT\}) = 0.$$

The event E is still the union of the same three simple events, but now

$$P(E) = \frac{1}{2} + \frac{1}{2} + 0 = 1.$$

These two solutions yield different values for the probability of the same event E. This should not be disturbing since, according to Definition 3.4, the probability of an event depends on the previous assignment of probabilities to the simple events. With different acceptable assignments of probabilities to the simple events, as in Solutions 1 and 2 of Example 3.7, we have no reason to be surprised if an event E turns out to have different probabilities. Which assignment of probabilities to the simple events *should* be made is not a mathematical question, but one that depends upon our assessment of the real-world situation to which the theory is to be applied. The assignment in Solution 1 is the natural one for unbiased, "fair" coins, but the assignment in Solution 2 is more sensible if one is sure that the first coin is loaded so as to turn up heads all the time.

We conclude with three remarks.

(1) Statements in the theory of probability, as in all of mathematics, are of the conditional form; i.e., "If such and such is assumed, then such and such follows." Ordinarily, when we select a card from a full deck and inquire, "What is the probability of selecting a spade?" we answer, as in Example 3.5, "The probability of selecting a spade is $\frac{1}{4}$." But a complete answer would read as follows: "*If* we choose as sample space the set D containing 52 elements, one for each card in the deck, and *if* we assign equal probabilities of $\frac{1}{52}$ to each of the 52 simple events of D, *then* the probability that a spade is selected is $\frac{1}{4}$." The antecedents of this conditional assertion are usually omitted and only the consequent is stated, when it is clear from the context which sample space and which assignment of probabilities to simple events have been chosen. But if two people do the

same problem and get different numbers for the probability of the same event, it becomes important to spell out the hypotheses each used. Each answer may logically follow from the hypothesis employed in its derivation, and the answers may differ because different sample spaces or different assignments of probabilities to simple events were made. However, if the same sample space is used and the simple events are assigned the same probabilities by both people, then two *different* values for the probability of the *same* event can only mean that at least one of the people has committed a logical error.

The situation here is analogous to that in plane geometry. The assertions "The sum of the angles of a triangle is 180 degrees" and "The sum of the angles of a triangle is less than 180 degrees" certainly differ, but we cannot say whether they are true or false, since neither is a complete mathematical assertion. The hypotheses must be explicitly stated or implicitly understood. Thus, the first statement is true if it is intended to read, "If the axioms of Euclidean geometry are accepted, then the sum of the angles of a triangle is 180 degrees." And the second statement is true if we expand it to read, "If the axioms of Lobachewskian geometry are accepted, then the sum of the angles of a triangle is less than 180 degrees." Both Euclidean and non-Euclidean geometries are fruitful mathematical theories, and since their premises differ, nobody is now disturbed when their conclusions also differ. Which geometry *should* be used in a particular context is not a mathematical question, but one that is of great interest to those (like physicists) who apply geometry to the real world.

(2) Our definitions are so framed that *events* and only events have probabilities. Some authors prefer to formulate the theory so that *statements* are assigned probabilities. Except for linguistic differences, the theories are equivalent.

(3) Often, a particular assignment of probabilities to the simple events of a sample space S is implied by the use of certain adjectives. For example, if we say a "fair" coin is tossed, we shall mean that the two simple events {H} and {T} are to receive equal probabilities (each $\frac{1}{2}$). When we say a pair of "fair" or "unbiased" dice are thrown, we mean to insist that each of the 36 simple events of the familiar sample space be assigned probability $\frac{1}{36}$. When we say that a number is selected "at random" from n different numbers, then we agree to assign probability $1/n$ to each of the n simple events of the

appropriate sample space. Further examples will be discussed later, but for the present we adopt this convention in the problems that follow.

PROBLEMS

3.1. Consider the dice experiment of Example 3.4 and make the "natural" assignment of probabilities to the 36 simple events, so that each has probability $\frac{1}{36}$. Find the probability of the following events.

(a) The sum of the numbers on the dice is less than 4.
(b) One die gives a 3 and the other die a number less than 3.
(c) The sum of the numbers on the dice is 2 or 12.

3.2. A letter of the alphabet is chosen at random. Find the probability of the event that the letter selected

(a) is a vowel.
(b) is a consonant.
(c) precedes u (in alphabetical order) and is a vowel.
(d) follows t and is a vowel.
(e) follows v and is a vowel.

3.3. A committee of three is selected from six people A, B, C, D, E, and F. (Cf. Problem 1.2.)

(a) Specify a suitable sample space S and make an acceptable assignment of probabilities to the simple events of S.
(b) Find the probability that A is selected.
(c) Find the probability that A and B are selected.
(d) Find the probability that A or B is selected.
(e) Find the probability that A is not selected.
(f) Find the probability that neither A nor B is selected.

3.4. Let the sample space $S = \{o_1, o_2, o_3, o_4\}$ be given. Probabilities are assigned to the simple events so that

$$P(\{o_1\}) = P(\{o_2\}), \qquad P(\{o_3\}) = P(\{o_4\}) = 2P(\{o_1\}).$$

Find $P(\{o_1, o_3\})$.

3.5. In each of the following, specify an appropriate sample space S, assign probabilities to the simple events of S, and *then* find the required probability.

(a) Find the probability of obtaining exactly two tails if three coins are tossed.

(b) Find the probability that one cell is empty when two distinguishable objects are distributed in two cells. [Cf. Problem 1.1(d).]

(c) Find the probability that one cell is empty when two indistinguishable objects are distributed in two cells. [Cf. Problem 1.1(e).]

(d) Find the probability of finding a family with no boys among families with two children; with three children; with r children. [Cf. Problem 1.1(f)–(h).]

(e) Find the probability of all coins falling heads when r coins are tossed. [Cf. Problem 1.1(i).]

(f) A die is loaded in such a way that the probability of the face marked j turning up is proportional to j for $j = 1, 2, \cdots, 6$. Find the probability that an odd number turns up when the die is rolled.

(g) A month of the year is randomly selected, and we note the day of the week on which the 13th day of the month falls. Find the probability that this 13th day falls on a Sunday.

3.6. Refer to Table 4 (p. 24) reporting the result of a survey of 321 union men. Let one man be selected at random from this group of 321 men. Decide on a suitable sample space and assignment of probabilities to its simple events, and then find the probability that the man selected

(a) answers "yes."

(b) answers "yes" and is in the union four or more years.

(c) answers "don't know" and is in the union less than four years.

(d) answers "don't know" and is in the union over 10 years.

3.7. Refer to Problem 2.3 and find the probability of the events $E \cap F$ and $E \cup F$. Make the same assignment of probabilities to simple events of S as was made in Example 3.6.

3.8. Find the probability that among r people, there will be at least one whose birthday is the same as yours. Use logarithms, or otherwise determine the smallest value of r for which this probability is at least $\frac{1}{2}$.

3.9. Three squares numbered 1, 2, and 3 are marked on a table. A deck of three cards numbered 1, 2, and 3 is shuffled and then dealt so that one card appears in each numbered square. This experiment can be thought of as resulting in a *permutation* of the numbers 1, 2, and 3.

(a) Enumerate the six permutations that make up the sample space.

(b) Assign equal probabilities to each simple event.

(c) If card number j is dealt so as to fall in square number j, we say that a *match* occurs in square j. Let E_j be the event that a match occurs in square j. (Of course, j can be equal to 1, 2, or 3.) Compute the probability of each of the following events: $E_1, E_2, E_3,$ $E_1 \cup E_2, E_1 \cap E_2, E_1 \cap E_2 \cap E_3, E_1 \cup E_2 \cup E_3.$

(d) Is $P(E_1 \cup E_2)$ equal to $P(E_1) + P(E_2)$? Why not? Can you write a correct formula for $P(E_1 \cup E_2)$?

3.10. Find the probability that the player wins in each of the following lotteries. For each lottery, first define a sample space and assign probabilities to its simple events.

(a) Two white and four black balls are placed in an urn and thoroughly stirred. The player draws one ball and wins if the ball he draws is black.

(b) Same as (a), except that the player tosses a coin with one face painted white and the other face painted black just before he draws the ball. He wins if the ball drawn is of the color he tossed with the coin.

4. Some probability theorems

We now derive some consequences of the definitions given in the preceding section. We assume throughout that a finite sample space S is given,

(4.1) $$S = \{o_1, o_2, \cdots, o_n\},$$

and that some acceptable assignment of probabilities to the simple events of S has been made.

Because it will be helpful in visualizing the results to be proved, we pause to reformulate the definitions of the preceding section in a more picturesque language. We use the basic idea of a Venn diagram to represent the sample space S and events (subsets) of S. But now we use dots to indicate elements of S. Each dot determines one simple event, namely, the simple event containing as its only member the element represented by the dot. We imagine a flag erected at each dot and on this flag we write the probability assigned to the simple event determined by this dot. The flag erected at the dot representing outcome o_1 flies the number $P(\{o_1\})$, the flag erected at the dot representing

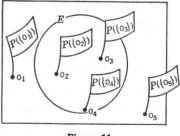

Figure 11

outcome o_2 flies the number $P(\{o_2\})$, etc. See Figure 11. Because of Definition 3.2, the number on each flag is nonnegative and the sum of the numbers on all the flags is 1. We now show that we can para-phrase our definitions as follows: *The probability of E is the sum of the numbers on the flags erected at elements of E.*

If $E = \emptyset$, then no dots, and therefore no flags, appear in the region representing E. In order to have $P(\emptyset) = 0$, as required by Definition 3.3, we shall agree to say that the sum of the numbers on the flags is 0 when there are in fact no flags. If E is a simple event, then exactly one dot appears in the region representing E and the flag erected at that dot carries precisely the number $P(E)$. If E is the union of two or more, say x, different simple events, then the region representing E contains exactly x dots and, when we add the numbers on the flags erected at these x dots, we are adding the probabilities of the simple events whose union is E. By Definition 3.4, we obtain $P(E)$. Thus, we have shown that the italicized phrase concluding the preceding paragraph is correct for all events E. We shall therefore feel free to use this picturesque "numbers on flags" language whenever it seems helpful.

Theorem 4.1. $P(S) = 1$; i.e., the probability of a certain event is 1.

Proof. The sample space S is the union of all n simple events,

$$S = \{o_1\} \cup \{o_2\} \cup \cdots \cup \{o_n\}.$$

Hence, by Definition 3.4,

$$P(S) = \sum_{j=1}^{n} P(\{o_j\})$$

and this sum is 1 by our agreement (in Definition 3.2) that the sum of the probabilities assigned to all simple events must be 1.

Using our picturesque language, the proof of Theorem 4.1 is equally easy. For to find $P(S)$ we must add the numbers on the flags erected at elements of S. But this means adding the numbers on *all* flags, and we know this sum is 1.

Theorem 4.2. If E and F are events such that $E \subseteq F$, then $P(E) \leq P(F)$; i.e., if E implies F, then the probability of E cannot exceed the probability of F.

Proof. By hypothesis, each element of E is also an element of F. Hence, each simple event among those whose union is E is also a simple event among those whose union is F. Since the probability of each simple event is nonnegative, it follows that the sum of the probabilities of the simple events of F is at least as large as the sum of the probabilities of the simple events of E. But this is precisely the required conclusion.

In a Venn diagram, all the points representing elements of E would also be in the region corresponding to F. To find $P(E)$ we add the numbers on the flags erected at points of E. All of these numbers as well as those, if any, that are erected at points in F but outside of E, are summed to get $P(F)$. Hence $P(E) \leq P(F)$, as before.

Theorem 4.2 says that if event F occurs whenever event E occurs, then the probability of F is at least as large as the probability of E.

Theorem 4.3. If E is any event, then $0 \leq P(E) \leq 1$.

Proof. We have $E \subseteq S$, since an event is by definition a subset of the sample space S. Hence by Theorems 4.1 and 4.2, we conclude that $P(E) \leq P(S) = 1$. Also $\emptyset \subseteq E$, so that Theorem 4.2 yields $P(\emptyset) \leq P(E)$. Since $P(\emptyset) = 0$, our proof is complete.

The extreme values 0 and 1 are worthy of special attention. We know that $P(\emptyset) = 0$ and $P(S) = 1$. Recalling the definition of impossible and certain events given in the glossary on p. 52, we can say that if an event is impossible, then it has probability 0, and if an event is certain, then it has probability 1. But the converse of each of these implications is false; i.e., if $P(E) = 0$ we *cannot* conclude that E is impossible, and if $P(E) = 1$ we *cannot* conclude that E is certain. For example, in Solution 2 of Example 3.7, the event that the first coin falls tails is {TH, TT}, certainly not the empty set. Yet, by our assignment of probabilities to the simple events, this event has probability 0. In that same example, the event {HH, HT} has probability 1, but is not certain since it is not the entire sample space.

The reason for this state of affairs is that we have allowed simple events to be assigned probability 0. If we insisted, as some authors do, that the probability of each simple event must be *positive*, then only the empty event would have probability 0, and only the whole sample space would have probability 1. However, it turns out to be the case in problems involving infinite sample spaces that there must exist events that are not impossible but yet have probability 0. Although we cannot pursue this matter here, our definitions are formulated in such a way that the reader need not be surprised by this fact when he goes on to study probabilities in infinite sample spaces.

Theorem 4.4. Let E and F be two events. Then

(4.2) $\qquad P(E \cup F) = P(E) + P(F) - P(E \cap F).$

In words, the probability that at least one of the events E and F occurs is obtained by adding the probability that E occurs and the probability that F occurs, and then subtracting the probability that both E and F occur.

Proof. First add the probabilities of all simple events containing elements of E. Their sum is $P(E)$. Then add the probabilities of all simple events containing elements of F. Their sum is $P(F)$. In the sum $P(E) + P(F)$ we have included $P(\{o_j\})$ if and only if $o_j \in E \cup F$. But we have added $P(\{o_j\})$ *twice* for every $o_j \in E \cap F$: once in the sum $P(E)$, and again in the sum $P(F)$. The sum of the probabilities of simple events that are counted twice is $P(E \cap F)$. We conclude that $P(E) + P(F) - P(E \cap F)$ is precisely the sum of the probabilities of all simple events in $E \cup F$, each counted once. Since this sum is $P(E \cup F)$, the theorem is proved.

The reader should draw a Venn diagram and test his understanding of this proof by formulating each step in the "numbers on flags" language. Before illustrating how Theorem 4.4 is used in a particular example, we deduce two more results.

Theorem 4.5. If E and F are *mutually exclusive* events, then

(4.3) $\qquad P(E \cup F) = P(E) + P(F).$

Proof. In the result of Theorem 4.4, we have only to note that now $P(E \cap F) = P(\emptyset) = 0.$

Theorem 4.5 says that the probability of the occurrence of at least one of two *mutually exclusive* events is the sum of their individual probabilities. Let us not forget the italicized hypothesis that must be true before using Formula (4.3). The use of Formula (4.2) requires no such caution, since it holds for *any* two events.

Theorem 4.6. Let E and E' be any *complementary* events. Then

(4.4) $\qquad P(E') = 1 - P(E).$

In words, the probability that E does *not* occur is obtained by subtracting from 1 the probability that E does occur.

Proof. E and E' are mutually exclusive events, since $E \cap E' = \emptyset$. Hence by (4.3),

$\qquad P(E \cup E') = P(E) + P(E').$

But $E \cup E' = S$, the entire sample space, and $P(S) = 1$ by Theorem 4.1. Hence

$$1 = P(E) + P(E'),$$

which is equivalent to (4.4).

In our less formal language, (4.4) is merely the result of noting that we obtain the sum of the numbers on all flags in S by adding the sum of the numbers on flags in E to the sum of the numbers on flags not in E.

The following examples illustrate how our formulas can be used to compute probabilities.

Example 4.1. Three coins are tossed. Find the probability of getting at least one head. We assign equal probabilities to the eight simple events of the sample space S defined in (2.1), p. 51. If E is the event "at least one head," then the complementary event E' is "no heads." By Theorem 4.6,

$$\begin{aligned} P(E) &= 1 - P(E') \\ &= 1 - P(\{\text{TTT}\}) \\ &= 1 - \tfrac{1}{8} = \tfrac{7}{8}. \end{aligned}$$

Note that we could have computed $P(E)$ directly by recognizing that E is itself the union of seven simple events. This example is so simple that either method is easy. But often in more complicated problems, the most efficient way to find the probability of an event is first to compute the probability of its complementary event and then use Formula (4.4). Recall that we followed this procedure in solving the birthday problem. Although interested in the event E (*at least* two people have the same birthday) we found it convenient (see Example 2.1) to first study the event E' (*no* two people have the same birthday).

Example 4.2. An integer is chosen at random from the first 200 positive integers. What is the probability that the integer chosen is divisible by 6 or by 8?

Let E be the event "integer selected is divisible by 6" and F the event "integer selected is divisible by 8." We are required to find $P(E \cup F)$. We define $S = \{1, 2, 3, \cdots, 200\}$ and assign probability

$\frac{1}{200}$ to each simple event of S. Now E contains $[\frac{200}{6}]^* = 33$ integers, and is therefore the union of 33 simple events, each with probability $\frac{1}{200}$. Hence $P(E) = \frac{33}{200}$. Similarly, F is the union of $[\frac{200}{8}] = 25$ simple events, so that $P(F) = \frac{25}{200}$. Since there are integers among the first 200 (like 24 and 48) that are divisible by both 6 and 8, the events E and F are *not* mutually exclusive. Hence we must compute $P(E \cap F)$. An integer is divisible by both 6 and 8 if and only if it is divisible by 24, the least common multiple of 6 and 8. There are $[\frac{200}{24}] = 8$ integers among the first 200 that are divisible by 24. Hence $P(E \cap F) = \frac{8}{200}$. By applying Formula (4.2), we find the required probability,

$$P(E \cup F) = \frac{33}{200} + \frac{25}{200} - \frac{8}{200} = \frac{1}{4}.$$

(For a generalization of this result, see Problem 4.11.)

We have seen that in many examples it is reasonable to assign the same probability to each simple event of the sample space. In this circumstance, there is a simple formula for the probability of an event.

Theorem 4.7. Suppose each of the n simple events of the sample space S in (4.1) is assigned the same probability. (This probability must then be $1/n$.) If E is an event containing f elements, then

(4.5) $$P(E) = \frac{f}{n}.$$

In other words, the probability of an event is the ratio of the number of elements in the event to the number of elements in the entire sample space.

Proof. Since E contains f elements, E is the union of f simple events of the sample space S. Hence, directly from the definition, $P(E)$ is the sum of f probabilities, each equal to $1/n$. But this sum is precisely f/n, so that our proof is complete.

If we call an outcome of the experiment "favorable" to E whenever E occurs, then Theorem 4.7 can be paraphrased as follows: If an experiment can result in n *equally likely* outcomes, then the probability of E is the ratio of the number of outcomes favorable to E to the total number of outcomes. This is the classic "definition" of probability

---

* The symbol $[x]$ stands for the greatest integer less than or equal to the number x. Thus,

$$[3.6] = 3, \qquad [\tfrac{4}{5}] = 0, \qquad [\tfrac{23}{4}] = 5, \qquad \text{etc.}$$

given by Laplace (1749–1827), one of the first and most important contributors to the mathematical theory of probability. Let us not forget that this rule for computing probabilities is applicable only when all simple events have been assigned the same probability. Thus, Formula (4.5) does not apply to the wide variety of important problems where it is not reasonable to make this special assignment of probabilities to simple events.

Ordinarily, to compute $P(E)$ we must first determine *which* elements of the sample space are in E, and then we add the probabilities of the corresponding simple events. But when Theorem 4.7 applies, we need only know *how many* elements are in E. It is therefore extremely useful to have effective techniques for counting the elements in sets specified by defining properties. In Example 3.6, for instance, the probability of the event that at least two people have the same birthday was easy to find because we had been able (in Example 2.1) to count the elements in this event by using the fundamental principle of counting. We discuss some other techniques for counting in the next chapter. Until then, our examples will be chosen so as to lead to events whose elements can be counted by explicit enumeration or by use of the fundamental principle.

We conclude this section with a brief discussion of the relation between the probability of an event and "odds" for the event.

Definition 4.1. Let E be any event. We say that *odds for E* are a to b if and only if

$$P(E) = \frac{a}{a + b}.$$

If odds for E are a to b, then *odds against E* are b to a.

Table 11 gives some common odds and corresponding probabilities.

TABLE 11

Odds for E	$P(E)$
1 to 1	$\frac{1}{2}$
2 to 1	$\frac{2}{3}$
3 to 1	$\frac{3}{4}$
3 to 5	$\frac{3}{8}$
1 to 2	$\frac{1}{3}$
12 to 5	$\frac{12}{17}$

Given the odds for E we have only to apply Definition 4.1 to find the probability of E. On the other hand, if $P(E)$ is given, we write it in the fractional form $a/(a + b)$ and then know that the odds for E are a to b. For example, if $P(E) = 0.7$, we first write

$$P(E) = \frac{7}{10} = \frac{7}{7 + 3}.$$

Hence, odds for E are 7 to 3. Since odds are often used to express probabilities of events, it is useful to be able to translate odds to probabilities and vice versa.

PROBLEMS

4.1. Two fair dice are rolled. Find the probability of the event E that the dots on the two uppermost faces do not add to 4. What are odds for E?

4.2. A card is drawn at random from a standard deck of playing cards. Let E be the event "card selected is an ace" and F the event "card selected is a spade."

  (a) Are E and F mutually exclusive events?
  (b) Find the probability that at least one of the events E and F occurs.
  (c) What are odds for the event $E \cup F$?

4.3. A fair die is rolled twice. What are odds for the event that at least one roll yields a number less than 3?

4.4. Odds a to b and c to d are said to be equal if $a{:}b = c{:}d$, i.e., if their ratios are equal. For example, odds of 10 to 5, 4 to 2, and 2 to 1 are equal.

  (a) Show that if odds for two events are equal, then the events have equal probabilities.
  (b) Show that odds against an event E are equal to odds for the complementary event E'.

4.5. Odds for event E are 2 to 1. Odds for $E \cup F$ are 3 to 1. Consistent with this information, what are the smallest and largest possible values for the probability of event F?

4.6. A card is drawn at random from an ordinary deck of 52 cards. This card is replaced, and then another card is selected at random from the full deck.

  (a) Define a suitable sample space for this experiment and assign probabilities to its simple events.

(b) Find the probability that at least one of the cards selected is the ace of spades.

(c) What are the odds for the event that neither card is the ace of spades?

4.7. Repeat the preceding problem, but now assume that the first card is *not* replaced before the second is drawn.

4.8. The output of a machine producing nails is known to contain 2% defectives, the other 98% meeting specifications. From the very large lot of nails produced by the machine, two nails are drawn at random and inspected.

(a) Define a suitable sample space for this experiment and make a reasonable assignment of probabilities to its simple events.

(b) Find the probability that at least one of the nails is defective.

4.9. A high school senior applies for admission to college A and college B. He estimates that the probability of being admitted to A is 0.7, that his application will be rejected at B with probability 0.5, and that the probability of at least one of his applications being rejected is 0.6. What is the probability that he will be admitted to at least one of the colleges?

4.10. If in Theorem 4.2 we make the hypothesis that E is a *proper* subset of F, i.e., that $E \subseteq F$ but $E \neq F$, does it then follow that $P(E) < P(F)$?

4.11. (a) An integer is chosen at random from the first 20 positive integers. What is the probability that the integer chosen is divisible by 6 or 8?

(b) An integer is chosen at random from the first 2000 positive integers. What is the probability that the integer chosen is divisible by 6 or 8?

(c) The result of Example 4.2 in the text together with the results of parts (a) and (b) should lead you to conjecture a general theorem of which these results are special cases. State such a theorem and try to prove it.

4.12. Prove that if E and F are any events, then
$$P(E \cap F) \leq P(E) \leq P(E \cup F) \leq P(E) + P(F).$$

4.13. Let E and F be any two events. Suppose the numbers $P(E)$, $P(F)$, and $P(E \cap F)$ are known. Find formulas in terms of these numbers for the following probabilities. In each case give a verbal description of the event whose probability you are finding.

(a) $P(E' \cup F')$ (b) $P(E' \cap F')$
(c) $P(E' \cup F)$ (d) $P(E' \cap F)$
(e) $P(E \cap F')$ (f) $P((E \cap F)')$

4.14. Generalize Theorem 4.4 by showing that the probability of the occurrence of at least one among three events E_1, E_2, and E_3 is given by

(4.6) $\quad P(E_1 \cup E_2 \cup E_3) = P(E_1) + P(E_2) + P(E_3) - P(E_1 \cap E_2)$
$\qquad\qquad - P(E_1 \cap E_3) - P(E_2 \cap E_3) + P(E_1 \cap E_2 \cap E_3).$

[*Note:* You will find a Venn diagram like the one in Figure 7 helpful in checking that the probability of each simple event making up $E_1 \cup E_2 \cup E_3$ is counted once and only once in the expression on the right in (4.6).]

4.15. From a standard deck we select one card at random. Use Formula (4.6) to find the probability that the card is a spade, an honor card, or a deuce.

4.16. Use Formula (4.6) to find the probability that a number selected at random from the first 200 positive integers is divisible by 6 or 8 or 10.

4.17. We make a definition and then state a theorem. Use Formula (4.6) to prove the theorem.

　　Definition 4.2 Let k be any integer greater than 1. Events E_1, E_2, $\cdots$, E_k are said to be *mutually exclusive in pairs* if and only if all possible pairs of events from E_1, E_2, $\cdots$, E_k are mutually exclusive, i.e., $E_i \cap E_j = \emptyset$ for all $i \neq j$ where i and j can assume the values $1, 2, \cdots, k$.

　　Theorem 4.8 If E_1, E_2, and E_3 are mutually exclusive in pairs, then

(4.7) $\quad P(E_1 \cup E_2 \cup E_3) = P(E_1) + P(E_2) + P(E_3).$

4.18. Suppose we assume only that $E_1 \cap E_2 \cap E_3 = \emptyset$. Show by example that (4.7) does *not* necessarily hold. (Cf. Problem I.4.6b.)

4.19. Prove the following generalization of Theorem 4.8 by mathematical induction.

　　Theorem 4.9 Let k be any integer greater than 1 and suppose the events E_1, E_2, $\cdots$, E_k are mutually exclusive in pairs. Then

(4.8) $\quad P(E_1 \cup E_2 \cup \cdots \cup E_k) = P(E_1) + P(E_2) + \cdots + P(E_k).$

4.20. Modify the hypothesis in Theorem 4.9 so that E_1, E_2, $\cdots$, E_k are *any* events, not necessarily mutually exclusive in pairs. With this weaker hypothesis, prove the following weaker result:

$\qquad P(E_1 \cup E_2 \cup \cdots \cup E_k) \leq P(E_1) + P(E_2) + \cdots + P(E_k).$

4.21. (a) Find the probability of at least one match when using a deck of three cards. (Cf. Problem 3.9.)

　　　　(b) Find the probability of at least one match using four numbered squares and four cards. (First define a sample space and make an acceptable assignment of probabilities to its simple events.)

(c) We want to find a formula for the probability of at least one match when using N squares and a deck of N cards (numbered $1, 2, \cdots, N$), where N is any positive integer. Define a suitable sample space for this experiment, determine the number of elements in this sample space, and then make an acceptable assignment of probabilities to its simple events. Note that we could find the probability of at least one match if we had a formula for $P(E_1 \cup E_2 \cup \cdots \cup E_N)$, where E_j denotes the event that a match occurs at card number j. Can you guess this formula by detecting a pattern in Formulas (4.2) and (4.6)? If not, then first use (4.2) and (4.6) to derive a formula for the special case $N = 4$, and then try guessing again. The proof of the correct general formula and its use to find the probability of at least one match require counting techniques that we have not yet discussed.* But even when we can't complete a problem, it is useful to think about it and try to see what we need to learn in order to be able to complete it. This problem is the famous *problem of rencontre* in probability theory and was originally discussed by the French mathematician Montmort (1678–1719).

5. Conditional probability and compound experiments

Suppose an experiment is performed and we are interested in the probability of some event E. But now assume that we are given additional information, namely, that another event F has occurred. In this section, we discuss how the computation of the probability of E is affected by the information that F is known to have occurred.

It is helpful first to take a close look at an example in which we can find reasonable answers on intuitive grounds. The methods we employ in this simple example will lead us to formulate precise definitions that will become part of our mathematical theory.

Example 5.1. A club with five male and five female charter members elects two women and three men to membership. From the total of 15 members, one person is selected at random. We are interested in two events:

$$E = \text{person selected is a male},$$
$$F = \text{person selected is a charter member}.$$

---

* See W. Feller, *An Introduction to Probability Theory and Its Applications*, 2nd edition, John Wiley and Sons, Inc., 1957, pp. 88–91. For another solution and interesting historical comments, see I. Todhunter, *A History of the Mathematical Theory of Probability*, Chelsea Publishing Co., 1949, pp. 91–93.

As sample space we take a set S of 15 elements, one for each club member. Since the selection is "at random" we assign probability $\frac{1}{15}$ to each simple event of S. Observing that E is the union of eight simple events, F the union of ten simple events, and $E \cap F$ the union of five simple events, we calculate

$$P(E) = \tfrac{8}{15}, \quad P(F) = \tfrac{10}{15} = \tfrac{2}{3}, \quad P(E \cap F) = \tfrac{5}{15} = \tfrac{1}{3}.$$

So far we have nothing new. But now suppose we are informed that the person selected is a charter member. What is the probability of E, now that this fact about the outcome of the experiment has been made known to us? Most people quickly answer that the revised probability of E should be $\frac{5}{10}$. They reason as follows: Since F is known to have occurred, we know that one of the ten charter members was selected. The event E occurs if one of the five *male* charter members is selected. Because the selection is at random, the probability of selecting one of the five males from the ten charter members is $\frac{5}{10}$. If we introduce the symbol $P(E|F)$ to denote this revised or conditional probability of E given F, then

$$P(E) = \tfrac{8}{15} \quad \text{and} \quad P(E|F) = \tfrac{1}{2}.$$

Thus, in this example the probability of E decreases due to the added information that event F has occurred.

This informal and intuitive reasoning can be described in another way. Ordinarily, given a sample space S and an acceptable assignment of probabilities to the simple events of S, we compute the probability of an event E by adding the probabilities of the simple events whose union is E. Since $P(S) = 1$ and $E \cap S = E$, we can write the identity

$$(5.1) \qquad\qquad P(E) = \frac{P(E \cap S)}{P(S)},$$

which shows that $P(E)$ is the ratio of the probability of that part of E included in S (which happens to be all of E) to the probability of S itself (which happens to be 1).

But if we are told that event F has occurred, then the outcomes corresponding to elements of F', the complement of F, are no longer possible. Hence, in the light of our added information about the outcome of the experiment, the event F replaces the sample space S as the set whose elements correspond to all *possible* outcomes of the experiment. With this in mind, observe how reasonable it appears to write, in analogy with (5.1),

(5.2)
$$P(E|F) = \frac{P(E \cap F)}{P(F)},$$

which says that $P(E|F)$, the conditional probability of E given F, is the ratio of the probability of that part of E included in F (which is $E \cap F$) to the probability of F itself.

Applied to the problem in Example 5.1, this ratio is

$$P(E|F) = \frac{\frac{1}{3}}{\frac{2}{3}} = \frac{1}{2},$$

as before.

Formula (5.2) is the basis of our formal definition of conditional probability.

Definition 5.1. Let E and F be two events of a sample space S. Suppose an acceptable assignment of probabilities has been made to the simple events of S in such a way that $P(F) > 0$. Then *the conditional probability of E given F*, denoted by $P(E|F)$, is defined by Equation (5.2). The conditional probability of E given F is undefined if $P(F) = 0$.

Formulas (5.1) and (5.2) show that the role of F in computing $P(E|F)$ is analogous to the role of S in computing $P(E)$. It is helpful to carry this analogy further. When we are told that F has occurred, then F can be considered as a new sample space, since all possible outcomes of the experiment must now correspond to elements of F. Then we must be sure to have the probabilities of the simple events of F add to 1, as they must for any sample space. But they actually add to $P(F)$. If $P(F) = 1$, then no changes are required. However, if $P(F) < 1$, we imagine the probabilities of all simple events of F increased *proportionately* by dividing each by the same number $P(F)$. We thus obtain new probabilities for the simple events of F. In view of the relation between original and new probabilities of simple events of F, Formula (5.2) can be paraphrased as follows: The conditional probability of E given F is the sum of the *new* probabilities of those simple events whose union is the event $E \cap F$, i.e., whose union is the part of E included in the new sample space F.

Thus $P(E|F)$ is simply a probability calculated for events considered as subsets of the new sample space F. It follows that the formulas we proved in Section 4 for probabilities relative to the sample space S apply without modification to conditional probabilities relative to the information that a fixed event F has occurred. (See Problem 5.16.)

Example 5.2. Three fair coins are tossed, one after the other. Let E be the event "at least two heads" and F the event "first coin falls heads." We define the usual sample space containing the eight outcomes HHH, HHT, $\cdots$, TTT and assign each simple event the probability $\frac{1}{8}$. Then E is the union of four simple events, F the union of four simple events, and $E \cap F$ the union of three simple events. Hence $P(E) = P(F) = \frac{4}{8}$ and $P(E \cap F) = \frac{3}{8}$. Thus, the conditional probability of E given F is

$$P(E|F) = \frac{P(E \cap F)}{P(F)} = \frac{\frac{3}{8}}{\frac{4}{8}} = \frac{3}{4}.$$

As expected, the added knowledge that the first coin falls heads *increases* the probability of getting at least two heads. Before this additional information is revealed, $P(E) = \frac{1}{2}$. Afterwards, $P(E|F) = \frac{3}{4}$.

Example 5.3. A person is selected at random from among 321 union men whose opinions were reported in Table 4 on p. 24. Let E denote "man answers yes" and F "man is in union less than one year." Then we compute (as in Problem 3.6),

$$P(E) = \tfrac{246}{321}, \quad P(F) = \tfrac{44}{321}, \quad P(E \cap F) = \tfrac{27}{321}.$$

Therefore,

$$P(E|F) = \frac{\frac{27}{321}}{\frac{44}{321}} = \tfrac{27}{44}.$$

Note that $P(E|F) < P(E)$; i.e., the knowledge that the man is in the union less than one year *decreases* the probability that he answers "yes."

Example 5.4. A card is selected at random from a standard deck. Let E denote "card is a spade" and F "card is an ace." Then

$$P(E) = \tfrac{1}{4}, \quad P(F) = \tfrac{1}{13}, \quad P(E \cap F) = \tfrac{1}{52},$$

and

$$P(E|F) = \frac{\frac{1}{52}}{\frac{1}{13}} = \tfrac{1}{4}.$$

Here we have a case in which $P(E)$ and $P(E|F)$ are *equal*: the knowledge that the card is an ace does not change the probability that it is a spade.

The important but special case when $P(E)$ and $P(E|F)$ are equal, as in Example 5.4, will be discussed in Section 7 where we introduce the concept of independent events. In the remainder of this section, we consider some consequences of Definition 5.1, as well as an application of conditional probabilities to so-called compound or composite experiments.

If $P(E) > 0$, the roles of E and F in (5.2) can be interchanged. Then the conditional probability of F given E is

(5.3) $$P(F|E) = \frac{P(F \cap E)}{P(E)} = \frac{P(E \cap F)}{P(E)},$$

the last equality following from the commutative law for the intersection of two sets.

By solving (5.2) and (5.3) for $P(E \cap F)$, we obtain the following result, sometimes referred to as the *theorem on compound probabilities:*

(5.4) $$P(E \cap F) = P(E)P(F|E) = P(F)P(E|F).$$

Formula (5.4) finds extensive use when we compute probabilities for events defined in terms of a compound experiment. For example, the experiment in which we toss a coin, toss it again and then toss it a third time is an example of a compound experiment with three trials. If we have two urns containing colored balls and we choose an urn and then a ball from that urn, we have performed a compound experiment with two trials. Many experiments are most conveniently described as a compounding of two or more trials: first, something is done (trial number 1); then, after the first trial is completed, something else is done (trial number 2); etc. An example will best serve to illustrate the use of conditional probabilities in such compound experiments.

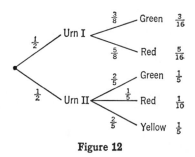

Figure 12

Example 5.5. Urn I contains three green and five red balls. Urn II contains two green, one red, and two yellow balls. We select an urn at random and then draw one ball at random from that urn. What is the probability that we obtain a green ball?

The data of the problem are conveniently summarized in the tree diagram of Figure 12. Since the urn is selected at random, we write probability $\frac{1}{2}$ on each branch leading from the starting point to an outcome of the first trial (number of the urn). We are also given conditional probabilities of drawing a ball of a specified color, given the urn selected. These *conditional* probabilities appear on each branch leading from an urn to an outcome of the second trial (color of ball). We give two solutions to the problem posed in this example.

Solution 1. The event "green ball selected" can occur in one of these two mutually exclusive ways: (1) select urn I and draw a green ball, or (2) select urn II and draw a green ball. Hence the event "green ball selected" is the union of the mutually exclusive events described in (1) and (2). By Formula (4.3), we obtain (with obvious shorthand notation for events),

$$P(\text{green}) = P(\text{urn I and green}) + P(\text{urn II and green}).$$

Each of the terms on the right is the probability of an intersection of two events. Applying Formula (5.4),

$$P(\text{green}) = P(\text{urn I})P(\text{green}|\text{urn I}) + P(\text{urn II})P(\text{green}|\text{urn II}),$$

and using the data summarized in Figure 12,

$$(5.5) \qquad P(\text{green}) = (\tfrac{1}{2})(\tfrac{3}{8}) + (\tfrac{1}{2})(\tfrac{2}{5}) = \tfrac{31}{80}.$$

Solution 2. We go back to first principles. Let us define as sample space for this compound experiment the set

$$S = \{\text{Ig, Ir, IIg, IIr, IIy}\}$$

whose elements are ordered pairs denoting the outcomes of the two trials making up the experiment. Thus, Ig denotes the outcome for which urn I is selected and then the green ball drawn, etc.

Each of the five simple events of S corresponds to one path from left to right through the tree in Figure 12. We assign to each simple event the probability given by the product of the numbers appearing

TABLE 12

Simple Event of S	Probability
{Ig}	$(\tfrac{1}{2})(\tfrac{3}{8}) = \tfrac{3}{16}$
{Ir}	$(\tfrac{1}{2})(\tfrac{5}{8}) = \tfrac{5}{16}$
{IIg}	$(\tfrac{1}{2})(\tfrac{2}{5}) = \tfrac{1}{5}$
{IIr}	$(\tfrac{1}{2})(\tfrac{1}{5}) = \tfrac{1}{10}$
{IIy}	$(\tfrac{1}{2})(\tfrac{2}{5}) = \tfrac{1}{5}$

in the tree along the path to which that simple event corresponds. Using this rule, we obtain the probabilities listed in Table 12 and appearing at the end of each path through the tree in Figure 12. Let us note that this is an *acceptable* assignment of probabilities to the simple events of S: each probability is nonnegative and their sum is 1.

Now the event "green ball selected" is the union of the two simple events {Ig} and {IIg}. Hence

(5.6) $P(\text{green}) = \frac{3}{16} + \frac{1}{5} = \frac{31}{80}$,

as in Solution 1.

The reader may object that in Solution 1 we have violated our rule requiring the designation of a sample space and an assignment of probabilities to its simple events *before* probabilities can be computed. Strictly speaking, this claim is correct. But by comparing (5.5) and (5.6) we observe that the sample space and assignment of probabilities in Solution 2 were implicit in Solution 1. Indeed, let us agree that a compound experiment of n trials will always have as sample space the set S of ordered n-tuples denoting possible outcomes of the experiment. If we are given enough data (in the form of certain probabilities and conditional probabilities) to fill in a tree diagram like the one in Figure 12, then an acceptable assignment of probabilities to simple events of S is made as in Solution 2: Each simple event corresponds to one path through the tree, and the product of the numbers appearing on the branches of a path is the probability assigned to its corresponding simple event. It can be shown (see Problem 5.17) that the resulting assignment of probabilities to the simple events of S is not only acceptable, but is the *only* assignment consistent with the data of the problem.

Solution 1 is shorter, more direct, and easier than Solution 2. It is typical of many problems involving compound experiments that we choose to compute unknown probabilities by using the data of the problem directly, and thus bypass the explicit construction of a sample space and assignment of probabilities to its simple events. We shall adopt the shorter direct solution from now on, but with the knowledge that we could, if called upon to do so, go back to first principles and complete the longer, less direct solution that underlies the shorter procedure.

The theorem on compound probabilities, as expressed in (5.4), is a special case of an extremely useful formula which we now prove.

Theorem 5.1. If n is any integer ($n \geq 2$) and E_1, E_2, $\cdots$, E_n are any n events for which $P(E_1 \cap E_2 \cap \cdots \cap E_{n-1}) \neq 0$, then

$$(5.7) \quad P(E_1 \cap E_2 \cap \cdots \cap E_n) = P(E_1)P(E_2|E_1)P(E_3|E_1 \cap E_2)$$
$$\cdots P(E_n|E_1 \cap E_2 \cap \cdots \cap E_{n-1}).$$

Proof. Denote by S_n the statement expressed by Equation (5.7) and let N denote the set of those integers n for which S_n is true. We use the method of mathematical induction to prove that N is the set of *all* integers greater than 1.

(i) $2 \in N$. For S_2 is the statement

$$P(E_1 \cap E_2) = P(E_1)P(E_2|E_1).$$

That S_2 is true follows from the definition of $P(E_2|E_1)$. Note that our hypothesis reduces to $P(E_1) \neq 0$ when $n = 2$, so that the conditional probability is defined.

(ii) Now assume $k \in N$, where k is any integer greater than 1. We want to prove that also $(k + 1) \in N$. But S_k is the statement

$$(5.8) \quad P(E_1 \cap E_2 \cap \cdots \cap E_k)$$
$$= P(E_1)P(E_2|E_1) \cdots P(E_k|E_1 \cap E_2 \cap \cdots \cap E_{k-1}).$$

We verify that by the definition of conditional probability (and using properties of set intersection),

$$(5.9) \quad \frac{P(E_1 \cap E_2 \cap \cdots \cap E_{k+1})}{P(E_1 \cap E_2 \cap \cdots \cap E_k)} = P(E_{k+1}|E_1 \cap E_2 \cap \cdots \cap E_k).$$

Multiplying corresponding sides of Equations (5.8) and (5.9) yields

$$P(E_1 \cap E_2 \cap \cdots \cap E_{k+1})$$
$$= P(E_1)P(E_2|E_1) \cdots P(E_{k+1}|E_1 \cap E_2 \cap \cdots \cap E_k),$$

which is precisely the statement S_{k+1}. Hence we have shown that if $k \in N$, then $(k + 1) \in N$ for every $k \geq 2$.

We conclude from (i) and (ii) that N is the set of all integers greater than or equal to 2, and thus Theorem 5.1 is proved.

Example 5.6. You are told that an urn contains x red and $5 - x$ green balls, but the value of x (which can be 0, 1, 2, 3, 4, or 5) is not disclosed to you. Mr. Y is to draw a ball at random from the five balls in the urn, and you must guess the color of the ball he draws. We shall say that you win if you guess correctly and lose otherwise.

Let us consider each of the following strategies for making your guess.

Strategy 1. Guess that Y will draw a red ball.

Strategy 2. Guess that Y will draw a green ball.

Strategy 3. First draw a ball from the urn. If it is red, then guess Y will draw a red ball. If it is green, then guess Y will draw a green ball.

Strategy 4. Draw a ball from the urn and replace it. Then draw another ball and replace it. If both balls are red, then guess Y will draw a red ball. If both balls are green, then guess Y will draw a green ball. If you draw one red and one green ball, then draw one more ball from the urn. If this ball is red, then guess Y will draw a red ball. If it is green, then guess Y will draw a green ball.

Strategy 5. Same as Strategy 4, except that the first ball is *not* replaced before the second is drawn. Also, if a third draw is required, it is done *without* replacing the first two balls.

We are interested in calculating the probability that you win, i.e., your guess is correct. This probability will depend on the unknown value of x (which determines the composition of the urn) and on the strategy you decide to use. For example, if you choose Strategy 1, then you guess red. Y draws a red ball from the urn with probability $x/5$. Putting $x = 0, 1, 2, 3, 4, 5$ in turn, we get the probabilities of winning listed in Table 13 in the column headed Strategy 1. In that

TABLE 13

Number of Red Balls in Urn	Probability of Winning with Strategy				
x	1	2	3	4	5
0	0	1	1	1	1
1	.20	.80	.68	.74	.80
2	.40	.60	.52	.53	.54
3	.60	.40	.52	.53	.54
4	.80	.20	.68	.74	.80
5	1	0	1	1	1

table, we list the probability of winning for each possible composition of the urn and for each of the five available strategies. To see how these probabilities are calculated, consider a few examples.

(a) Suppose $x = 2$ and you adopt Strategy 3. Then you win whenever you and Y both draw red balls or both draw green balls. In order to simplify the notation, let us write R_1 to denote the event that the first ball you draw is red, R_Y the event that Y draws a red ball, G_2 the event that the second ball you draw is green, etc. Since the events $R_1 \cap R_Y$ and $G_1 \cap G_Y$ are mutually exclusive,

$$P(\text{win}) = P(R_1 \cap R_Y) + P(G_1 \cap G_Y)$$
$$= P(R_1)P(R_Y|R_1) + P(G_1)P(G_Y|G_1),$$

the last equality following from Formula (5.4). But since Y draws from the full urn which we are assuming contains two red and three green balls,

$$P(R_Y|R_1) = P(R_Y) = .4, \qquad P(G_Y|G_1) = P(G_Y) = .6.$$

Also $\qquad\qquad P(R_1) = .4, \qquad P(G_1) = .6,$

so we compute

$$P(\text{win}) = (.4)(.4) + (.6)(.6) = .52.$$

This probability appears in Table 13 in the row labeled $x = 2$ and column headed Strategy 3. In Figure 13, we have drawn the tree

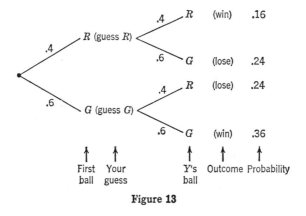

Figure 13

diagram for this experiment when you use Strategy 3. We have written next to each branch the probability that applies when $x = 2$. The reader would do well to follow on the diagram the computation we have just completed.

(b) Suppose $x = 2$ and you adopt Strategy 5. The tree diagram for Strategy 5 is much more complicated, and is drawn in Figure 14. Here too, we have written next to each branch the appropriate probabilities for the case $x = 2$. We note that there are now six mutually exclusive ways of winning. For example, you win if the experiment

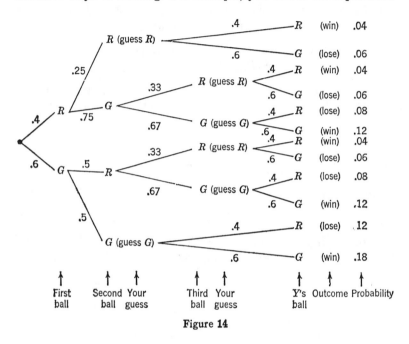

First	Second	Your	Third	Your
ball	ball	guess	ball	guess

Figure 14

results in the event $R_1 \cap G_2 \cap R_3 \cap R_Y$ in which you first draw a red, then a green, then another red (and thus guess red) and then Y draws a red ball. This event corresponds to the third path through the tree in Figure 14. We find the probability of this event by applying Formula (5.7) with $n = 4$.

(5.10) $P(R_1 \cap G_2 \cap R_3 \cap R_Y)$
$$= P(R_1)P(G_2|R_1)P(R_3|R_1 \cap G_2)P(R_Y|R_1 \cap G_2 \cap G_3).$$

But again

$$P(R_Y|R_1 \cap G_2 \cap G_3) = P(R_Y) = .4 = P(R_1).$$

Given that you drew a red ball first, since you do *not* replace this ball, there remain four balls in the urn, of which three are green. Hence

$$P(G_2|R_1) = .75.$$

If you get a red and a green, then the third ball is selected from an urn containing one red and two green balls. Hence

$$P(R_3|R_1 \cap G_2) = \tfrac{1}{3} = .33, \text{ approximately.}$$

Putting these values in (5.10) we find

$$P(R_1 \cap G_2 \cap R_3 \cap R_Y) = (.4)(.75)(.33)(.4) = .04.$$

This probability appears at the end of the third path through the tree in Figure 14 and is just the product of the probabilities for the branches of that path. Adding the probabilities of all events (paths) for which you win, we find that when $x = 2$ and you use Strategy 5 your probability of winning is .54. This number appears in the appropriate row and column in Table 13. In this way all the entries in Table 13 are computed.

All other things being equal, you prefer the strategy which gives you the highest probability of winning. Thus, referring to Table 13, since the probabilities in Column 4 are at least as large as those in Column 3 for all possible compositions of the urn, you prefer Strategy 4 to Strategy 3. For the same reason, Strategy 5 is preferred to Strategy 4. However, if for some reason, you are sure that the urn contains more red than green balls (i.e., $x = 3$, 4, or 5), then you might reasonably prefer Strategy 1 over any of the other strategies. A complete analysis is not possible here, but it is clear that the strategy you prefer will depend upon a number of factors that we have omitted from our discussion. For example, there may be a prize for winning and a penalty for losing. You may have to pay for the information you get by drawing one or more balls from the urn before making your guess. Thus, Strategy 3 may cost you more than Strategy 1 or 2, and Strategies 4 and 5 may cost still more than Strategy 3. The strategy you prefer will depend upon all of these factors, as well as on your belief about the composition of the urn. References 2 and 10 in the supplementary reading list at the end of this chapter may be consulted for discussions of how to evaluate strategies and why this is of great importance in statistical decision theory.

PROBLEMS

5.1. A green and red die are rolled.

(a) Find the conditional probability of obtaining a sum greater than 10, given that the red die resulted in a 5.

(b) Find the conditional probability of obtaining a sum less than 6, given that the red die resulted in a 2.

(c) Find the conditional probability of obtaining sum 7, given that the red die resulted in a number less than 4.

(d) In parts (a)–(c), how does the given information affect the probability of the event in question?

5.2. A fair coin is tossed three successive times. Find the odds for obtaining three heads. How do the odds change if it is given that the second toss resulted in a head?

5.3. Three indistinguishable objects are distributed in three cells. Find the conditional probability that all three occupy the same cell, given that at least two of them are in the same cell.

5.4. A committee of three is selected from six people A, B, C, D, E, and F. [Cf. Problem 3.3.] Find the conditional probability of A and B being selected, given that neither C nor D were selected.

5.5. Two people are selected (one after the other) at random from the 321 union men whose opinions are recorded in Table 4 on page 24. Find the probability that both men answered "yes."

5.6. Students in a summer school program took two courses: Chemistry and History. The registrar reports that 4 percent failed Chemistry, 3 percent failed History, and 1 percent failed both Chemistry and History.

(a) What percentage passed Chemistry and failed History?

(b) Among those who failed Chemistry, what percentage also failed History?

(c) Among those who failed History, what percentage also failed Chemistry?

5.7. (a) A fair coin is tossed three successive times.

    (i) Find the probability that the third toss results in heads.

    (ii) Find the conditional probability that the third toss results in heads, given that the first two tosses result in heads.

(b) A fair coin is tossed N successive times, where N is a positive integer.

    (i) Define a suitable sample space and make the appropriate assignment of probabilities to its simple events.

    (ii) Find the probability that the Nth toss results in heads.

    (iii) Find the conditional probability that the Nth toss results in heads, given that all preceding tosses result in heads. (*Question:* Does the coin have a memory?)

5.8. The manager of a retail grocery store advertises the following promotional scheme in the newspaper. During a specified week, each customer purchasing at least $10 worth of groceries at one time will receive a numbered ticket. At the same time, the cashier places a duplicate ticket in a large bowl. Tickets are numbered serially, starting with number 1. At the end of a week, the tickets in the bowl are thoroughly stirred, one ticket is chosen, and its number determines the winner of a previously announced cash prize. Let us suppose that 200 tickets are distributed during the week.

(a) What is the probability that the first digit of the winning number is 1?

(b) What is the conditional probability that the first digit of the winning number is 1, given that the winning number is greater than 100?

(c) What is the probability that the first digit of the winning number is 9?

(d) What is the probability that the first digit of the winning number is 9, if it is known that the winning number is greater than 100?

(e) Suppose the number of tickets distributed is a positive integer, say N. We want each of the nine digits $(1, 2, 3, \cdots, 9)$ to have the same probability of being the first digit of the winning number. What are all the possible values of N?

5.9. Two defective radio tubes get mixed up with two good ones. You start testing the tubes, one by one, until you have discovered both defectives.

(a) Construct a tree diagram for this experiment.

(b) What is the probability that the second defective tube will be the second tested? the third tested? the fourth tested? What is the sum of the three probabilities you computed? Is this a sum that could have been expected *before* doing the computations?

5.10. An urn contains g green and r red balls. One ball is drawn at random. It is replaced and c more balls of the same color are added to the urn, where c is some positive integer. Another ball is drawn at random from the urn and this ball, together with c more of the same color are again added to the urn. This procedure can be repeated any number of times and supplies a model (first studied by G. Polya) in which the drawing of a ball of either color increases the probability of the same color in the next drawing. (Polya drew the analogy with contagious diseases, where each case of a disease increases the probability of further cases.)

(a) Find the conditional probability that the second ball is red, **given** that the first ball is red.

(b) Find the probability that the first three drawings all result in red balls.

(c) Find the conditional probability that the first ball is red, given that the second ball is red.

5.11. A drawer contains four black, six brown, and two blue socks. Two socks are taken at random from the drawer, one after the other. What is the probability that both socks will be of the same color?

5.12. Use Formula (5.7) to find the probability that r people selected at random will all have different birthdays. Then find the probability that at least two people among the r will have the same birthday, and compare with the answer in (3.6).

5.13. Refer to Table 14, which is a fragment from the American Men Mortality Table published in 1918 by the Actuarial Society of America.

TABLE 14

Rate of Mortality Per 1000						
Age at Issue of Policy	Duration of Policy in Years					
	0	1	2	3	4	5
20	2.73	3.59	3.80	3.96	4.13	4.31
21	2.78	3.66	3.86	4.01	4.18	4.35
22	2.83	3.72	3.91	4.06	4.21	4.39

[For example, the entry 3.96 at age 20, duration of policy 3 years, means that 0.00396 is the probability that a person now aged 23 who was issued insurance at age 20 will die before attaining age 24. Similarly, the entry 3.91 at age 22, duration 2 years, means that 0.00391 is the probability that a person now aged 24 who was issued insurance at age 22 will die before attaining age 25.] Calculate the probability that a man now aged 21 who was issued insurance a year ago will die (a) between ages 21 and 22, (b) between ages 22 and 23, (c) between ages 23 and 24.

5.14. (a) Let n be a positive integer and define $_np_x$ as the probability that a person aged x years will survive n years. Put $p_x = {}_1p_x$ and show that

$$_np_x = p_x p_{x+1} \cdots p_{x+n-1}.$$

(b) Let l_0 be an arbitrary positive integer (an observed number of newborn babies) and define for all $x > 0$,

$$l_x = l_0({}_xp_0), \quad d_x = l_x - l_{x+1}.$$

Give interpretations for l_x and d_x and show that

$$_np_x = \frac{l_{x+n}}{l_x}.$$

(c) Let $q_x = 1 - p_x$. Show that

$$q_x = \frac{d_x}{l_x}.$$

The probability q_x is known in actuarial mathematics as the rate of mortality at age x.

5.15. Let E and F be events and suppose $P(F)$ is neither 0 nor 1. Show that

if $P(E|F) > P(E)$,   then $P(E|F') < P(E)$.

State why this result is intuitively reasonable.

5.16. Prove the following laws, in each case assuming the conditional probabilities are defined.

(a) $P(F|F) = 1$.
(b) $P(\emptyset|F) = 0$.
(c) If $E_1 \subseteq E_2$, then $P(E_1|F) \leq P(E_2|F)$.
(d) $P(E'|F) = 1 - P(E|F)$.
(e) $P(E_1 \cup E_2|F) = P(E_1|F) + P(E_2|F) - P(E_1 \cap E_2|F)$.
(f) $P(E|F') = \dfrac{P(E) - P(E \cap F)}{1 - P(F)}$.
(g) If $P(F) = 1$, then $P(E|F) = P(E)$.
(h) If $P(F) > 0$ and E and F are mutually exclusive events, then $P(E|F) = 0$.

5.17. In Table 12, let the probabilities of the five simple events be $a, b, c, d,$ and e. We know that the sum of these numbers must be 1. Also, these numbers must be consistent with the probabilities associated with the branches of the tree in Figure 12 since these probabilities are given in the statement of Example 5.5. Show that a, b, c, d, e must have the values given in Table 12.

5.18. Refer to Example 5.6 of the text.

(a) By drawing tree diagrams or otherwise, verify the probabilities given in Table 13.
(b) The tree diagram for Strategy 5 is drawn in Figure 14. Note that the tree diagram for Strategy 4 is the same diagram. But the probabilities associated with the branches of the tree are not the same. For the case $x = 2$, what are the branch probabilities for Strategy 4?

(c) Suppose you adopt the following strategy: You draw two balls from the urn, not replacing the first before drawing the second. If you get two red, then you guess Y will draw a red ball. Otherwise you guess Y will draw a green ball. Draw a tree diagram for this strategy and calculate the probability that you win for each possible composition of the urn.

5.19. A seller of rebuilt television tubes and a buyer get together to draw up a contract. The seller will supply tubes in lots of 100 tubes. The buyer, when a lot is offered to him, wants to protect himself against the possibility that the lot contains too many defective tubes. The contract therefore provides that out of each lot, two tubes will be selected at random and tested. The buyer considers the following alternative plans as guides for making his decision in the light of the experimental evidence.

Plan 1 If both of the tubes tested are satisfactory, then accept the whole lot. Otherwise, reject the lot.

Plan 2 If both of the tubes tested are defective, then reject the whole lot. Otherwise, accept the lot.

Plan 3 If both of the tubes tested are satisfactory, then accept the lot. If both are defective, then reject the lot. If one of the tubes is satisfactory and the other defective, then select a third tube at random from the remaining 98 tubes in the lot and accept or reject the lot according as this tube is satisfactory or defective.

Denote by E the event that the buyer accepts the lot and let x be the (unknown) number of defective tubes in a lot offered to the buyer. Clearly, $P(E)$ depends upon both the plan adopted by the buyer and the quality of the lot as determined by the value of x.

(a) Obtain a formula for $P(E)$ in terms of x for each of the three plans.

(b) Substitute the values $x = 0, 5, 10, 20, 30, 40, 50$ in the formulas obtained in (a) and make three graphs plotting (for each plan) the value of x along the horizontal axis and the value of $P(E)$ along the vertical axis. A graph of this kind is called an *operating characteristic curve* (OC-curve) for a plan.

(c) Which rule is most favorable to the buyer? to the seller? (*Note:* these questions become more interesting and more difficult when such things as the utilities resulting from the desirable actions of accepting good lots and rejecting poor lots, the disutilities from the undesirable actions of accepting poor lots and rejecting good lots, and the costs of testing tubes are brought into the analysis.)

6. Bayes' formula

In this section, as an application of conditional probabilities, we derive a famous formula first used by Thomas Bayes in a paper published posthumously in 1763. To prepare the way, we make a definition and prove a preliminary result.

Definition 6.1. A *partition* of a set E is a set $\{E_1, E_2, \cdots, E_n\}$ with the following properties:

(i) $E_j \subseteq E$ $(j = 1, 2, \cdots, n)$

(ii) $E_j \cap E_k = \emptyset$ $(j = 1, 2, \cdots, n; k = 1, 2, \cdots, n; j \neq k)$

(iii) $E_1 \cup E_2 \cup \cdots \cup E_n = E$

In words, a partition of a set E is a set of subsets of E [property (i)] that are disjoint [property (ii)] and exhaustive [property (iii)]. Every element of E is a member of one and only one of the subsets in the partition.

We are already acquainted with partitions of sets. Two complementary events F and F' form the partition $\{F, F'\}$ of the sample space S. For F and F' are certainly subsets of S, and we have $F \cap F' = \emptyset$ and $F \cup F' = S$, as required by Definition 6.1.

From a Venn diagram, we see immediately that $\{E \cap F', E \cap F\}$ is a partition of the set E, $\{E \cap F', E \cap F, E' \cap F\}$ is a partition of the set $E \cup F$, and $\{E \cap F', E \cap F, E' \cap F, E' \cap F'\}$ is a partition of the entire sample space S.

Two more examples of partitions should suffice to make the notion clear. In the sample space S of 52 elements, each denoting one outcome of the experiment in which a card is selected from a standard deck, let E_s, E_h, E_d, and E_c, denote the events that the card selected is a spade, heart, diamond, and club respectively. Then $\{E_s, E_h, E_d, E_c\}$ is a partition of S, since the four subsets are clearly mutually exclusive in pairs and exhaustive. Another partition of S is the set of all 52 simple events of the sample space S.

Theorem 6.1. Let $\{E_1, E_2, \cdots, E_n\}$ be a partition of the sample space S, and suppose each of the events $E_1, E_2, \cdots, E_n$ has nonzero probability. Let E be any event. Then

$$P(E) = P(E_1)P(E|E_1) + P(E_2)P(E|E_2) + \cdots + P(E_n)P(E|E_n)$$

or, using the summation symbol,

(6.1) $$P(E) = \sum_{j=1}^{n} P(E_j)P(E|E_j).$$

Proof. From the hypothesis that $\{E_1, E_2, \cdots, E_n\}$ is a partition of S, it can be shown (see Problem 6.13) that $\{E \cap E_1, E \cap E_2, \cdots, E \cap E_n\}$ is a partition of the event E. Hence

$$E = (E \cap E_1) \cup (E \cap E_2) \cup \cdots \cup (E \cap E_n)$$

expresses E as the union of n *mutually exclusive* events. Applying Formula (4.8) in Problem 4.19 yields the equation

(6.2) $P(E) = P(E \cap E_1) + P(E \cap E_2) + \cdots + P(E \cap E_n).$

But, directly from the definition of conditional probability, we have for $j = 1, 2, \cdots, n$,

$$P(E \cap E_j) = P(E_j)P(E|E_j).$$

Making this substitution in (6.2) proves the theorem. Note that we have guaranteed the existence of the conditional probabilities in (6.1) by our assumption that the events $E_1, E_2, \cdots, E_n$ do not have zero probability.

As the following examples show, Formula (6.1) is useful because an evaluation of the probabilities $P(E_j)$ and conditional probabilities $P(E|E_j)$ is often easier than a direct calculation of $P(E)$.

Example 6.1. Freshmen account for 30%, sophomores 25%, juniors 25%, and seniors 20% of the members of a college fraternity. Fifty percent of the freshmen, 30% of the sophomores, 10% of the juniors, and 2% of the seniors are enrolled in a mathematics course. A member of the fraternity being chosen at random, what is the probability that he is enrolled in a mathematics course?

We let E denote "member selected is enrolled in a mathematics course," and E_1, E_2, E_3, and E_4 denote the events that the member selected is a freshman, sophomore, junior, and senior respectively. Then $\{E_1, E_2, E_3, E_4\}$ is a partition of the sample space S consisting of all ordered pairs the first object of which identifies the class and the second object the presence or absence of the student in a mathematics course. In fact, the data of the problem specify all the probabilities needed to apply Formula (6.1) with $n = 4$. We find

$$\begin{aligned} P(E) &= (.30)(.50) + (.25)(.30) + (.25)(.10) + (.20)(.02) \\ &= .254, \end{aligned}$$

or slightly more than 25 percent of the fraternity are enrolled in a mathematics course.

Example 6.2. Find the probability that in a well-shuffled deck of cards, the ace of spades is next to the king of spades. Here, as in the preceding example, it seems sensible to break the problem up into cases, i.e., to consider first the event in which the ace of spades is the top card of the deck, then the event that it is the bottom card of the deck, and finally the remaining event in which the ace of spades is somewhere within the deck. Let E_1, E_2, and E_3 denote these events in the order stated. We choose as sample space S the set of ordered 52-tuples denoting all possible orderings of the 52-card deck. Then $\{E_1, E_2, E_3\}$ is a partition of S. If E denotes the event that the ace and king of spades are neighboring cards, then noting that only one card is next to the top or bottom cards but two cards are next to a card within the deck, we find

$$P(E_1) = P(E_2) = \tfrac{1}{52}, \quad P(E_3) = \tfrac{50}{52} = \tfrac{25}{26},$$
$$P(E|E_1) = P(E|E_2) = \tfrac{1}{51}, \quad P(E|E_3) = \tfrac{2}{51}.$$

Hence, by Formula (6.1) with $n = 3$,

$$P(E) = (\tfrac{1}{52})(\tfrac{1}{51}) + (\tfrac{1}{52})(\tfrac{1}{51}) + (\tfrac{25}{26})(\tfrac{2}{51}) = \tfrac{1}{26}.$$

From Theorem 6.1 it is only a short step to Bayes' formula.

Theorem 6.2. Let $\{E_1, E_2, \cdots, E_n\}$ be a partition of the sample space S, and suppose each of the events E_1, E_2, $\cdots$, E_n has nonzero probability. Let E be any event for which $P(E) > 0$. Then for each integer k $(1 \leq k \leq n)$, we have *Bayes' formula:*

$$(6.3) \qquad P(E_k|E) = \frac{P(E_k)P(E|E_k)}{\sum\limits_{j=1}^{n} P(E_j)P(E|E_j)}.$$

Proof. Applying the definition of conditional probability twice, we find

$$P(E_k|E) = \frac{P(E \cap E_k)}{P(E)} = \frac{P(E_k)P(E|E_k)}{P(E)}.$$

The theorem is proved by rewriting $P(E)$ according to Formula (6.1).

The following example illustrates the use of Bayes' formula.

Example 6.3. Suppose that the reliability of a chest X-ray test for the detection of tuberculosis is specified as follows: of people with tuberculosis, 90% of the X-ray examinations detect the disease but 10% go undetected. Of people free of tuberculosis, 99% of the X rays are judged free of the disease, but 1% are diagnosed as showing tuberculosis. From a large population of which only 0.1% have tuberculosis, one person is selected at random, given a chest X ray, and the radiologist reports the presence of tuberculosis. What is the probability that the person has tuberculosis?

We let E_1 denote the event that the person selected has tuberculosis, and E the event that the person's X ray is diagnosed as positive, i.e., as showing tuberculosis. We seek $P(E_1|E)$. Now $\{E_1, E_1'\}$ is a partition of the sample space of all people in the population. We are given the following probabilities:

$$P(E_1) = .001, \quad P(E_1') = .999, \quad P(E|E_1) = .9, \quad P(E|E_1') = .01.$$

From Bayes' formula, we find

$$(6.4) \qquad P(E_1|E) = \frac{P(E_1)P(E|E_1)}{P(E_1)P(E|E_1) + P(E_1')P(E|E_1')}$$

$$= \frac{(.001)(.9)}{(.001)(.9) + (.999)(.01)} = .083, \text{ approximately.}$$

Note that although the X-ray test is fairly reliable, we have found that only slightly more than 8% of those with positive X rays turn out to have tuberculosis. The results of such calculations must be taken into account when large-scale medical diagnostic tests are planned.

We note here the terminology often used when Bayes' formula is applied. The events $E_1, E_2, \cdots, E_n$ are called *hypotheses*, and they are assumed to be disjoint and exhaustive. The probability $P(E_k)$ is called the *a priori probability* of hypothesis E_k. The conditional probability $P(E_k|E)$ is called the *a posteriori probability* of the hypothesis E_k, given the observed event E. Thus, in Example 6.3, the events E_1 (person has tuberculosis) and E_1' (person does not have tuberculosis) are the hypotheses. The a priori probability of a person having tu-

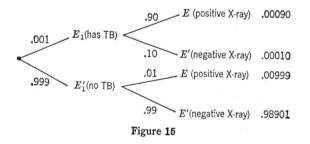

Figure 15

berculosis is $P(E_1) = .001$. But the a posteriori probability of a person having tuberculosis, given that his X ray is positive, is $P(E_1|E) = .083$.

In Figure 15, we have the tree diagram for Example 6.3 in which a person is first classified according to whether or not he has tuberculosis and then according to whether his X ray is positive or negative. Probabilities given in the data of the example are written on the appropriate branches of the tree. To the right of each of the four possible paths from left to right through the tree we have written the probability associated with that path. For example,

$$P(\text{has TB and X ray positive}) = P(\text{has TB})P(\text{X ray positive}|\text{has TB})$$
$$= (.001)(.9) = .0009$$

is the probability for the topmost path through the tree in Figure 15.

These path probabilities can also be recorded in tabular form, as in Table 15, where the entry in each cell is the probability of the intersection of the events given by the row and column in which the cell appears. If we add the entries in any row or column, then by (6.2), we obtain the probability of the event defining that row or column. These probabilities appear in the margins of the table.

TABLE 15

	E (X ray positive)	E' (X ray negative)	
E_1 (has TB)	$P(E_1 \cap E)$ .00090	$P(E_1 \cap E')$ .00010	$P(E_1)$ .001
E_1' (no TB)	$P(E_1' \cap E)$ .00999	$P(E_1' \cap E')$ .98901	$P(E_1')$ .999
	$P(E)$ .01089	$P(E')$ .98911	Total 1

From the entries in Table 15, we can derive all possible conditional probabilities. In particular,

$$P(E_1|E) = \frac{P(E_1 \cap E)}{P(E)} = \frac{.00090}{.01089} = .083, \text{ approximately,}$$

in agreement with (6.4).

Computing probabilities from the Table or by Bayes' formula, we can construct the tree diagram in Figure 16 which differs from the diagram in Figure 15 because the order of events has been reversed. In Figure 16, we think of a person *first* being classified according to whether his X ray is positive or negative and *then* according to whether or not he has tuberculosis. (Probabilities in Figure 16 are

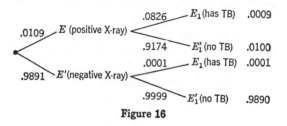

Figure 16

rounded to four decimal place accuracy.) Using the language associated with Bayes' formula, we have in Figure 15 the conditional probabilities of the possible observed events given the various hypotheses, whereas in Figure 16 we have conditional probabilities of the possible hypotheses given the various observed events.

We conclude with two more examples in which Bayes' formula proves useful.

Example 6.4. Three urns contain colored balls as specified in Table

TABLE 16

Urn	Red	White	Blue
1	3	4	1
2	1	2	3
3	4	3	2

16. One urn is chosen at random and a ball is withdrawn. It happens to be red. What is the probability that it came from urn 2?

We let E denote the event "ball selected is red." To account for the occurrence of E we have three hypotheses: E_1 (urn 1 selected), E_2 (urn 2 selected), E_3 (urn 3 selected). Since the urn is chosen at random,

$$P(E_1) = P(E_2) = P(E_3) = \tfrac{1}{3}.$$

We also are given the conditional probabilities,

$$P(E|E_1) = \tfrac{3}{8}, \quad P(E|E_2) = \tfrac{1}{6}, \quad P(E|E_3) = \tfrac{4}{9}.$$

Since $\{E_1, E_2, E_3\}$ is a partition of the sample space for this compound experiment, Bayes' formula is applicable. Putting $k = 2$, $n = 3$ in (6.3) we find

$$P(E_2|E) = \frac{(\tfrac{1}{3})(\tfrac{1}{6})}{(\tfrac{1}{3})(\tfrac{3}{8}) + (\tfrac{1}{3})(\tfrac{1}{6}) + (\tfrac{1}{3})(\tfrac{4}{9})} = \frac{12}{71}.$$

The reader may find it helpful to construct tree diagrams like those in Figure 15 and 16 for this example.

Example 6.5. After a severe flood, a warehouse finds itself stocked with boxes of flashbulbs from which identification labels have been washed off. There are three kinds of bulbs, each packed in units of 100 in identical boxes: low quality, medium quality, and high quality. It is known that in the entire warehouse, the proportions of boxes with low, medium, and high quality bulbs are .25, .25, and .50, respectively.

Since testing a flashbulb means destroying it, exhaustive testing of the bulbs is impractical. Instead, the distributor orders that two bulbs from each box be tested. The manufacturer, on the basis of past experience, estimates the conditional probabilities given in Table 17.

TABLE 17

Conditional Probabilities of Finding x Defectives Given That Two Bulbs Tested Were from Box of Known Quality			
Number of Defectives x	Quality of Box		
	Low	Medium	High
0	.49	.64	.81
1	.42	.32	.18
2	.09	.04	.01

Suppose two bulbs are selected from a box, tested, and both are found to fire satisfactorily. What is the probability that the box contains high quality bulbs? Our *hypotheses* are the three events L, M, H that the box contains low, medium, and high quality bulbs, respectively. If we let E denote the observed event that neither of the two bulbs tested was defective, then by Bayes' formula we find

$$P(\mathrm{H}|E) = \frac{P(\mathrm{H})P(E|\mathrm{H})}{P(\mathrm{L})P(E|\mathrm{L}) + P(\mathrm{M})P(E|\mathrm{M}) + P(\mathrm{H})P(E|\mathrm{H})}$$

$$= \frac{(.50)(.81)}{(.25)(.49) + (.25)(.64) + (.50)(.81)} = .59, \text{ approximately.}$$

Proceeding in this way, we compute the a posteriori probabilities of the three hypotheses, and so obtain Table 18.

TABLE 18

Quality of Box	A Priori Probability	A Posteriori Probability Given That We Observe		
		0 Defectives	1 Defective	2 Defectives
Low	.25	.18	.38	.60
Medium	.25	.23	.29	.27
High	.50	.59	.33	.13

We see that the most probable hypothesis in the event neither of the two bulbs tested is defective is that the bulbs come from a high quality box. But if one or both of the two tested bulbs are defective, then the most probable hypothesis is that the bulbs come from a low quality box.

Calculations of this sort using Bayes' formula are quite common in statistical decision theory. The reader interested in further details can consult References 2 or 10 listed at the end of this chapter.

PROBLEMS

6.1. From a group of four boys and two girls, first one child is selected at random and then, from the remaining five children, another child is selected at random. Find the probability that the second child selected will be a girl (a) from first principles, i.e., by defining a suitable sample space, assigning probabilities to its simple events, etc. (b) by use of Theorem 6.1.

6.2. Refer to Example 5.5 and construct a tree diagram in which the selected ball is identified first by its color and then by the urn from which it was drawn. Find probabilities associated with each branch of the tree, as well as for each path through the tree. Compare with Figure 12.

6.3. Refer to Example 6.1 of the text. Construct a tree diagram in which a fraternity member is first classified according to whether he is enrolled in a mathematics course and then according to his class. Find probabilities associated with each branch of the tree as well as for each path through the tree.

6.4. In the Polya urn model of Problem 5.10, find
   (a) The probability that the first ball is green.
   (b) The probability that the second ball is green.
   (c) The probability that the third ball is green. [In view of the answers in (a), (b), and (c), are you willing to make a conjecture? Try proving it!]

6.5. This problem should be done from first principles and also by using Bayes' formula. Three identical boxes each contain two coins. In one box both are pennies, in one both are nickels, and in the third there is one penny and one nickel. A man chooses a box at random and takes out a coin. If the coin is a penny, what is the probability that the other coin in the box is also a penny?

6.6. (a) Bolts are made by two machines A and B, but A produces twice as many bolts as B in a given time. A is known to produce two percent defectives and B one percent defectives. A bolt is examined and found to be defective. What are the probabilities a priori and a posteriori that the bolt was produced by A?
   (b) Suppose n_1 to n_2 is the ratio of the number of bolts produced by A to the number produced by B. Let p_1 and p_2 denote the proportion of defectives produced by A and B, respectively. Suppose a bolt is tested and found to be defective. Show that if $n_1 p_1 > n_2 p_2$, then the a posteriori probability that the bolt was produced by A is greater than the a posteriori probability that the bolt was produced by B.

6.7. Mr. Smith, having lived in his city many years, estimates the a priori probability that today's weather will be inclement is .2. (He thinks today will be fair with probability .8.) Mr. Smith listens to an early morning weather forecast to get some information on the day's weather. The forecaster makes one of three predictions: fair weather, inclement weather, uncertain weather. Mr. Smith has made estimates of conditional probabilities of the different predictions given the day's weather, as shown in Table 19. For example, he believes that of the fair days

TABLE 19

Day's Weather	Forecast		
	Fair	Inclement	Uncertain
Fair	.7	.2	.1
Inclement	.3	.6	.1

70% are correctly forecast, 20% are forecast as inclement and 10% as uncertain.

Suppose Mr. Smith hears the forecaster predict fair weather. What is the a posteriori probability of fair weather?

6.8. In a T-maze, a laboratory animal is given a choice of going to the left and getting food or going to the right and receiving a mild electric shock. Before any conditioning (in trial number 1) animals are equally likely to go to the left or right in the maze. Having received food on any trial, the probabilities of going to the left and right become .6 and .4, respectively, on the following trial. Having received the electric shock on any trial, the probabilities of going to the left and right on the next trial are .8 and .2, respectively.

    (a) What is the probability that the animal will turn left on trial number 2? On trial number 3?

    (b) Let us denote by p_n the probability that the animal will turn left on trial number n. Derive an equation relating p_n and p_{n-1} and use this equation to find a general formula for p_n in terms of p_1 and n.

6.9. Refer to Problem 5.11 and suppose the selected socks are of the same color. What is the probability that they are black?

6.10. A multiple-choice test question lists five alternative answers, of which just one is correct. If a student has done his homework, then he is certain to identify the correct answer; otherwise, he chooses an answer at random. Let p denote the probability of the event E that a student does his homework, and let F be the event that he answers the question correctly.

    (a) Find a formula for $P(E|F)$ in terms of p.

    (b) Show that $P(E|F) \geq P(E)$ for all values of p. When does the equality hold?

    (c) Suppose the test lists n alternative answers of which only one is correct. Now find $P(E|F)$ in terms of n and p, and show that if p is fixed but unequal to 0 or 1, then $P(E|F)$ increases as n increases. Is this result reasonable?

6.11. Of the freshmen in a certain college, it is known that 40% attended private secondary schools and 60% attended public schools. The registrar reports that 30% of all students who attended private secondary schools but only 20% of those who attended public schools attain A averages in their freshman year. At the end of the year, one student is chosen at random from the freshman class and he has an A average. What is the conditional probability that the student attended public schools?

6.12. You know that urn A contains two green and one red ball and urn B contains three green and two red balls. One of these urns is selected at random, but you don't know which one is selected. You may perform one of the following experiments before guessing which urn was selected.

    (i) Take one ball out of the selected urn and observe its color.

    (ii) Take two balls out of the selected urn, replacing the first before drawing the second, and observe their colors.

    (iii) Same as (ii), except that you do not replace the first ball before drawing the second.

Whichever experiment you choose, as soon as its outcome is known you compute the a posteriori probabilities of urns A and B being selected, given the observed outcome. You then guess the urn whose a posteriori probability is larger.

    (a) For each of the three experiments, determine which urn you guess for each possible experimental outcome.

    (b) For each of the three experiments, calculate the probability that you actually guess correctly which urn was selected. Which experiment leads to the highest probability of guessing correctly? [It is interesting to observe that most people, when offered a choice of one of the three experiments, prefer experiment (iii).]

6.13. Let $\{E_1, E_2, \cdots, E_n\}$ be a partition of a sample space S and let E be any subset of S. Show that

$$\{E \cap E_1, E \cap E_2, \cdots, E \cap E_n\}$$

is a partition of E.

6.14. Let a universal set $\mathfrak{U}$ of people be given. Let M, F, C, A, H, and S denote the subsets of male, female, child, adult, healthy, and sick people respectively. Then we can form the following partitions of $\mathfrak{U}$:

$$P_1 = \{M, F\} \qquad P_2 = \{C, A\} \qquad P_3 = \{H, S\}.$$

From P_1 and P_2 we can form a new partition, namely

$$P_4 = \{M \cap C, M \cap A, F \cap C, F \cap A\},$$

in which people are classified both according to sex *and* age. P_4 is called the *cross-partition* of P_1 and P_2. Analogously we can classify people according to their health in addition to sex and age. We thus are led to the partition

$$P_5 = \{M \cap C \cap H, M \cap A \cap H, F \cap C \cap H, F \cap A \cap H,$$
$$M \cap C \cap S, M \cap A \cap S, F \cap C \cap S, F \cap A \cap S\},$$

which is called the *cross-partition* of P_3 and P_4.

From these illustrative examples, formulate a reasonable definition for the *cross-partition* of any two partitions of an arbitrary set E. Can you prove that a cross-partition of E is a partition of E?

7. Independent events

As we have seen, the probability of E and the conditional probability of E given F are generally unequal, although they can be equal. The case of equality,

$$(7.1) \qquad P(E|F) = P(E),$$

is especially important, for (7.1) expresses the fact that knowing F has occurred does not change the probability of E having occurred. If (7.1) holds, we shall say that E *is independent of* F. Let us note that this relation between the events E and F is defined only if F has positive probability, i.e., only if $P(E|F)$ is meaningful.

Assuming $P(E) > 0$ and $P(F) > 0$, we rewrite (5.4) here,

$$(7.2) \qquad P(E \cap F) = P(E)P(F|E) = P(F)P(E|F),$$

and from the equality on the right deduce that if (7.1) is true, then

$$(7.3) \qquad P(F|E) = P(F)$$

is also true. We have thereby proved the following result.

Theorem 7.1. Let E and F be events with positive probability. If E is independent of F, then F is independent of E.

In words, if knowledge that F occurs does not change the probability of E, then knowledge that E occurs does not change the probability of F. Thus, if $P(E) > 0$ and $P(F) > 0$ so that the conditional probabilities in (7.1) and (7.3) are defined, then these equations must both be true or both be false. When either is true, we find from (7.2) that

$$(7.4) \qquad P(E \cap F) = P(E)P(F).$$

An important definition is based on equation (7.4).

Definition 7.1. Two events E and F are said to be *independent events* if and only if Equation (7.4) holds; i.e., the probability that both E and F occur is the product of the probability that E occurs and the probability that F occurs. Two events that are not independent are said to be *dependent events*. We shall refer to Equation (7.4) as the *multiplication rule* for the events E and F.

In the literature, one often finds two independent events referred to as "mutually independent," "stochastically independent," "independent in the sense of probability," or "statistically independent." We shall use the simpler language of Definition 7.1.

There is a reason for using Equation (7.4), rather than (7.1) or (7.3), as a means of defining independent events. With the latter equations, events E and F with zero probability would be excluded from our definition. No such restriction is involved when Equation (7.4) is used. In fact, it is easy to see from Definition 7.1 that if $P(E) = 0$ and F is *any* event, then E and F are independent. For $E \cap F$ is a subset of E, and so $P(E \cap F) \leq P(E)$ by Theorem 4.2. Since we are assuming $P(E) = 0$, it follows that $P(E \cap F) = 0$. Thus Equation (7.4) holds, and this proves that E and F are independent events, as claimed.

Whether or not two events E and F are independent is a question that we can answer in our present state of knowledge *only* by showing that Equation (7.4) does or does not hold. Although we will often have the intuitive feeling that two specified events E and F are independent, our intuition must be checked by computing $P(E)$, $P(F)$, and $P(E \cap F)$, and then verifying that the multiplication rule (7.4) is true. We give three examples.

Example 7.1. A green and red die are rolled. Let E be the event "six on green die" and F the event "five on red die." We choose the familiar sample space containing 36 outcomes and assign probability $\frac{1}{36}$ to each simple event. Then

$$P(E) = \tfrac{6}{36} = \tfrac{1}{6}, \quad P(F) = \tfrac{6}{36} = \tfrac{1}{6}, \quad P(E \cap F) = \tfrac{1}{36}.$$

Equation (7.4) holds and, therefore, E and F are independent events.

Example 7.2. Two fair coins are tossed. Let E be the event "not more than one head" and F the event "at least one of each face." Define the sample space $S = \{HH, HT, TH, TT\}$ and assign probability $\frac{1}{4}$ to each simple event. Then

$$P(E) = \tfrac{3}{4}, \quad P(F) = \tfrac{2}{4}, \quad \text{and} \quad P(E \cap F) = \tfrac{2}{4}.$$

Hence

$$P(E \cap F) \neq P(E)P(F),$$

and so E and F are dependent events.

Example 7.3. Three fair coins are tossed. Let E and F be events as described in Example 7.2. Now the sample space S contains the familiar eight elements HHH, HHT, $\cdots$, TTT. We assign probability $\frac{1}{8}$ to each simple event of S. Then

$$P(E) = \tfrac{4}{8}, \quad P(F) = \tfrac{6}{8}, \quad P(E \cap F) = P(\{\text{HTT, THT, TTH}\}) = \tfrac{3}{8}.$$

Now the multiplication rule (7.4) holds, and E and F are independent events.

It is not unusual for students to feel that the events E and F should either be independent in both Example 7.2 and 7.3, or dependent in both. But our intuition is not to be trusted—Equation (7.4) must be relied upon to determine if two events are independent or dependent. [See Problem 7.3.]

If the probability of E is unchanged by the knowledge that F occurred, then it seems reasonable that it should also be unchanged by the knowledge that F did not occur. This and related observations are made precise in the following result.

Theorem 7.2. Let E and F be independent events. Then the following pairs of events are also independent: (i) E and F', (ii) E' and F, (iii) E' and F'.

Proof. We prove (i) and leave (ii) and (iii) for the reader. (See Problem 7.4.) In view of Definition 7.1, to prove E and F' are independent events, we must prove that the multiplication rule (7.4) holds for E and F'; i.e.,

(7.5) $$P(E \cap F') = P(E)P(F').$$

Now, by the result of Problem 4.13(e),

$$P(E \cap F') = P(E) - P(E \cap F) = P(E) - P(E)P(F),$$

since the multiplication rule holds for E and F by hypothesis. Hence,

$$P(E \cap F') = P(E)[1 - P(F)] = P(E)P(F'),$$

by Theorem 4.6. Thus, Equation (7.5) holds and the proof of (i) is complete.

Example 7.4. Suppose A has probability p_A of surviving one year and B has probability p_B of surviving one year. If we assume that event E (A survives one year) and event F (B survives one year) are independent, then we have the following possibilities and their probabilities:

Event	Verbal Description	Probability
$E \cap F$	Both A and B survive 1 year	$p_A p_B$
$E \cap F'$	A survives 1 year but B does not	$p_A(1 - p_B)$
$E' \cap F$	A does not survive 1 year but B does	$(1 - p_A)p_B$
$E' \cap F'$	Neither A nor B survives 1 year	$(1 - p_A)(1 - p_B)$

PROBLEMS

7.1. A card is drawn at random from a standard deck of 52 cards. Let E be the event that the card is a spade, F the event that the card is a deuce, and G the event that the card is a deuce or a trey. Determine which of the following pairs of events are independent.

(a) E and F (b) E and G (c) F and G.

7.2. Refer to the mortality table in Problem 5.13. Mr. Smith is now aged 21 and Mr. Jones is now aged 23. Each man was issued insurance one year ago. Assuming that the events "Smith survives to age 22" and "Jones survives to age 24" are independent, calculate the probability that at least one of the men dies within one year.

7.3. (a) Four coins are tossed. Let E and F be the events described in Example 7.2. Show that E and F are dependent.
(b) Let n coins be tossed, where n is any positive integer greater than 1. Let E and F be the events described in Example 7.2. Show that E and F are independent events *if and only if $n = 3$*.

7.4. Complete the proof of Theorem 7.2 by proving (ii) and (iii).

7.5. Consider the data in Table 20 on the smoking habits of a sample of females in the United States.

TABLE 20

Income	Number of Persons	Distribution of Females 18–24 Years of Age, by Current Amount of Smoking and by Income, February 1955						
		Percentage Distribution						
		Not Regular Smokers			Regular Smokers			
		Nonsmokers		Occasional Smokers	Av. Daily No. of Cigarettes			Total
		Never	Discontinued		1–9	10–20	21–40	
		%	%	%	%	%	%	%
None	3335	64.1	2.5	4.1	13.4	15.1	.8	100
Under $1000	1677	65.1	2.9	6.2	14.1	11.2	.5	100
$1000–1999	1117	64.5	3.0	4.0	14.5	11.0	3.0	100
$2000–2999	956	59.3	0.8	11.6	12.2	15.5	.6	100
$3000–	375	40.5	6.5	13.1	10.2	27.6	2.1	100
Total:	7460	62.6	2.6	6.0	13.4	14.3	1.1	100

Source: *Tobacco Smoking in the United States in Relation to Income*, Marketing Research Report No. 189, U.S. Dept. of Agriculture, Washington D. C., July 1957, page 110.

Suppose that a single female is selected at random from the 7460 females making up this sample.

(a) Define the sample space S for this experiment. What probability is to be assigned to each simple event of S?

(b) Find the probability that the female selected has an income of at least $3000.

(c) Find the probability that the female selected smokes between 10 and 20 cigarettes per day on the average.

(d) Find the conditional probability that the female selected smokes between 10 and 20 cigarettes per day on the average, given that she has an income of at least $3000.

(e) Find the probability that the female selected has an income of at least $3000 and also smokes between 10 and 20 cigarettes per day on the average.

(f) Are the events "female selected has an income of at least $3000" and "female selected smokes between 10 and 20 cigarettes daily on the average" dependent or independent events?

7.6. One student is selected at random from the summer school students of Problem 5.6. Are the events "student failed Chemistry" and "student failed History" independent or dependent?

7.7. From a pack of playing cards, two cards are drawn successively, the first being replaced before the second is drawn. Let E be the event "first card is a spade," F the event "second card is not a king," and G the event "first card is an ace or a king." Determine which (if any) of the three pairs of events E and F, F and G, E and G are independent.

7.8. Repeat Problem 7.7, but now assume that the first card is *not* replaced before the second is drawn.

7.9. Show that if E is any event and $P(F) = 1$, then E and F are independent.

7.10. Of what events E can it be said that the events E and E are independent?

7.11. Let E, F, and G be three events. We are told that E and F are independent events, and that F and G are independent events. Does it follow that E and G are independent events? Defend your answer.

7.12. Of the three events E, F, and G, we know that E and F are independent and $G \subseteq E$. Does it follow that G and F are independent? .Defend your answer.

7.13. The 1957–1958 *Combined Membership List* of the American Mathematical Society (S), the Mathematical Association of America (A), and the Society for Industrial and Applied Mathematics (1) gives the following information for the 46 members listed on page 1.

Memberships	Number
S only	16
A only	15
I only	7
S and A	6
A and I	1
S and I	0
S and A and I	1

One person is selected at random from this group of 46 people.

(a) Show that the events "person belongs to the American Mathematical Society" and "person belongs to the Mathematical Association of America" are dependent.

(b) Assuming everyone else maintains their memberships, how many of the 16 members of only the American Mathematical Society must also become members of the Mathematical Association of America in order that the events in (a) be independent?

7.14. Two partitions of S, say $\{E_1, E_2, \cdots, E_n\}$ and $\{F_1, F_2, \cdots, F_m\}$ are defined to be independent if

$$P(E_i \cap F_j) = P(E_i)P(F_j)$$

for $i = 1, 2, \cdots, n$ and $j = 1, 2, \cdots, m$, i.e., if the multiplication rule (7.4) holds for every pair of events formed by taking one event from each partition. Show that the events E and F are independent if and only if the partitions $\{E, E'\}$ and $\{F, F'\}$ are independent.

8. Independence of several events

In this section, we generalize the notion of independence to an arbitrary (but finite) number of events. Let us first consider the special case of three events E_1, E_2, and E_3.

Definition 8.1. The events E_1, E_2, and E_3 are *pairwise independent* (or *independent in pairs*) if all of the possible pairs of events (i.e., E_1 and E_2, E_1 and E_3, E_2 and E_3) are independent.

Thus, if E_1, E_2, and E_3 are pairwise independent, then the multiplication rule (7.4) holds for each pair of events:

$$(8.1) \quad \left\{ \begin{array}{l} P(E_1 \cap E_2) = P(E_1)P(E_2) \\ P(E_1 \cap E_3) = P(E_1)P(E_3) \\ P(E_2 \cap E_3) = P(E_2)P(E_3). \end{array} \right.$$

Let the reader note that we have defined what we mean by the *pairwise* independence of three events, but we have *not* yet said what is meant by the phrase, "E_1, E_2, and E_3 are independent events." Nevertheless, we do think of reasonable consequences that such a definition should entail. For example, we would like to be able to show that if E_1, E_2, and E_3 are independent events, then the two events $(E_1 \cap E_2)$ and E_3 are also independent. Does this result follow if we assume only that E_1, E_2, and E_3 are pairwise independent? This question amounts to asking if the equations in (8.1) imply that the multiplication rule holds for $(E_1 \cap E_2)$ and E_3, i.e., whether or not

$$(8.2) \qquad P((E_1 \cap E_2) \cap E_3) = P(E_1 \cap E_2)P(E_3)$$

follows from (8.1).

But $(E_1 \cap E_2) \cap E_3 = E_1 \cap E_2 \cap E_3$ and if we use (8.1) to simplify $P(E_1 \cap E_2)$, then (8.2) becomes

$$(8.3) \qquad P(E_1 \cap E_2 \cap E_3) = P(E_1)P(E_2)P(E_3).$$

We are thus led to inquire whether Equations (8.1) imply Equation (8.3). That the assumption of pairwise independence of E_1, E_2, and E_3 does *not* imply the desirable consequence that $(E_1 \cap E_2)$ and E_3 are independent is shown by the following example where three events are defined for which (8.1) is true but (8.3) is false.

Example 8.1. To control the quality of a manufacturing process, each unit produced passes through three inspections. Of four units A, B, C, and D it is known that A passed only inspection 1, B passed only inspection 2, C passed only inspection 3, and D passed all 3 inspections. One of the four units is selected at random. Let $E_1 =$ "unit passed inspection 1," $E_2 =$ "unit passed inspection 2," and $E_3 =$ "unit passed inspection 3." Then

$$P(E_1) = P(E_2) = P(E_3) = \tfrac{2}{4} = \tfrac{1}{2},$$
$$P(E_1 \cap E_2) = P(E_1 \cap E_3) = P(E_2 \cap E_3) = \tfrac{1}{4},$$

so that all three equations in (8.1) hold. Thus the events E_1, E_2, and E_3 are pairwise independent. But (8.3) does not hold, since

$$P(E_1 \cap E_2 \cap E_3) = \tfrac{1}{4} \neq P(E_1)P(E_2)P(E_3) = \tfrac{1}{8}.$$

We conclude that the pairwise independence of E_1, E_2, and E_3 does not imply the independence of $(E_1 \cap E_2)$ and E_3.

From this example (see also Problems 8.1–8.3), it is clear that the definition of independence for more than two events requires care. For three events, a suitable definition is obtained by demanding that Equation (8.3) hold in addition to the three equations in (8.1). We shall find it convenient to refer to Equation (8.3) as the *multiplication rule* for the events E_1, E_2, and E_3.

Definition 8.2. Three events E_1, E_2, and E_3 are said to be *independent* if and only if the multiplication rule holds for all combinations of two or more of the events.

The three equations in (8.1) express the multiplication rule for the three pairs of events obtainable from E_1, E_2, and E_3. Equation (8.3) is the multiplication rule for all three events. To say that E_1, E_2, and E_3 are independent is to say that all four of these equations are true.

It is now possible to prove that certain expected consequences do indeed follow from Definition 8.2.

Theorem 8.1. Let E_1, E_2, and E_3 be independent events. Then the following events are also independent:

(a) E_1 and $(E_2 \cap E_3)$ (b) E_2 and $(E_1 \cup E_3)$
(c) E_1' and $(E_2 \cap E_3')$ (d) E_1', E_2, and E_3'

More generally, E_1 and any event expressible in terms of E_2 and E_3 are independent, E_2 and any event expressible in terms of E_1 and E_3 are independent, etc.

Proof. We prove (c) here, and leave (a), (b), and (d) for the reader. The general result can be proved by considering these and all other similar combinations of the three events E_1, E_2, and E_3.

To prove (c) requires (by Definition 7.1) that we prove

$$(8.4) \qquad P(E_1' \cap (E_2 \cap E_3')) = P(E_1')P(E_2 \cap E_3').$$

We are given that E_1, E_2, and E_3 are independent, so that Equations (8.1) and (8.3) are true by hypothesis.

By drawing an appropriate Venn diagram and resorting to the fundamental definition of the probability of an event, let the reader verify that

$$(8.5) \quad P(E_1' \cap (E_2 \cap E_3')) = P(E_2 \cap E_3') - P(E_1 \cap E_2)$$
$$+ P(E_1 \cap E_2 \cap E_3).$$

(Using the "numbers of flags" language introduced in Section 4, the proof of (8.5) amounts to noting that one obtains the sum of the numbers on all flags in $E_1' \cap E_2 \cap E_3'$, each counted once, by writing the sum of the numbers on all flags in $E_2 \cap E_3'$ and in $E_1 \cap E_2 \cap E_3$, and then subtracting the numbers that have been added twice, their sum being $P(E_1 \cap E_2)$.)

By using Equations (8.1) and (8.3), we find

$$(8.6) \quad P(E_1' \cap (E_2 \cap E_3'))$$
$$= P(E_2 \cap E_3') - P(E_1)P(E_2) + P(E_1)P(E_2)P(E_3)$$
$$= P(E_2 \cap E_3') - P(E_1)P(E_2)[1 - P(E_3)]$$
$$= P(E_2 \cap E_3') - P(E_1)P(E_2)P(E_3').$$

But since E_2 and E_3 are independent by hypothesis, it follows from Theorem 7.2 that E_2 and E_3' are also independent, and so the multiplication rule holds for E_2 and E_3'. Hence, continuing from (8.6),

$$P(E_1' \cap (E_2 \cap E_3')) = P(E_2 \cap E_3') - P(E_1)P(E_2 \cap E_3')$$
$$= [1 - P(E_1)]P(E_2 \cap E_3')$$
$$= P(E_1')P(E_2 \cap E_3').$$

This completes the proof of part (c) of Theorem 8.1.

The following example shows how this theorem is applied.

Example 8.2. One shot is fired from each of three guns. Let E_1, E_2, E_3 denote the events that the target is hit by the first, second and third gun, respectively. Suppose

$$P(E_1) = 0.5, \quad P(E_2) = 0.6, \quad P(E_3) = 0.8.$$

Assuming E_1, E_2, E_3 are independent events, what is the probability that exactly one hit is registered?

Since the one hit can be made by any gun, the required probability is given by

$$P(E_1 \cap E_2' \cap E_3') + P(E_1' \cap E_2 \cap E_3') + P(E_1' \cap E_2' \cap E_3).$$

By the independence assumption and Theorem 8.1, each of these probabilities can easily be evaluated. For example,

$$P(E_1 \cap E_2' \cap E_3') = P(E_1)P(E_2')P(E_3')$$
$$= (0.5)(0.4)(0.2) = 0.04.$$

In this way, we find the probability of exactly one hit is 0.26. (See Problem 8.5.)

Definition 8.2 has been so formulated that it can be used as a definition of independence for any finite number of events.

Definition 8.3. The n events E_1, E_2, $\cdots$, E_n ($n \geq 2$) are said to be *independent* if and only if the multiplication rule holds for all combinations of two or more of the events, i.e., if and only if we have

$$(8.7) \begin{cases} P(E_i \cap E_j) = P(E_i)P(E_j) \\ \qquad\qquad (1 \leq i < j \leq n) \\ P(E_i \cap E_j \cap E_k) = P(E_i)P(E_j)P(E_k) \\ \qquad\qquad (1 \leq i < j < k \leq n) \\ P(E_i \cap E_j \cap E_k \cap E_l) = P(E_i)P(E_j)P(E_k)P(E_l) \\ \qquad\qquad (1 \leq i < j < k < l \leq n) \\ \cdots\cdots\cdots\cdots\cdots\cdots\cdots\cdots \\ \cdots\cdots\cdots\cdots\cdots\cdots\cdots\cdots \\ P(E_1 \cap E_2 \cap \cdots \cap E_n) = P(E_1)P(E_2) \cdots P(E_n). \end{cases}$$

How many defining conditions must be checked if n events are to be proved independent? Let us think of the set $\mathfrak{U}$ containing as elements the n events E_1, E_2, $\cdots$, E_n. The set $\mathfrak{U}$ has exactly 2^n subsets. The multiplication rule is required to hold for all subsets of $\mathfrak{U}$ containing at least two events. There is one null subset and there are n unit subsets for which multiplication rules are not required. Hence, there are $2^n - n - 1$ equations summarized in (8.7).

Let us observe that Definition 8.3 implies that if n events are independent, then any smaller number of events taken from these n are also independent.

In our later work, we shall find many applications of the important idea of independence of events. Most of these involve "independent trials of an experiment" or "experiments repeated independently under identical conditions." In the next section, we present a mathematical formulation of these important concepts.

PROBLEMS

8.1. A green and a red die are rolled. Let $E_1 = $ "6 on red die," $E_2 = $ "6 on green die," and $E_3 = $ "sum of numbers on two dice is odd." Show that E_1, E_2, and E_3 are pairwise independent but are not independent.

8.2. A card is selected at random from a standard deck. Let $E_1 = $ "card is a spade or a club," $E_2 = $ "card is a spade," and $E_3 = $ "card is ace, king, $\cdots$, 8 of diamonds or the ace of spades." Show that Equation (8.3) holds, but that none of the three equations in (8.1) is true.

8.3. Suppose of three events E_1, E_2, and E_3 it is known that E_1 and E_2 are independent and that the multiplication rule given in Equation (8.3) applies to all three events. Prove that $E_1 \cap E_2$ and E_3 are independent, but show by example that E_1 and E_3 need not be independent.

8.4. A coin is tossed three times in succession. We define the customary sample space

$$S = \{\text{HHH, HHT}, \cdots, \text{TTT}\},$$

and let E_1, E_2, and E_3 denote the events that a head is tossed on the first, second, and third toss, respectively. Suppose we require

$$P(E_1) = P(E_2) = P(E_3) = p,$$

where $0 \le p \le 1$, and also that E_1, E_2, and E_3 are independent. Show that there is one and only one acceptable assignment of probabilities to the simple events of S that is consistent with these assumptions.

8.5. Refer to Example 8.2 and find the most probable number of times that the target is hit.

8.6. Let all pairs of events from E_1, E_2, and E_3 be mutually exclusive.

   (a) Are E_1, E_2, and E_3 pairwise independent?
   (b) What additional hypotheses are required to make "yes" the correct answer in (a)?
   (c) With the hypotheses added in (b), are the events E_1, E_2, and E_3 independent?

8.7. Let E_1 and E_2 be independent events and suppose E_3 has probability zero or one. Show that E_1, E_2, and E_3 are independent events.

8.8. Complete the proof of Theorem 8.1 by proving (a), (b), and (d).

8.9. Suppose the events E_1, E_2, $\cdots$, E_n are independent and that $P(E_k) = 1/(k + 1)$ for $1 \le k \le n$. Find the probability that none of the n events occurs, justifying each step in your calculation.

8.10. Let p be the probability that a man aged x will die in a year. There are four men (A, B, C, and D) each aged x. We assume that events E_1, E_2, E_3, E_4 are independent whenever E_1 is defined in terms of only A's life, E_2 in terms of only B's life, etc. Find the probability that

   (a) A will die within the year.
   (b) A and B will die but C and D will not die in the year.
   (c) Only A will die within the year.

8.11. The president of a company must decide which of two actions to take, say whether to rent or buy expensive machinery. His vice-president is likely to make a faulty analysis and thus recommend the wrong decision with probability .05. The president hires two consultants who separately study the problem and make their recommendations. After watching them at work, the president estimates that one consultant is likely to recommend the wrong decision with probability .05, the other with probability .10. He decides to take the action recommended by a majority of the three reports he receives. What is the probability that he will make a wrong decision? Does the assumption of independence you have made seem reasonable for this problem?

9. Independent trials

The notion of "experiments repeated independently under identical conditions" is central to empirical science and, as such, is worthy of precise formulation. This we do in the present section. Since we shall make extensive use of Cartesian product sets, the reader may find it helpful to review Section 5 of Chapter 1 at this time.

Suppose an experiment is under consideration. As we know, we think instead of its mathematical counterpart, the sample space S, where

$$(9.1) \qquad S = \{o_1, o_2, \cdots, o_n\}.$$

We assume that an acceptable assignment of probabilities has been made to the simple events of S; i.e., to each $\{o_j\}$ there is assigned a nonnegative number $P(\{o_j\})$ in such a way that

$$(9.2) \qquad \sum_{j=1}^{n} P(\{o_j\}) = 1.$$

Now let us think of performing this experiment and then performing it again. The succession of two experiments is a new experiment that we want to describe mathematically. In order to avoid confusing references to original experiments and this new experiment, it is convenient to refer to the original experiments as *trials* and to describe the new experiment as made up of two trials, each represented by (or corresponding to) the sample space S. This new experiment is mathematically defined, as are all experiments, by a sample space. The elements (outcomes) of this new sample space are all the ordered pairs (o_j, o_k) denoting the occurrence of outcome o_j at the first trial and outcome o_k at the second trial. Thus the sample space for the experiment is the Cartesian product set $S \times S$. Since the sample space S for each of the two trials making up the experiment has n elements, there are n^2 ordered pairs in $S \times S$.

Before probability questions can be answered for the experiment, we must make some acceptable assignment of probabilities to the n^2 simple events of $S \times S$; i.e., we must assign a nonnegative number to $\{(o_j, o_k)\}$ for each j and k in such a way that the sum of all n^2 numbers is 1. As we know, there are infinitely many ways of doing this. *But if we say that the two trials are independent, then by definition there is one and only one way that we must use: the assignment must be made so that*

$$(9.3) \qquad P(\{(o_j, o_k)\}) = P(\{o_j\})P(\{o_k\})$$

for $j = 1, 2, \cdots, n$ and $k = 1, 2, \cdots, n$.

Formula (9.3) expresses the probability of the simple event $\{(o_j, o_k)\}$ of $S \times S$ as the *product* of the probabilities of the simple events $\{o_j\}$ and $\{o_k\}$ of S. Before discussing the significance of this rule, we first demonstrate that (9.3) provides an *acceptable* assignment of probabilities to the simple events of $S \times S$.

The number $P(\{(o_j, o_k)\})$ is certainly nonnegative, since it is the product of two nonnegative numbers. Now to find the sum of the probabilities of all simple events of $S \times S$, we first write them in rows and columns as follows:

$$
\begin{array}{cccc}
P(\{(o_1, o_1)\}) & P(\{(o_1, o_2)\}) & \cdots & P(\{(o_1, o_n)\}) \\
P(\{(o_2, o_1)\}) & P(\{(o_2, o_2)\}) & \cdots & P(\{(o_2, o_n)\}) \\
\cdot & \cdot & & \cdot \\
\cdot & \cdot & & \cdot \\
\cdot & \cdot & & \cdot \\
P(\{(o_n, o_1)\}) & P(\{(o_n, o_2)\}) & \cdots & P(\{(o_n, o_n)\}).
\end{array}
$$

The sum of the probabilities in the first column is

$$\sum_{j=1}^{n} P(\{(o_j, o_1)\}) = \sum_{j=1}^{n} P(\{o_j\})P(\{o_1\}), \text{ by (9.3)},$$

$$= P(\{o_1\}) \sum_{j=1}^{n} P(\{o_j\})$$

$$= P(\{o_1\}), \text{ by (9.2)}.$$

Similarly, the sum of the probabilities in the kth column is $P(\{o_k\})$ for $k = 1, 2, \cdots, n$. The sum of the probabilities of all n^2 simple events is the sum of the column totals,

$$\sum_{k=1}^{n} P(\{o_k\}) = 1,$$

and the assignment specified by (9.3) is acceptable, as claimed.

We summarize in the following formal definition.

Definition 9.1. Let S be a sample space with elements $o_1, o_2, \cdots, o_n$ and let $P(\{o_j\})$ be the probability of the simple event $\{o_j\}$ for $j = 1, 2, \cdots, n$. By the experiment consisting of two *independent trials* corresponding to S, we mean the sample space $S \times S$ (the Cartesian product set of S with itself) whose elements are the n^2 ordered pairs (o_j, o_k) and whose simple events $\{(o_j, o_k)\}$ are assigned probabilities in accordance with the product rule in Equation (9.3).

Example 9.1. A fair coin is tossed once and then tossed again. Each toss is a trial represented by the sample space $S = \{H, T\}$ whose two simple events are each assigned probability $\frac{1}{2}$. The experiment made up of the two trials is defined by the sample space $S \times S$, where

$$S \times S = \{HH, HT, TH, TT\}.$$

There are infinitely many acceptable assignments of probabilities to the four simple events of $S \times S$ (see the discussion in Example 3.7), but if the two tosses are said to be *independent*, then each simple event *must* be assigned probability $\frac{1}{4}$, in accordance with Equation (9.3). We remind the reader of our agreement to write HH rather than (H, H), HT rather than (H, T), etc. It is customary to write ordered pairs using parentheses with the two objects of the pair separated by a comma, but when no confusion can arise we shall continue to use the less cumbersome notation.

Example 9.2. From a population of n people, one person is selected at random. Another person is then selected at random from the full group; i.e., we allow the same person to be selected at both trials. Each selection (trial) is defined by the sample space $S = \{1, 2, \cdots, n\}$, where each person is identified by a positive integer. Each of the n simple events of S is assigned probability $1/n$; i.e., $P(\{j\}) = 1/n$ for $j = 1, 2, \cdots, n$. The experiment made up of these two trials is called *selecting a sample of two with replacement from the population* and is represented by the Cartesian product set $S \times S$ given by

$$S \times S = \{(j, k) \mid j \in S, k \in S\}.$$

To say that the two trials (i.e., the selection of the first person and the selection of the second person) are *independent* is to *require* the assignment of probabilities to the simple events of $S \times S$ in accordance with Equation (9.3). Since

$$P(\{(j, k)\}) = P(\{j\})P(\{k\}) = \left(\frac{1}{n}\right)\left(\frac{1}{n}\right) = \frac{1}{n^2},$$

the independence of the trials means that each simple event of $S \times S$ is assigned the same probability $1/n^2$. Thus we have the formal mathematical counterpart of our intuitive feeling that selecting a random sample of size two with replacement can be considered as a succession of two independent selections. (See Problem 9.9.)

Example 9.3. Consider the experiment consisting of two independent rolls of a fair die. Since each roll corresponds to the sample space $S = \{1, 2, \cdots, 6\}$, each simple event of which is assigned probability $\frac{1}{6}$, Definition 9.1 demands that the experiment be defined by the familiar sample space

$S \times S = \{(1, 1), (1, 2), \cdots, (1, 6), \cdots, (6, 1), (6, 2), \cdots, (6, 6)\}$

for which each simple event has probability $\frac{1}{36}$. Let $E_1 =$ "first roll results in a 6" and $E_2 =$ "second roll results in an even number." We intuitively expect that the independence of the *trials* will have as a consequence that E_1 and E_2 are independent *events*. That this is the case is easy to verify, since

$$P(E_1 \cap E_2) = \tfrac{3}{36}, \quad P(E_1) = \tfrac{6}{36}, \quad \text{and} \quad P(E_2) = \tfrac{18}{36},$$

so that the multiplication rule

$$P(E_1 \cap E_2) = P(E_1)P(E_2)$$

does hold, and the events E_1 and E_2 are independent, as expected.

We feel this result is reasonable because of the special nature of the events E_1 and E_2. Of the two independent rolls of the die, the first roll determines whether or not E_1 occurs, and the second roll determines whether or not E_2 occurs. More generally, given any two independent trials, it seems reasonable to expect that, if the first trial determines whether or not an event E_1 occurs and the second trial determines whether or not an event E_2 occurs, then E_1 and E_2 will be independent events. We want to prove this result is generally true.

But first we must define precisely what we mean when we say that the first trial (or the second trial) *determines* whether an event E occurs. Let us refer to Example 9.3 and note that, in the experiment consisting of two independent rolls of a fair die, we have

E_1 = "first roll results in 6"

    = $\{(6, 1), (6, 2), (6, 3), (6, 4), (6, 5), (6, 6)\}$

    = $\{6\} \times \{1, 2, 3, 4, 5, 6\}$

    = $\{6\} \times S.$

Similarly,

E_2 = "second roll results in an even number"

    = $\{(1, 2), (2, 2), (3, 2), (4, 2), (5, 2), (6, 2), (1, 4), (2, 4), (3, 4),$

         $(4, 4), (5, 4), (6, 4), (1, 6), (2, 6), (3, 6), (4, 6), (5, 6), (6, 6)\}$

    = $\{1, 2, 3, 4, 5, 6\} \times \{2, 4, 6\}$

    = $S \times \{2, 4, 6\}.$

Generally, our event E is of course a subset of $S \times S$ and, as such, is a set of ordered pairs. To say that E is determined by the first trial means that the first member of the ordered pair is restricted by the requirement that E occurs, but the second member is unrestricted. Similarly, to say that E is determined by the second trial means that the first member of the ordered pair is unrestricted, but the second member is restricted by the condition that E occurs. We make the following formal definition.

Definition 9.2. Consider an experiment consisting of two independent trials, each trial defined by the sample space S. (The two-trial experiment therefore has as sample space the Cartesian product set $S \times S$.) An event E (subset of $S \times S$) is said to be *determined by the first trial* if and only if there is some subset C_1 of S such that

$$E = C_1 \times S.$$

Similarly, an event F is said to be *determined by the second trial* if and only if there is some subset C_2 of S such that

$$F = S \times C_2.$$

We are now able to state and prove our main result.

Theorem 9.1. Consider the experiment consisting of two independent trials corresponding to the sample space

$$S = \{o_1, o_2, \cdots, o_n\}.$$

Let E_1 and E_2 be any two events of $S \times S$ such that E_1 is determined by the first trial and E_2 is determined by the second trial. Then E_1 and E_2 are independent events.

The following lemma is the essential result needed to prove Theorem 9.1.

Lemma. Let $S \times S$ be the sample space defining an experiment consisting of two independent trials, each trial corresponding to the sample space S. Let $C_1 \subseteq S$ and $C_2 \subseteq S$. Then

$$(9.4) \qquad P(C_1 \times C_2) = P(C_1)P(C_2).$$

Proof of Lemma. Let the reader first note that (9.4) reduces to (9.3) in the special case when C_1 and C_2 are *simple* events of S. And, of course, (9.4) is also true if C_1 or C_2 is the null event. Now let us suppose that the subsets C_1 and C_2 are given as follows:

$$C_1 = \{o_{j_1}, o_{j_2}, \cdots, o_{j_r}\}, \qquad C_2 = \{o_{k_1}, o_{k_2}, \cdots, o_{k_s}\}.$$

The event $C_1 \times C_2$ is the union of all those simple events $\{(o_j, o_k)\}$ of $S \times S$ for which $\{o_j\} \; \epsilon \; C_1$ and $\{o_k\} \; \epsilon \; C_2$. We arrange these events in r rows and s columns and write

$$C_1 \times C_2 = \left\{ \begin{array}{l} \{(o_{j_1}, o_{k_1})\} \cup \{(o_{j_1}, o_{k_2})\} \cup \cdots \cup \{(o_{j_1}, o_{k_s})\} \cup \\ \cdot \;\; \cdot \;\; \cdot \;\; \cdot \;\; \cdot \;\; \cdot \;\; \cdot \;\; \cdot \;\; \cdot \;\; \cdot \;\; \cdot \;\; \cdot \;\; \cdot \;\; \cdot \;\; \cdot \;\; \cdot \\ \{(o_{j_r}, o_{k_1})\} \cup \{(o_{j_r}, o_{k_2})\} \cup \cdots \cup \{(o_{j_r}, o_{k_s})\}. \end{array} \right.$$

Now to compute $P(C_1 \times C_2)$ we must add the probabilities of all these simple events. If we add the probabilities of the simple events in the first row, we find from the assumed independence of the trials,

$$\sum_{v=1}^{s} P(\{(o_{j_1}, o_{k_v})\}) = \sum_{v=1}^{s} P(\{o_{j_1}\})P(\{o_{k_v}\})$$

$$= P(\{o_{j_1}\}) \sum_{v=1}^{s} P(\{o_{k_v}\})$$

$$= P(\{o_{j_1}\})P(C_2),$$

the last equality following from the definition of $P(C_2)$ as the sum of the probabilities of the simple events whose union is C_2. The sum of the probabilities in any other row is obtained in the same way, the sum for the uth row ($u = 1, 2, \cdots, r$) being $P(\{o_{j_u}\})P(C_2)$. We obtain the sum of the probabilities of all simple events of $C_1 \times C_2$ by adding these row sums. Thus,

$$P(C_1 \times C_2) = \sum_{u=1}^{r} P(\{o_{j_u}\})P(C_2)$$

$$= P(C_2) \sum_{u=1}^{r} P(\{o_{j_u}\})$$

$$= P(C_1)P(C_2),$$

and the proof of the lemma is complete.

Proof of Theorem 9.1. Our hypothesis concerning E_1 and E_2, in view of Definition 9.2, implies the existence of sets C_1 and C_2, each subsets of S, such that

$$E_1 = C_1 \times S, \qquad E_2 = S \times C_2.$$

To prove the theorem, it suffices to prove that the multiplication rule holds for E_1 and E_2, i.e.,

(9.5) $$P(E_1 \cap E_2) = P(E_1)P(E_2).$$

We first note (cf. Problem I.5.5) that

(9.6) $$E_1 \cap E_2 = (C_1 \times S) \cap (S \times C_2) = C_1 \times C_2.$$

Now apply the lemma three times to obtain

(9.7) $\qquad P(E_1 \cap E_2) = P(C_1 \times C_2) = P(C_1)P(C_2),$
(9.8) $\qquad P(E_1) = P(C_1 \times S) = P(C_1)P(S) = P(C_1),$
(9.9) $\qquad P(E_2) = P(S \times C_2) = P(S)P(C_2) = P(C_2),$

where we have used the fact that $P(S) = 1$. Thus we see that the multiplication rule (9.5) holds, and so E_1 and E_2 are independent. The proof of Theorem 9.1 is now complete.

It is important to note the significance of our results. Referring back to the dice-rolling experiment in Example 9.3, we recall that the phrase "first roll results in 6" was interpreted as the description of the event $E_1 = \{6\} \times S$, a subset of $S \times S$. But this phrase also describes an event, namely $\{6\}$, which is a subset of S. In general, the event $E_1 = C_1 \times S$ and the event C_1, although events of *different* sample spaces, are both determined by the same first trial of the experiment. We expect that these events should therefore have the same probability. Formulas (9.8) and (9.9) guarantee that such expectations are realized.

Note also that when E_1 and E_2 satisfy the hypotheses of Theorem 9.1, then, as shown by (9.7), we can calculate $P(E_1 \cap E_2)$ as a product of probabilities of events (subsets) of the sample space S, and we do not have to do any computations relative to the sample space $S \times S$ of the two-trial experiment.

Our definitions and results can be generalized to any finite number of repetitions of the same experiment or, still more generally, to any number of successive experiments whether like or unlike. We omit proofs.

Definition 9.3. Suppose N is a positive integer and let S_j (for $j = 1, 2, \cdots, N$) be a sample space with outcomes $o_1^{(j)}, o_2^{(j)}, \cdots, o_{n_j}^{(j)}$. By the experiment consisting of the succession of N trials, the first corresponding to S_1, the second to S_2, etc., we mean the sample space $S_1 \times S_2 \times \cdots \times S_N$ (the Cartesian product set of $S_1, S_2, \cdots, S_N$) whose elements are all the $n_1 n_2 \cdots n_N$ ordered N-tuples $(o^{(1)}, o^{(2)}, \cdots, o^{(N)})$ where $o^{(1)} \in S_1$, $o^{(2)} \in S_2$, $\cdots$, $o^{(N)} \in S_N$. For each sample space S_j let there be an acceptable assignment of probabilities to its simple events. To say that the N trials are *independent* is to define the probabilities of all simple events in $S_1 \times S_2 \times \cdots \times S_N$ by the product rule

$$P(\{(o^{(1)}, o^{(2)}, \cdots, o^{(N)})\}) = P(\{o^{(1)}\})P(\{o^{(2)}\}) \cdots P(\{o^{(N)}\}).$$

Theorem 9.2. Consider the experiment consisting of N independent successive trials corresponding to the sample spaces $S_1, S_2, \cdots, S_N$, in that order. Let events $E_1, E_2, \cdots, E_N$ be such that E_j is determined by the jth trial for $j = 1, 2, \cdots, N$. [To say, for example, that E_1 is determined by the first trial means that there is a subset C_1 of S_1 such that

$$E_1 = C_1 \times S_2 \times S_3 \times \cdots \times S_N;$$

to say that E_2 is determined by the second trial means that there is a subset C_2 of S_2 such that

$$E_2 = S_1 \times C_2 \times S_3 \times \cdots \times S_N;$$

and so on.] Then the events $E_1, E_2, \cdots, E_N$ are independent.

If $N = 2$ and $S_1 = S_2 = S$, then Definition 9.3 and Theorem 9.2 reduce to Definition 9.1 and Theorem 9.1 respectively.

Example 9.4. A fair coin is tossed, then a symmetric die is rolled, and finally a card is selected at random from a standard deck. We assume that these three trials are independent. Introducing obvious notation, we note that the event "head" is determined by the first trial, "even number" is determined by the second trial, and "spade" is determined by the third trial. Applying Theorem 9.2, we conclude that "head," "even number," and "spade" are independent events. Hence

$$P(\text{head, even number, spade}) = P(\text{head})P(\text{even number})P(\text{spade})$$
$$= (\tfrac{1}{2})(\tfrac{3}{6})(\tfrac{13}{52}) = \tfrac{1}{16}.$$

Example 9.5. A quiz has four questions of multiple-choice type. There are three possible answers for each question, but only one answer is right. Assuming a student guesses at random for his answer to each question and that his successive guesses are independent, what is the probability that he gets more right than wrong answers?

The sample space for each trial (answering a question) is $S = \{R, W\}$, where R denotes a right answer, W a wrong answer. We are given that

$$P(\{R\}) = \tfrac{1}{3}, \qquad P(\{W\}) = \tfrac{2}{3}.$$

For the four-question test, the sample space is $S \times S \times S \times S$. The event "3 or 4 right" is the subset

$$\{RRRR, RRRW, RRWR, RWRR, WRRR\}$$

Because the trials are independent,

$$P(\{RRRR\}) = (\tfrac{1}{3})^4, \qquad P(\{RRRW\}) = (\tfrac{1}{3})^3(\tfrac{2}{3}),$$

and the probability of the other simple events for which exactly three answers are right is also $(\tfrac{1}{3})^3(\tfrac{2}{3})$. Hence

$$P(3 \text{ or } 4 \text{ right}) = (\tfrac{1}{3})^4 + 4(\tfrac{1}{3})^3(\tfrac{2}{3}) = \tfrac{1}{9}.$$

PROBLEMS

9.1. (a) A fair coin is tossed three independent times. Determine a suitable sample space for this three-trial experiment and make the required assignment of probabilities to its simple events.

(b) Repeat part (a), but now assume that the coin is constructed so that the probability of head on any toss is p $(0 \leq p \leq 1)$ and the probability of tail is $q = 1 - p$. (Cf. Problem 8.4.)

9.2. A random sample of five is selected with replacement from a population of which 40 percent are female and 60 percent are male. Define the sample space for this experiment and assign probabilities to its simple events. Find the probability that the sample contains

(a) no males   (b) at least one male

(c) exactly one male   (d) all males

9.3. A test has ten questions of multiple-choice type. There are six choices for each answer, but only one is correct. Suppose a student guesses his answer to each question. (For example, he can toss a fair die and let his answer be determined by the number that turns up.) Assuming his guesses are independent, define a sample space for this experiment and assign probabilities to its simple events. Find the probability that he gets nine or ten correct answers.

9.4. At a busy street intersection, it is estimated that a jaywalker will be hit by a car with probability .01. Assuming individual trips form independent trials, find the probability of a jaywalker remaining unhit if he crosses the street twice per day for 30 days.

9.5. A football team wins its weekly game with probability .7, loses with probability .2, and ties with probability .1. Consider the games played on three consecutive weekends as a three-trial experiment in which the trials are independent. Find the probability that the number of wins exceeds the sum of the number of losses and ties.

9.6. A baseball player approximates his chances at bat as follows: probability .3 of getting a hit, .1 of getting a base on balls, and .6 of being out. Consider the four times the player is at bat in a game as four independent trials and compute the probability that he gets (a) one walk and three hits, (b) one walk, one hit, and is put out twice.

9.7. Suppose a missile has probability $\frac{1}{2}$ of destroying its target and probability $\frac{1}{2}$ of missing it. Assuming the missile firings form independent trials, determine the number of missiles that should be fired at a target in order to make the probability of destroying the target at least .99.

9.8. (a) Each of two urns contains three identical balls numbered from 1 to 3. One ball is drawn from each urn, and we assume these drawings are independent. What is the probability that 2 is the greatest number drawn?

(b) Each of k urns contains n identical balls numbered from 1 to n. One ball is drawn from each urn, and we assume these drawings are independent. What is the probability that m is the greatest number drawn?

9.9. From a population of n people, one person is selected at random. A second person is then selected at random from the remaining $(n - 1)$ people. Imagine the n people lined up in order in positions numbered $1, 2, \cdots, n$. The first trial (selecting the first person) amounts to selecting one of the n positions, and so can be represented by the sample space $S_n = \{1, 2, \cdots, n\}$, in which each simple event is assigned probability $1/n$. After this first person is selected, imagine the remaining $(n - 1)$ people keeping their relative positions but closing ranks so that they are lined up in positions numbered 1, 2, $\cdots$, $n - 1$. The second trial (selecting the second person) amounts to selecting one of these $(n - 1)$ positions, and so can be represented by the sample space $S_{n-1} = \{1, 2, \cdots, n - 1\}$, in which each simple event is given probability $1/(n - 1)$. If these two trials are assumed to be independent, then the experiment consisting of the two independent trials is called *selecting a random sample of two without replacement from the population.*

(a) What sample space are we thus led to for the experiment of selecting a sample of two without replacement from the population? What probability is assigned to each simple event of this sample space?

(b) Suppose $n = 26$ and these 26 people are named A, B, C, $\cdots$, X, Y, Z. You select a random sample of two without replacement from this population and report the outcome (2, 3). Which two people were selected?

(c) Generalize our discussion and show that the selection of a random sample of N *without* replacement from a population of n people can be considered as a succession of N *independent* selections. (Of course, $N \leq n$.)

(d) With $n = 26$ as in (b), suppose $N = 4$ people were selected without replacement and the outcome reported as the 4-tuple (2, 3, 1, 22). Which four people were selected?

9.10. We have an n-trial experiment, each trial of which corresponds to the sample space S. Show that the null event $\emptyset$ and the entire sample space for the n-trial experiment are determined by every trial of the experiment. Does this seem reasonable?

10. A probability model in genetics

Probability concepts have come to play an increasingly important role, not only as the foundation of mathematical statistics, but also in formulating mathematical models for phenomena in all the sciences, biological, physical, and social. In one brief section, we can hardly hope to do more than illustrate the latter kind of application. We shall use the theory developed to this point, especially the ideas of conditional probability and independent trials, to consider (in greatly oversimplified form) some important questions arising in population genetics and involving the factors influencing evolution.

Incidentally, we are also able to illustrate the use in probability of the method of difference equations.*

We restrict our attention to a single gene which has only two forms: recessive (r) and dominant (D). We assume that each individual in the population under consideration has two such genes in his chromosomes and therefore can be classified as one of the following types: (1) pure dominant, DD, in which both genes are of dominant form; (2) hybrid, rD, in which one gene is recessive and the other dominant; (3) pure recessive, rr, in which both genes are recessive. (Biologists refer to these classes as *genotypes;* DD individuals are called *homozygous dominant,* rr individuals are *homozygous recessive,* and rD individuals are *heterozygous.*)

The genetic make-up (with respect to this particular gene) of each generation is described by the proportions of individuals of this generation in the three genotypes. If we think of drawing a sample of one individual at random from this generation, then the proportion of any genotype will be the probability of the individual being of that genotype. Thus we introduce symbols as follows:

n = generation number $(0, 1, 2, \cdots)$,

u_n = probability that individual selected from nth generation is DD,

$2v_n$ = probability that individual selected from nth generation is rD,

w_n = probability that individual selected from nth generation is rr.

It is clear that

$$(10.1) \qquad u_n + 2v_n + w_n = 1 \qquad (n = 0, 1, 2, \cdots).$$

The general problem of population genetics can be formulated in the following manner: Given the initial $(n = 0)$ probabilities u_0, $2v_0$, w_0 and a set of assumptions describing the dependence of future generations on this initial one (i.e., assumptions concerning the mating system, gene mutations, forces of natural selection, etc.), find the genotype probabilities for $n \geq 1$.

---

* Additional material on probability methods in genetics can be found in Chapter 3 of the book by Neyman listed in the references at the end of this chapter. For a systematic exposition of the construction and application of a probability model in psychology, see R. R. Bush and F. Mosteller, *Stochastic Models for Learning,* John Wiley and Sons, Inc., 1955. A number of articles containing probability models appear in P. F. Lazarsfeld (Ed.), *Mathematical Thinking in the Social Sciences,* The Free Press, 1954. The method of difference equations used in this section is expounded in the author's *Introduction to Difference Equations,* John Wiley and Sons, Inc., 1958, Dover Publications, Inc., 1986.

We shall consider a problem of this sort in which the system is one of random (Mendelian) mating, sometimes called *panmixia*, modified by both selection and mutation forces. The salient features of this model are described in the context of the following rules for obtaining an individual of any generation (say the $(n + 1)$st) if the genotype probabilities u_n, $2v_n$, w_n are known for the preceding generation (the nth).

(i) A male parent is selected at random from this population; i.e., the probabilities are u_n, $2v_n$, w_n for such a parent to be DD, rD, rr respectively. (We assume that genotypes occur among males and females with the same probabilities as in the whole population.) A single gene is then selected at random from the two genes that the male parent carries. For example, if the male parent is DD, then the gene D is transmitted with probability 1; if the parent is rD, then genes r and D are each selected with probability $\frac{1}{2}$, etc.

(ii) A female parent is selected at random from the population, as in (i). A single gene is then selected at random from the two genes carried by the female parent.

The genotype of the new individual of the $(n + 1)$st generation is determined by the union of the male and the female genes selected in (i) and (ii). We shall speak of steps (i) and (ii) as *trials* of the experiment in which an individual of one generation is formed from individuals of the preceding generation.

Random Mendelian mating (*panmixia*) with respect to the single gene under study is characterized by the assumption that the selections involved in trials (i) and (ii) are carried out at random and that these trials are independent.

As an example, let us calculate the probability of genotype rr in the $(n + 1)$st generation of a population undergoing random Mendelian mating. An rr individual can arise only when the genes selected from both the male and female parent are recessive. Now the event E_1 (male gene is r) can occur in two mutually exclusive ways: (a) the male parent is rr and then an r gene is transmitted, or (b) the male parent is rD and then an r gene is transmitted. We thus find that

(10.2) $$P(E_1) = (w_n)(1) + (2v_n)(\tfrac{1}{2}) = v_n + w_n.$$

Similarly, we observe that if E_2 is the event "female gene is r," then

$$P(E_2) = v_n + w_n.$$

But E_1 is determined by trial (i) and E_2 is determined by trial (ii). Since the trials are assumed independent, we conclude by Theorem 9.1 that E_1 and E_2 are independent events. Hence

$$P(E_1 \cap E_2) = P(E_1)P(E_2).$$

Since $E_1 \cap E_2$ is the event "individual of $(n + 1)$st generation is rr," we have

$$P(E_1 \cap E_2) = w_{n+1},$$

and therefore

(10.3) $$w_{n+1} = (v_n + w_n)^2.$$

Although we can continue and similarly derive the other genotype probabilities in the $(n + 1)$st generation in terms of those in the preceding generation, we turn instead to a more general model in which the assumptions of random mating are modified by mutation and selection forces.

We first assume that the dominant gene D *mutates* to the recessive form r with probability α ($0 \leq \alpha \leq 1$), this mutation being independent of the source (male or female) of the gene. Let us think of the mutation occurring, if at all, after the male and female genes are selected, but before their union. To illustrate, we recalculate the probability of the event E_1 as follows. E_1 can now occur in four mutually exclusive ways: (a) male parent is DD, gene selected is D, this gene mutates to r; (b) male parent is rD, gene selected is r; (c) male parent is rD, gene selected is D, this gene mutates to r; (d) male parent is rr, gene selected is r. Thus we find

(10.4) $$\begin{aligned} P(E_1) &= (u_n)(1)(\alpha) + (2v_n)(\tfrac{1}{2}) + (2v_n)(\tfrac{1}{2})\alpha + (w_n)(1) \\ &= (v_n + w_n) + \alpha(u_n + v_n). \end{aligned}$$

Note that (10.4) reduces to (10.2) in case there is no mutation ($\alpha = 0$).

We also add a *selection* force which affects the participation of pure recessive individuals in the mating process. Up to this point, we have assumed that all genes are viable. Now let us suppose that the fertility of rr individuals is impaired so that a proportion β of these individuals (whether male or female) do not have viable genes to transmit. To illustrate the impact of this assumption, we again calculate the probability of E_1 (male gene is r) but now *on the condition F that the gene is viable;* i.e., we calculate $P(E_1|F)$. For the moment, we neglect the mutation effect. Until now $P(F)$ has been 1, but with the addition of fertility differences, we note that the male gene is viable if and only if the male parent is *not* one of those whose fertility is impaired. Hence

(10.5) $$P(F) = 1 - \beta w_n.$$

Also we calculate

$$P(E_1 \cap F) = (2v_n)(\tfrac{1}{2}) + (w_n - \beta w_n)(1)$$
$$= v_n + w_n - \beta w_n.$$

Hence

(10.6) $$P(E_1|F) = \frac{v_n + w_n - \beta w_n}{1 - \beta w_n}.$$

Note that (10.6) reduces to (10.2) when $\beta = 0$ (no selection force or all genes viable.)

We should observe that the mutation and selection forces are opposite in effect: the mutation force is directed toward an *increase* of the recessive gene in the population, whereas the selection force tends to *decrease* the relative frequency of this gene.

Our problem can now be summarized as follows: Suppose that our system of mating is panmixia modified by the above mutation and selection forces. Let f_n equal the probability (proportion) of the recessive gene r among the genes of parents in the nth generation. The genotype probabilities u_0, $2v_0$, w_0 are given for the initial ($n = 0$) generation of parents. How does f_n depend upon n? Is there an equilibrium value of f_n that is approached as n gets larger and larger and, if such an equilibrium proportion exists, how quickly is it reached and what is its dependence on the initial genotypic composition of the population?

We have in (10.6) calculated the probability $P(E_1|F)$ that a viable gene produced by a parent of the nth generation is recessive, but we neglected the mutation force. Similarly we find that the probability, say p_n, that a viable gene produced by a parent of the nth generation is dominant (before mutation process occurs) is given by

(10.7) $$p_n = \frac{u_n + v_n}{1 - \beta w_n}.$$

It follows that the probability that this viable gene is dominant after mutation is $p_n(1 - \alpha)$, and so we find

(10.8) $$f_n = 1 - (1 - \alpha)p_n.$$

Knowing f_n, we obtain the genotype probabilities in generation $(n + 1)$:

(10.9) $$\begin{cases} u_{n+1} = (1 - f_n)^2 \\ 2v_{n+1} = 2f_n(1 - f_n) \\ w_{n+1} = f_n^2, \end{cases}$$

and using these together with (10.7) and (10.8) we derive (Cf. Problem 10.4) a *difference equation* for the probabilities f_n:

(10.10) $\quad f_{n+1} = 1 - (1 - \alpha) \dfrac{1 - f_n}{1 - \beta f_n^2} \quad (n = 0, 1, 2, \cdots)$.

Our problem can now be formulated analytically: Given numbers f_0, α, β, each between 0 and 1 inclusive, find the dependence of f_n (the proportion of recessive genes among parents of generation number n) on the generation number n and the prescribed parameters f_0 (the proportion of recessive genes among parents of the initial generation $n = 0$), α (the probability with which a dominant gene mutates to the recessive form), and β (the proportion of rr parents in each generation who do not have viable genes).

This problem is quite difficult to solve in all generality, but there are some important special cases that are fairly easy.

Case 1: *Panmixia* ($\alpha = 0$, $\beta = 0$; neither mutation nor selection.)

In this case, the difference equation (10.10) reduces to

(10.11) $\qquad\qquad f_{n+1} = f_n \qquad (n = 0, 1, 2, \cdots)$

and we immediately conclude that $f_n = f_0$ for all n. In view of Equations (10.9), it follows that

$$u_1 = (1 - f_0)^2 = u_2 = u_3 = \cdots$$
$$2v_1 = 2f_0(1 - f_0) = 2v_2 = 2v_3 = \cdots$$
$$w_1 = f_0^2 = w_2 = w_3 = \cdots.$$

We have in this way demonstrated the so-called *Hardy-Weinberg law: With repeated mating under panmixia, the distribution of the three genotypes (with respect to a single gene) is fixed after one generation.* Thus we see that the Mendelian laws are conservative in effect, and one may regard evolution as the study of those forces (mutation, selection, assortative mating, etc.) which tend to disturb this unchanging equilibrium of genotype proportions.

Case 2: $\alpha = 0$ and $\beta = 1$; i.e., no mutation and all pure recessives completely sterile.

Now Equation (10.10) becomes

(10.12) $f_{n+1} = 1 - \dfrac{1}{1 + f_n}$ $(n = 0, 1, 2, \cdots)$.

If we make the substitution

(10.13) $f_n = \dfrac{1}{g_n}$,

then (10.12) takes on the simpler form

(10.14) $g_{n+1} = g_n + 1$ $(n = 0, 1, 2, \cdots)$.

From (10.14), with $n = 0$, we find $g_1 = g_0 + 1$. Then putting $n = 1$ in (10.14) we see that $g_2 = g_1 + 1 = g_0 + 2$. Writing (10.14) with $n = 3$ and then using our newly found expression for g_2, we find $g_3 = g_2 + 1 = g_0 + 3$. By mathematical induction, we can prove

(10.15) $g_n = g_0 + n$ $(n = 0, 1, 2, \cdots)$.

In view of (10.13) we thus find that

(10.16) $f_n = \dfrac{f_0}{1 + nf_0}$.

Equation (10.16) describes the elimination of a single gene under complete sterilization of pure recessives. Although f_n decreases and approaches zero as n gets larger, this decrease is quite slow. For example, let us compute the number of generations required in order that, with complete sterility of all rr individuals, the proportion of the r gene decreases to half its initial value. We put $f_n = (\tfrac{1}{2})f_0$ in (10.16) and solve to find $n = 1/f_0$. Thus if $f_0 = .001$, then 1000 generations are required to reduce the proportion of the recessive gene to .0005. Quantitative considerations of this kind are clearly of importance in eugenics.

Other conclusions that follow from Equation (10.10) are included in the problems.

PROBLEMS

10.1. Find the proportions of the three genotypes in the first $(n = 1)$ and second $(n = 2)$ generation of random Mendelian mating if

    (a) $u_0 = 0, 2v_0 = \tfrac{1}{2}, w_0 = \tfrac{1}{2}$
    (b) $u_0 = w_0 = 0, 2v_0 = 1$
    (c) $u_0 = 1, 2v_0 = w_0 = 0$
    (d) $u_0 = 2v_0 = 0, w_0 = 1$

10.2. Redo the preceding problem with a mutation force added. Assume $\alpha = 0.1$ is the probability that the dominant gene D mutates to the recessive form r.

10.3. Redo Problem 10.1 assuming both mutation and selection forces present. Use Equations (10.7)–(10.9) and assume $\alpha = 0.1$, $\beta = 0.2$. In each case write the first three terms of the sequence $f_0, f_1, f_2, \cdots$.

10.4. To derive the difference equation (10.10), proceed as follows. (a) Use (10.8) to write f_{n+1} in terms of p_{n+1}. (b) In the equation found in (a) replace p_{n+1} by an equivalent expression obtained by using (10.7) and involving $u_{n+1}, v_{n+1}, w_{n+1}$. (c) In the equation obtained in (b), substitute for $u_{n+1}, v_{n+1}, w_{n+1}$ the expressions given in (10.9) and simplify.

10.5. Write out the details of the derivation of Equation (10.14) in the text.

10.6. Show by mathematical induction that (10.15) is the solution of the difference equation (10.14) for all $n = 0, 1, 2, \cdots$.

10.7. Consider the case in which all pure recessives are completely sterile (i.e., $\beta = 1$), but the supply of the recessive gene r is replenished by the mutation D → r; i.e., $\alpha > 0$.

(a) Show that Equation (10.10) becomes

$$(10.17) \qquad f_{n+1} = \frac{\alpha + f_n}{1 + f_n} \qquad n = 0, 1, 2, \cdots.$$

(b) Suppose $f_0 = \sqrt{\alpha}$. Show that then $f_n = \sqrt{\alpha}$ for $n = 1, 2, \cdots$.
(c) Suppose $\alpha = 1$. Find the terms of the sequence $f_0, f_1, f_2, \cdots$.
(d) Suppose $f_0 \neq \sqrt{\alpha}$ and $\alpha \neq 1$. Let $f_n = \sqrt{\alpha} + 1/g_n$ and show that g_n satisfies the difference equation

$$(10.18) \qquad g_{n+1} = Ag_n + B \qquad (n = 0, 1, 2, \cdots),$$

where

$$A = \frac{1 + \sqrt{\alpha}}{1 - \sqrt{\alpha}}, \qquad B = \frac{1}{1 - \sqrt{\alpha}}.$$

(e) Show that Equation (10.18) is satisfied for all $n = 0, 1, 2, \cdots$ by

$$g_n = \left(g_0 - \frac{B}{1 - A} \right) A^n + \frac{B}{1 - A}.$$

(f) Conclude that if $f_0 \neq \sqrt{\alpha}$ and $\alpha \neq 1$, then Equation (10.17) implies

$$(10.19) \qquad f_n = \sqrt{\alpha} + \cfrac{1}{\left(\dfrac{1}{2\sqrt{\alpha}} + \dfrac{1}{f_0 - \sqrt{\alpha}} \right)\left(\dfrac{1 + \sqrt{\alpha}}{1 - \sqrt{\alpha}} \right)^n - \dfrac{1}{2\sqrt{\alpha}}}$$

$$(n = 0, 1, 2, \cdots).$$

(g) From (10.19) note that since $A > 1$, A^n gets larger and larger as n increases. Conclude that f_n approaches $\sqrt{\alpha}$ as n increases without bound. Thus the population approaches a balanced (equilibrium) state in which the recessive gene occurs with proportion $\sqrt{\alpha}$.

SUPPLEMENTARY READING

1. Bizley, M. T. L., *Probability: An Intermediate Textbook*, Cambridge Univ. Press, 1957.

2. Chernoff, H. and L. E. Moses, *Elementary Decision Theory*, John Wiley and Sons, Inc., 1959, Dover Publications, Inc., 1986.

3. Commission on Mathematics, College Entrance Examination Board, *Introductory Probability and Statistical Inference*, An Experimental Course, Revised Preliminary Edition, 1959.

4. Cramér, H., *The Elements of Probability Theory and Some of its Applications*, John Wiley and Sons, Inc., 1955.

5. Feller, W., *An Introduction to Probability Theory and its Applications*, 2nd edition, John Wiley and Sons, Inc., 1957.

6. Kemeny, J. G., H. Mirkil, J. L. Snell, G. L. Thompson, *Finite Mathematical Structures*, Prentice-Hall, Inc., 1959.

7. Munroe, M. E., *Theory of Probability*, McGraw-Hill Book Company, Inc., 1951.

8. Neyman, J., *First Course in Probability and Statistics*, Henry Holt and Company, Inc., 1950.

9. Parzen, E., *Modern Probability Theory and its Applications*, John Wiley and Sons, Inc., 1960.

10. Schlaifer, R., *Probability and Statistics for Business Decisions*, McGraw-Hill Book Company, Inc., 1959.

11. Todhunter, I., *A History of the Mathematical Theory of Probability*, Chelsea Publishing Company, 1949.

12. Uspensky, J. V., *Introduction to Mathematical Probability*, McGraw-Hill Book Company, Inc., 1937.

Chapter 3

SOPHISTICATED COUNTING

---

1. Counting techniques and probability problems

Up to this point, we have illustrated our theory with examples that require only very direct and elementary counting procedures. We accomplished this by restricting ourselves either to experiments leading to sample spaces with a small number of elements (where we were able to count by direct enumeration) or, if the experiment had a large number of possible outcomes, to events whose elements were able to be counted by a direct application of the fundamental principle of counting. These restrictions were intentional, since our aim was to present the basic theory of probability unencumbered by difficulties due to incidental counting problems. But now we must face the fact that many interesting and important probability problems require more sophisticated counting techniques. We develop a few of these techniques in this section. Since we shall be using the fundamental principle of counting time and again, the reader may find it helpful to review the discussion on pp. 9–11.

The following counting problems are solved in this section. We suppose that a nonempty set A with n (distinct) elements is given.

Problem 1. For any positive integer $r \leq n$, find the number of ordered r-tuples the objects of which are different elements of A. Each such ordered r-tuple specifies an ordered arrangement or *permu-*

tation of the r objects taken from the n elements of A. Because of this fact, the required number is denoted by $P(n, r)$.

Problem 2. For any nonnegative integer $r \leq n$, find the number of subsets of A that have exactly r elements. (Such a subset will be called an r-subset of A.) The number of r-subsets of a set with n elements is denoted by $\binom{n}{r}$. Each r-subset specifies a selection *without regard to order* of r elements from the n elements in A.

Problem 3. We are given k numbered cells and k nonnegative integers $n_1, n_2, \cdots, n_k$ whose sum is n; i.e., for $k \geq 1$,

$$n_1 + n_2 + \cdots + n_k = n.$$

Find the number of ways of putting the n elements of A into these k cells so that n_1 elements are in the first cell, n_2 elements are in the second cell, $\cdots$, n_k elements are in the kth cell. This number is denoted by

$$\binom{n}{n_1, n_2, \cdots, n_k}.$$

A simple example will help to clarify these problems and the special notation we have introduced.

Example 1.1. Let $A = \{a, b, c, d, e\}$ be a set of $n = 5$ elements. We list the following ordered pairs (2-tuples) formed by selecting two elements of A and paying attention to the order in which they are selected:

(1.1)
$$
\begin{array}{cccccccccc}
ab & ac & ad & ae & bc & bd & be & cd & ce & de \\
ba & ca & da & ea & cb & db & eb & dc & ec & ed.
\end{array}
$$

These are the 20 ordered pairs or 20 permutations of two elements from the five elements in A. In symbols, $P(5, 2) = 20$. If a chairman and a secretary must be elected from among five men on a committee, there are $P(5, 2) = 20$ different possible results of the election. Note that we correctly count as different the results leading to chairman = a, secretary = b on the one hand and chairman = b, secretary = a on the other.

Now, however, suppose we want to elect a subcommittee of two men from the five committee members. The order in which the choices are made is now irrelevant; we care only *which* two men are elected. Thus we want the number of 2-subsets of the set A. The pairs of elements in the ten possible 2-subsets are enumerated in the first row of (1.1). Although ab and ba are *different* permutations, they determine the *same* 2-subset of A since $\{a, b\} = \{b, a\}$. We thus find that there are ten 2-subsets of A. In symbols, $\binom{5}{2} = 10$.

Finally, suppose four members of the committee are arranging rides to the funeral of the fifth member, say e. Three cars are available and can take 2, 1, and 1 passenger respectively. We list the possible assignments of men to cars:

car 1	ab	ab	ac	ac	ad	ad	bc	bc	bd	bd	cd	cd
car 2	c	d	b	d	b	c	a	d	a	c	a	b
car 3	d	c	d	b	c	b	d	a	c	a	b	a

Thus we find that there are 12 ways of placing the four objects (committeemen a, b, c, d) into three numbered cells (the three cars) so that two objects are in the first cell, 1 object in the second cell, and one object in the third cell. In symbols, we have computed $\binom{4}{2, 1, 1} = 12$. (We killed off committeeman e only to create a counting problem with an answer small enough for us to easily list all the possible assignments to cars. Let the reader show that if all five committee members are being driven to a happy occasion, then there are 30 ways of assigning the men to three cars so that two ride in the first car, two in the second car, and one rides in the third car. In symbols, $\binom{5}{2, 2, 1} = 30$.)

We turn now to derivations of general formulas that solve the three problems we have stated. But first a definition to simplify our notation.

Definition 1.1. If n is a positive integer, then the product of the integers from 1 to n is called "n factorial" and is denoted by $n!$. By special convention, we agree to put $0! = 1$.

For example:

$1! = 1$	$5! = 5 \cdot 4 \cdot 3 \cdot 2 \cdot 1 = 120$
$2! = 2 \cdot 1 = 2$	$6! = 6 \cdot 5! = 720$
$3! = 3 \cdot 2 \cdot 1 = 6$	$7! = 7 \cdot 6! = 5040$
$4! = 4 \cdot 3 \cdot 2 \cdot 1 = 24$	$8! = 8 \cdot 7! = 40,320$

Note that in computing $6!$, $7!$, and $8!$, we used the fact that

$$(1.2) \qquad (n+1)! = (n+1)(n!) \quad \text{for } n = 0, 1, 2, \cdots$$

Observe also that when $n = 0$, Formula (1.2) reads $1! = (1)(0!)$ and hence is correct by virtue of our convention that $0! = 1$. Although it may seem artificial now, we assure the reader that defining $0!$ in this way will turn out to be very convenient in the formulas that follow.

Theorem 1.1. With the notation introduced in the statements of Problems 1–3, we have

$$(1.3) \qquad P(n, r) = \frac{n!}{(n-r)!}$$

> The number of ordered r-tuples or permutations of r objects from a set of n objects.

$$(1.4) \qquad \binom{n}{r} = \frac{n!}{r!(n-r)!}$$

> The number of r-subsets (subsets with exactly r elements) of a set of n elements.

$$(1.5) \qquad \binom{n}{n_1, n_2, \cdots, n_k} = \frac{n!}{n_1! n_2! \cdots n_k!}$$

> The number of ways of placing n distinct objects into k cells so that n_i objects are in cell i for $i = 1, 2, \cdots, k$. $(n_1 + n_2 + \cdots + n_k = n.)$

Proof. The fundamental principle of counting is the key tool in proving these formulas. To prove (1.3) note that to form an r-tuple $(a_1, a_2, \cdots, a_r)$ from the n given objects we must choose a_1 (task 1) from the n objects, then choose a_2 (task 2) from the remaining $n-1$ objects, and so on until we choose a_r (task r) from the remaining $n - r + 1$ objects. Hence $P(n, r)$, the number of ordered r-tuples from a set of n objects, is equal to the number of ways of completing these r tasks in the stated order. By the fundamental principle we find

$$(1.6) \qquad P(n, r) = n(n-1) \cdots (n-r+1).$$

Multiplying and dividing by $(n-r)!$ we obtain the alternative form

$$P(n, r) = \frac{n(n-1) \cdots (n-r+1)(n-r)!}{(n-r)!},$$

from which (1.3) follows by observing that the numerator is indeed the product of all the positive integers from 1 to n. Note that when $r = n$, (1.6) becomes

$$(1.7) \qquad P(n, n) = n(n-1) \cdots 1 = n!.$$

Formula (1.3) also gives this result due to our convention that $0! = 1$. Hence (1.3) is true for all positive integers $r \leq n$, as claimed.

To prove (1.4) we observe that there are as many ways of writing an ordered r-tuple as there are ways of completing the following tasks in the stated order: (1) choose an r-subset from the given set of n elements and thus determine *which* r objects will be used to form the r-tuple, and then (2) arrange these r objects in some order so that there is a first, a second, $\cdots$, an rth object specified. It follows that $P(n, r)$, the number of ordered r-tuples, must be precisely the number of ways of completing these two tasks. The first task can be done in $\binom{n}{r}$ ways by definition of this symbol. The second task can be done in $P(r, r) = r!$ ways by virtue of the meaning of $P(r, r)$ together with Formula (1.7). Hence

$$(1.8) \qquad P(n, r) = \binom{n}{r} r!$$

or, using (1.3),

$$\binom{n}{r} = \frac{P(n, r)}{r!} = \frac{n!}{r!(n-r)!},$$

as claimed in (1.4). Note that this argument fails when $r = 0$, since $P(n, 0)$ has not been defined. But we can easily check that (1.4) is correct when $r = 0$. For a set with n elements has exactly one 0-subset (the null set $\emptyset$) and Formula (1.4) yields this answer, since when $r = 0$, (1.4) becomes

$$(1.9) \qquad \binom{n}{0} = \frac{n!}{0!(n-0)!} = 1.$$

(By now the reader should be convinced that putting $0! = 1$ is indeed sensible.)

Finally, to prove (1.5) we use (1.4) in conjunction with the fundamental principle of counting. To determine the n_1 objects that go into the first cell we choose an n_1-subset from the available n objects. We can therefore allocate n_1 objects to the first cell in $\binom{n}{n_1}$ ways. To determine the n_2 objects that are put in the second cell, we choose an n_2-subset from the remaining $n - n_1$ objects. Hence we can put n_2 objects into this second cell in $\binom{n - n_1}{n_2}$ ways. Continuing in this way, we see that to determine the n_k objects that go into the last cell we must choose an n_k-subset from the remaining $n - (n_1 + n_2 + \cdots + n_{k-1})$ objects, and this can be done in

$$\binom{n - n_1 - \cdots - n_{k-1}}{n_k}$$

ways. By the fundamental principle of counting, we conclude that

$$\binom{n}{n_1, n_2, \cdots, n_k} = \binom{n}{n_1}\binom{n - n_1}{n_2} \cdots \binom{n - n_1 - \cdots - n_{k-1}}{n_k}.$$

Now we use (1.4) to simplify the product on the right. The product of the first two terms is

$$\binom{n}{n_1}\binom{n - n_1}{n_2} = \frac{n!}{n_1!(n - n_1)!} \frac{(n - n_1)!}{n_2!(n - n_1 - n_2)!}$$

$$= \frac{n!}{n_1! n_2!(n - n_1 - n_2)!}.$$

The product of the first three factors is

$$\binom{n}{n_1}\binom{n - n_1}{n_2}\binom{n - n_1 - n_2}{n_3}$$

$$= \frac{n!}{n_1! n_2!(n - n_1 - n_2)!} \frac{(n - n_1 - n_2)!}{n_3!(n - n_1 - n_2 - n_3)!}$$

$$= \frac{n!}{n_1! n_2! n_3!(n - n_1 - n_2 - n_3)!}.$$

When all k factors are multiplied, we similarly find that

$$\binom{n}{n_1, n_2, \cdots, n_k} = \frac{n!}{n_1! n_2! \cdots n_k!(n - n_1 - n_2 - \cdots - n_k)!}.$$

But $(n - n_1 - n_2 - \cdots - n_k)! = 0! = 1$, and so the proof of (1.5) is complete.

Two special cases and an important alternative interpretation of Formula (1.5) are worth noting here. If we have n objects and n cells, then there are clearly just as many ways of putting one object in each cell as there are ordered n-tuples of these n objects. Hence we expect that

$$\binom{n}{1, 1, \cdots, 1} = P(n, n) = n!$$

and indeed, Formula (1.5) yields this expected result if we put $k = n$ and $n_1 = n_2 = \cdots = n_n = 1$.

Also, if we put $k = 2$ and $n_1 = r$ (and therefore $n_2 = n - r$), then (1.5) becomes

$$(1.10) \qquad \binom{n}{r, \, n - r} = \frac{n!}{r!(n - r)!} = \binom{n}{r}.$$

This equation merely expresses the fact that there are as many ways of placing n objects into two cells with r objects in one cell and $(n - r)$ in the other cell as there are different r-subsets of a set with n elements. For in determining the r elements in the r-subset (cell 1), we automatically place the remaining $n - r$ elements in the complement of the r-subset (cell 2).

A second interpretation of Formula (1.5) is introduced in the following examples.

Example 1.2. We know that there are $P(5, 5) = 5! = 120$ permutations of five *distinct* letters. But now suppose the five letters are two a's and three b's. How many different permutations are there now? We find that there are only ten:

> aabbb   ababb   abbab   abbba   baabb
> babab   babba   bbaab   bbaba   bbbaa.

To see how to obtain the number 10 without explicitly enumerating the permutations, think not of the letters themselves but rather of the *positions* they occupy in the permutation. Counting from left to right, a permutation contains five positions and is uniquely determined as soon as we specify the two positions for the a's and the three positions for the b's. Thus there are just as many permutations as there are ways of putting the five positions into two cells, the first containing two positions for the a's and the second containing three positions for the b's. But this number is what we denoted by $\binom{5}{2, \, 3}$ and by (1.5) is seen to be 10, as expected.

Example 1.3. To determine the number of distinguishable arrangements on one shelf of four different books, for each of which there are two copies, we note that each arrangement contains eight distinct positions and is determined as soon as we specify the two positions for the first book, the two positions for the second book, etc. The eight positions can be placed in four cells, each containing two positions, in

$$\binom{8}{2, 2, 2, 2} = \frac{8!}{(2!)^4} = 2520$$

different ways by (1.5). Thus there are 2520 distinguishable arrangements of the eight books.

The arguments used in the preceding examples are readily generalized to prove the following theorem.

Theorem 1.2. If we have n objects, n_1 of which are of one kind, n_2 of a second kind, $\cdots$, n_k of a kth kind (where $n_1 + n_2 + \cdots + n_k = n$), then the number of distinguishable permutations of the n objects is given by

$$\binom{n}{n_1, n_2, \cdots, n_k} = \frac{n!}{n_1! n_2! \cdots n_k!}.$$

Proof. We have only to observe that each permutation contains n positions and is uniquely determined as soon as we specify the n_1 positions for the objects of the first kind, the n_2 positions for the objects of the second kind, $\cdots$, the n_k positions for the objects of the kth kind. Thus there are as many permutations as there are ways of putting the n positions into k cells, the ith cell containing the n_i positions for the objects of the ith kind for $i = 1, 2, \cdots, k$. But this number of ways is given in (1.5), and so the proof is complete.

We turn now to some examples illustrating the use of these counting techniques in computing probabilities.

Example 1.4. Find the probability when a bridge game is dealt that each player has exactly one ace.

There are as many ways of dealing four bridge hands as there are ways of placing 52 objects (the cards of the full deck) into four cells (the North, East, South, and West hands) so that each cell contains 13 cards. Thus, by (1.5), there are

$$N = \binom{52}{13, 13, 13, 13} = \frac{52!}{(13!)^4}$$

different deals in bridge. Let us choose as our sample space S a set with N elements, each denoting a different deal, and let us assign to each simple event of S the same probability $1/N$. (N is actually equal to a number larger than 53 billion billion billion, but we do not need to know the precise value of N to complete this problem.)

Now the event "each player has one ace" is the union of as many simple events of S as there are different deals for which North, East, South, and West each have one ace. If we call this number x, then the required probability is x/N by Theorem II.4.7. To determine x, note that there are as many deals for which each player has one ace as there are ways of completing the following tasks in the stated order: (1) deal the four aces, one to each player, (2) deal the remaining 48 cards, 12 to each player. By (1.5), task 1 can be done in

$$\binom{4}{1,\,1,\,1,\,1} = \frac{4!}{(1!)^4} \quad \text{ways}$$

and task 2 in

$$\binom{48}{12,\,12,\,12,\,12} = \frac{48!}{(12!)^4} \quad \text{ways.}$$

By the fundamental principle of counting, the number of deals for which each player has one ace is

$$x = \frac{4!48!}{(12!)^4}$$

and the required probability, say p, is given by

$$p = \frac{x}{N} = \frac{4!48!(13!)^4}{(12!)^4 52!}.$$

To compute p, we first use Table 21 to find the logarithm of p:

$$\begin{aligned}
\log p &= \log 4! + \log 48! + 4(\log 13!) - 4(\log 12!) - \log 52! \\
&= 1.3802 + 61.0939 + 4(9.7943) - 4(8.6803) - 67.9067 \\
&= -.9766 \\
&= 9.0234 - 10.
\end{aligned}$$

Now we use Table 22 to estimate the value of p. Since the mantissa .0234 is between .0000 and .0414, we know that p is between .10 and .11. Thus, odds against the event "each player has one ace" are approximately 9 to 1.

Example 1.5. From five married couples, four people are selected. What is the probability that two men and two women are chosen?

There are $\binom{10}{4}$ ways of selecting a subset of four people from all ten people. Since

$$\binom{10}{4} = \frac{10!}{4!6!} = \frac{10 \cdot 9 \cdot 8 \cdot 7}{4 \cdot 3 \cdot 2 \cdot 1} = 210,$$

TABLE 21.   COMMON LOGARITHMS OF FACTORIALS

n	$\log n!$	n	$\log n!$	n	$\log n!$
1	0.0000	26	26.6056	51	66.1906
2	0.3010	27	28.0370	52	67.9067
3	0.7782	28	29.4841	53	69.6309
4	1.3802	29	30.9465	54	71.3633
5	2.0792	30	32.4237	55	73.1037
6	2.8573	31	33.9150	56	74.8519
7	3.7024	32	35.4202	57	76.6077
8	4.6055	33	36.9387	58	78.3712
9	5.5598	34	38.4702	59	80.1420
10	6.5598	35	40.0142	60	81.9202
11	7.6012	36	41.5705	61	83.7055
12	8.6803	37	43.1387	62	85.4979
13	9.7943	38	44.7185	63	87.2972
14	10.9404	39	46.3096	64	89.1034
15	12.1165	40	47.9117	65	90.9163
16	13.3206	41	49.5244	66	92.7359
17	14.5511	42	51.1477	67	94.5620
18	15.8063	43	52.7812	68	96.3945
19	17.0851	44	54.4246	69	98.2333
20	18.3861	45	56.0778	70	100.0784
21	19.7083	46	57.7406	71	101.9297
22	21.0508	47	59.4127	72	103.7870
23	22.4125	48	61.0939	73	105.6503
24	23.7927	49	62.7841	74	107.5196
25	25.1907	50	64.4831	75	109.3946

TABLE 22.   COMMON LOGARITHMS*

	0	1	2	3	4	5	6	7	8	9
1	0000	0414	0792	1139	1461	1761	2041	2304	2553	2788
2	3010	3222	3424	3617	3802	3979	4150	4314	4472	4624
3	4771	4914	5051	5185	5315	5441	5563	5682	5798	5911
4	6021	6128	6232	6335	6435	6532	6628	6721	6812	6902
5	6990	7076	7160	7243	7324	7404	7482	7559	7634	7709
6	7782	7853	7924	7993	8062	8129	8195	8261	8325	8388
7	8451	8513	8573	8633	8692	8751	8808	8865	8921	8976
8	9031	9085	9138	9191	9243	9294	9345	9395	9445	9494
9	9542	9590	9638	9685	9731	9777	9823	9868	9912	9956

* For example, $\log 1.2 = .0792$, $\log .12 = 9.0792 - 10 = -.9208$, $\log .76 = 9.8808 - 10 = -.1192$.

we take as sample space a set S with 210 elements and assign each simple event of S the same probability $\frac{1}{210}$.

Now we can select two men from the five available men in $\binom{5}{2} = 10$ ways. Similarly, there are ten ways of selecting two women. By the fundamental principle of counting, there are $10 \cdot 10 = 100$ ways of selecting two men and two women. Hence the required probability is $\frac{100}{210}$ or $\frac{10}{21}$.

Example 1.6. A coin is tossed five independent times. What is the probability of the event E that we get exactly three heads?

We define $S_1 = \{H, T\}$ as sample space for a single toss and, since we have no information about the coin, let us put

$$P(\{H\}) = p, \qquad P(\{T\}) = q = 1 - p,$$

where p is some number between 0 and 1 inclusive. For the five-toss experiment, we must define as sample space S the Cartesian product

$$S = S_1 \times S_1 \times S_1 \times S_1 \times S_1.$$

Because of the assumed independence of the tosses, the probability of each simple event of S is determined by the product rule given in Section II.9. For example,

$$P(\{HHHTT\}) = P(\{H\})P(\{H\})P(\{H\})P(\{T\})P(\{T\}) = p^3 q^2$$

and similarly

$$P(\{HHTHT\}) = p \cdot p \cdot q \cdot p \cdot q = p^3 q^2.$$

In fact, any simple event whose sole element corresponds to an outcome resulting in three heads and two tails will have probability $p^3 q^2$. The number of such simple events is the same as the number of 5-tuples containing exactly three H's and two T's. Such a 5-tuple is uniquely determined as soon as we select the positions (i.e., numbers identifying which are the tosses) that resulted in heads. We can select three positions from the available five in $\binom{5}{3} = 10$ ways. Since the event E is the union of these ten simple events, each with probability $p^3 q^2$, we have

$$P(E) = 10 p^3 q^2.$$

If the coin is fair, then

$$p = q = \tfrac{1}{2} \quad \text{and} \quad P(E) = \tfrac{10}{32} = .31, \quad \text{approximately.}$$

But if the coin is biased so that, let us say, $p = \frac{1}{3}$ and $q = \frac{2}{3}$, then

$$P(E) = \frac{40}{243} = .16, \quad \text{approximately.}$$

This coin-tossing example is typical of an important class of problems that we will study in Chapter 5. Note especially that we have here an example in which (assuming $p \neq q$) the simple events of S are *not* assigned the same probability.

Example 1.7. What is the probability that a poker hand will have one pair?

A poker hand is a 5-subset of the set of 52 cards in the full deck and so there are $\binom{52}{5}$ different poker hands. If for the moment, we call this number N, then our sample space S has N elements and each simple event of S is assigned probability $1/N$. Using Formula (1.4) and some arithmetic, we find that $N = 2{,}598{,}960$.

Now a poker hand with one pair has two cards of the same face value (i.e., two aces, two kings, etc.) and three cards whose face values are all different and different from that of the pair. We obtain a unique poker hand with one pair by completing the following tasks in order: (1) choose the face value for the pair from the 13 available face values. This can be done in $\binom{13}{1} = 13$ ways. (2) Choose two cards with the face value selected in (1). This can be done in $\binom{4}{2} = 6$ ways. (3) Choose the three face values for the other three cards in the hand. Since there are 12 face values from which to choose, this can be done in $\binom{12}{3} = 220$ ways. (4) Choose one card (from the four available) of each face value chosen in (3). This can be done in $4^3 = 64$ ways. By the fundamental principle of counting, there are

$$13 \cdot 6 \cdot 220 \cdot 64 = 1{,}098{,}240$$

poker hands with one pair. Hence the required probability is

$$\frac{1{,}098{,}240}{2{,}598{,}960} = .42, \quad \text{approximately.}$$

As poker players know, and as this answer shows, it is not at all unusual to have a one-pair hand. Only hands with no pair at all have a higher probability of occurring. (See Problem 1.25.)

Let us conclude by outlining the procedure we followed in answering the probability questions posed in the preceding problems.

(1) Define a sample space S for the experiment and assign probabilities to its simple events. This may involve counting the elements of S (as in Examples 1.4, 1.5, and 1.7) or it may require other noncounting considerations (as in Example 1.6).

(2) Determine $P(E)$ by calculating the sum of the probabilities of the simple events whose union is the event E. In the problems under discussion here, this requires counting the number of elements of E. To do this, it is often helpful first to construct a sequence of tasks having the following property: each way of completing the tasks in the specified sequential order produces an experimental outcome corresponding to exactly one element of E and conversely, each element of E corresponds to an outcome produced by completing the tasks in exactly one way. (We did this in Example 1.4 where a bridge deal for which each player has one ace was produced by completing two tasks, and also in Example 1.7 where each poker hand with one pair was thought of as produced by completing four tasks in order.) The problem of counting the number of elements in E is thereby reduced to that of counting the number of ways of completing these tasks.

(3) Now use one or more of the formulas discussed in this section to count the number of ways of completing each task. Then invoke the fundamental principle of counting to determine the number of ways of completing all the tasks in the stated order. This number is the number of elements in E, and $P(E)$ can thus be evaluated.

The problems that follow will help the reader develop his ability to count by means of the formulas presented in this section. He thus becomes able to find probabilities in a wide class of more complicated but more interesting experiments than heretofore considered.

PROBLEMS

1.1. Evaluate:

(a) $\dfrac{8!}{3!5!}$ (b) $\dfrac{52!}{50!}$ (c) $\dbinom{9}{5}$

(d) $\dbinom{10}{6}$ (e) $\dbinom{9}{4,\,3,\,2}$ (f) $\dbinom{10}{7,\,0,\,3}$

1.2. How large must n be before $n!$ exceeds (a) a thousand? (b) a million? (c) a billion? (d) a trillion? [*Hint:* Use Table 21.]

1.3. Compute to two decimal place accuracy (using logarithms) the value of

(a) $\dfrac{\dbinom{60}{12}}{\dbinom{70}{12}}$ (b) $\dfrac{\dbinom{10}{4}\dbinom{60}{8}}{\dbinom{70}{12}}$ (c) $\dbinom{12}{3}\left(\dfrac{1}{7}\right)^3\left(\dfrac{6}{7}\right)^9$

1.4. Let r be a positive integer. For any number x, let

$$(x)_r = x(x-1)(x-2)\cdots(x-r+1).$$

Show that

(a) if x is an integer and $x \geq r$, then $(x)_r = P(x, r)$.

(b) $(-1)_r = (-1)^r r!$

(c) $(-2)_r = (-1)^r (r+1)!$

(d) $(-\tfrac{1}{2})_r = (-1)^r r! 2^{-2r} \dbinom{2r}{r}$.

1.5. (a) One straight line is determined by two points in a plane. Three lines are determined by three noncolinear points. Six lines are determined by four points, no three of which are colinear. How many lines are determined by n points, no three of which are colinear?

(b) A triangle has no diagonals; a quadrilateral has two diagonals; a pentagon has five diagonals. How many diagonals does a polygon of n sides have?

1.6. The 11 digits 1, 2, 2, 2, 3, 3, 3, 3, 4, 4, 4 are permuted in all distinguishable ways. How many permutations (a) begin with 22? (b) begin with 343?

1.7. Each permutation of the digits 1, 2, 3, 4, 5, 6 determines a six-digit number. If the numbers corresponding to all possible permutations are listed in order of increasing magnitude, which is the 417th?

1.8. The six digits 1, 1, 1, 2, 3, 3 are permuted and, as in the preceding problem, we list the corresponding six-digit numbers in order of increasing magnitude.

(a) How many numbers start with the digit 2?

(b) How far down in the list is the number 321,311?

1.9. We have two each of n different objects, $2n$ objects altogether. How many distinguishable selections of four objects are there for which (a) all four objects are different? (b) two are alike and two different? (c) two are alike and the other two are also alike? (d) The *total* number of distinguishable selections of four objects from these $2n$ objects is equal to six times the number of 4-subsets of the set of n different objects. Find the value of n.

1.10. A group of ten boys and ten girls is divided into two groups of ten each. Find the probability that each group contains as many boys as girls.

1.11. A bookstore clerk has ten books, five each of two titles, to place on a bookshelf. If he places them at random so that all distinguishable arrangements are equally likely, what is the probability that (a) five copies of one title follow five copies of the other title on the shelf? (b) the two titles alternate on the shelf?

1.12. You need four eggs to make omelets for breakfast. You find a dozen eggs in the refrigerator but do not realize that two of these eggs are rotten. What is the probability that of the four eggs you choose (a) none are rotten? (b) exactly one is rotten? (c) exactly two are rotten?

1.13. In the preceding problem, suppose you break the four eggs into a saucer and discover that you have chosen at least one rotten egg.

(a) What is the conditional probability that both rotten eggs are in the saucer?

(b) If you choose four other eggs from the remaining eight eggs, what is the conditional probability that they will all be good?

1.14. Refer to Example 1.6 of the text and find the probability that the coin falls heads exactly k times, where $k = 0, 1, 2, 3, 4, 5$.

1.15. Baseball team A plays team B ten times in a given month. Assume that team A is better than team B and has probability $\frac{3}{5}$ of winning and probability $\frac{2}{5}$ of losing each game. If the games are considered as ten independent trials, find the probability team A wins (a) exactly six games, (b) exactly seven games, (c) a majority of the games.

1.16. A pack of ten cards consists of three aces, two kings, two queens, and three jacks. We shuffle the deck and pick one card. Let this trial be performed eight independent times. What is the probability that an ace is selected twice, a king three times, and a jack three times?

1.17. Find the probability that in eight independent rolls of a fair die, the numbers 1, 3, and 5 turn up two, three, and three times, respectively.

1.18. From a panel of 20 seniors, 15 juniors, ten sophomores, and five freshmen, a committee of five is selected at random. What is the probability that the committee consists of two seniors and one from each of the other classes?

1.19. In the preceding problem, suppose you know that the committee contains exactly one freshman. What is the conditional probability that there are also two seniors, one sophomore, and one junior on the committee?

1.20. There are ten defective and 60 good transistors in a lot from which you select a sample (without replacement) of 12. Calculate with two decimal place accuracy the probability that the sample contains (a) no defectives. (b) exactly one defective. (c) exactly two defectives. (d) exactly three defectives. (e) exactly four defectives. (f) exactly five defectives.

1.21. In the preceding problem, suppose the sample of 12 was selected *with* replacement. Find the probability that the sample contains three defectives and compare with the corresponding answer for sampling without replacement. Do the same for 0, 1, 2, 4, and 5 defectives.

1.22. We scramble the letters of the word "Muhammadan" and then arrange them in some order.

(a) What is the probability that the three a's will be consecutive letters?

(b) What is the probability that the three a's will be consecutive and the three m's will also be consecutive?

(c) What is the probability that no three consecutive letters are alike?

1.23. The 11 letters of the word "Mississippi" are scrambled and then arranged in some order.

(a) What is the probability that the four i's are consecutive letters in the resulting arrangement?

(b) What is the conditional probability that the four i's are consecutive, given that the arrangement starts with "M" and ends with "s"?

(c) What is the conditional probability that the four i's are consecutive, given that the arrangement ends with four consecutive esses?

1.24. A poker player holds a pair of aces and a king, queen, and jack. He discards three cards, holding his pair, and draws three more cards from the deck of 47 cards. What is the probability that his hand contains (a) three aces after the draw? (b) two pairs, aces high, after the draw? (*Note:* When a hand is spoken of as containing a pair of aces, three aces, etc., we will mean that it contains no higher count. Thus, to say a hand contains three aces means that it contains exactly three aces and two different cards. To say a hand contains two pairs, aces high, means that it contains a pair of aces, a different pair, and a fifth card of a still different face value.)

1.25. In Example 1.7 we found the probability of a one-pair poker hand. Now find the probability of the following poker hands:

(a) no pair (five different face values, not in sequence, not same suit.)

    (b) two pairs (one pair of each of two different face values plus a card of a third face value.)

    (c) three of a kind (exactly three cards of one face value plus two different cards.)

    (d) straight (five cards in sequence, but not all of the same suit.)

    (e) flush (five cards of the same suit but not in sequence.)

    (f) full house (three cards of one face value, and two cards of another face value.)

    (g) four of a kind (four cards of one face value). See Example I.2.3.

    (h) straight flush (five cards in sequence and of the same suit.)

1.26. In seven-card stud poker, your first three cards are of the same suit. What is the probability that you will find at least two more cards of the same suit among the other four cards in your hand?

1.27. Let North be dealt 13 cards from a bridge deck and suppose S_1 is the sample space corresponding to this deal (trial).

    (a) Show that dealing hands to all four players in a bridge game can be considered as a four-trial experiment in which the trials are identical but *not* independent. What is the sample space S for the four-trial experiment?

    (b) Calculate the probability of the event E that North has exactly one ace, considering E as a subset of S.

    (c) Calculate $P(E)$, but now considering E as a subset of S_1.

    (d) Give a precisely worded and complete explanation of why your answers in (b) and (c) are the same. (*Hint:* Refer to Section II.9.)

1.28. In a bridge game, what is the probability that you and your partner together have exactly k aces, where $k = 0, 1, 2, 3, 4$?

1.29. When a bridge hand is dealt, what is the probability of (a) a 5–4–3–1 distribution, i.e., of a hand containing five cards of one suit, four of another, etc.? (b) a 4–4–3–2 distribution? (c) a 4–3–3–3 distribution?

1.30. You and your partner in bridge are declarers and hold nine spades, including the ace and king. The defenders hold four spades, including the queen. What is the probability that the distribution of the four spades in the opposing hands is (a) four in one hand, none in the other? (b) three in one hand, one in the other? (c) two in one hand, two in the other?

1.31. Continuing the preceding problem, you know that the queen will fall when you lead the ace and king if the four spades are divided equally between the opposing hands or if the queen is the only spade in one of them. Show that odds for the queen's falling on the lead of the ace and king are approximately 1.13 to 1.

2. Binomial coefficients

The numbers $\binom{n}{r}$ introduced in the preceding section have many interesting and important properties that we will need to know for our later work. Although it is a slight digression at this point, we pause here to develop some of these properties.

Our first task is to explain the title of this section. The reader is familiar with the formulas

(2.1) $(x + y)^2 = x^2 + 2xy + y^2$
(2.2) $(x + y)^3 = x^3 + 3x^2y + 3xy^2 + y^3$
(2.3) $(x + y)^4 = x^4 + 4x^3y + 6x^2y^2 + 4xy^3 + y^4.$

For any positive integer n, $(x + y)^n$ is the product of n equal factors:

(2.4) $(x + y)^n = (x + y)(x + y) \cdots (x + y).$

In seeking a general formula for $(x + y)^n$, it is helpful first to identify the coefficients in the special cases (2.1)–(2.3). Indeed, the reader should check that these can be written as follows, thus suggesting our first theorem:

$$(x + y)^2 = \binom{2}{0}x^2 + \binom{2}{1}xy + \binom{2}{2}y^2 = \sum_{r=0}^{2}\binom{2}{r}x^{2-r}y^r;$$

$$(x + y)^3 = \binom{3}{0}x^3 + \binom{3}{1}x^2y + \binom{3}{2}xy^2 + \binom{3}{3}y^3$$

$$= \sum_{r=0}^{3}\binom{3}{r}x^{3-r}y^r;$$

$$(x + y)^4 = \binom{4}{0}x^4 + \binom{4}{1}x^3y + \binom{4}{2}x^2y^2 + \binom{4}{3}xy^3 + \binom{4}{4}y^4$$

$$= \sum_{r=0}^{4}\binom{4}{r}x^{4-r}y^r.$$

Theorem 2.1. (The binomial theorem.) If n is any positive integer and x and y are any numbers, then

$$(x + y)^n = \binom{n}{0}x^n + \binom{n}{1}x^{n-1}y + \binom{n}{2}x^{n-2}y^2 + \cdots + \binom{n}{n}y^n$$

or, more concisely,

(2.5) $$(x + y)^n = \sum_{r=0}^{n}\binom{n}{r}x^{n-r}y^r.$$

Proof. To compute $(x + y)^n$, we choose either the letter x or the letter y from each of the n factors in (2.4) and multiply these n choices. If we do this for all possible choices of x's and y's and add the results, we obtain $(x + y)^n$. For example, we get the product x^n by choosing x from each factor, we get the product $x^{n-1}y$ whenever we choose x from all but one factor, we get $x^{n-2}y^2$ whenever we choose x from all but two factors, etc.

For a given integer r $(0 \leq r \leq n)$, the product $x^{n-r}y^r$ is obtained whenever we choose exactly r y's (and therefore $n - r$ x's.) To determine our choice uniquely, we have only to decide from which r of the n factors we select y's. Hence there are $\binom{n}{r}$ choices, each leading to the product $x^{n-r}y^r$ and so the term $\binom{n}{r} x^{n-r}y^r$ appears in the expansion of $(x + y)^n$. Since r is any integer from 0 to n inclusive, the theorem is proved.

Because the numbers $\binom{n}{r}$ appear as coefficients in the expansion of a power of a binomial, they are called *binomial coefficients*.

Example 2.1. To expand $(1 + t)^n$ we put $x = 1$, $y = t$ in (2.5) and find

$$(2.6) \qquad (1 + t)^n = \sum_{r=0}^{n} \binom{n}{r} t^r = \binom{n}{0} + \binom{n}{1} t + \cdots + \binom{n}{n} t^n.$$

If we now put $t = 1$, there follows the interesting identity

$$(2.7) \qquad\qquad 2^n = \binom{n}{0} + \binom{n}{1} + \cdots + \binom{n}{n}.$$

Since $\binom{n}{r}$ is the number of r-subsets of a set with n elements, we see that the right-hand side of (2.7) is the number of 0-subsets (which is 1, there being only one null set) plus the number of 1-subsets plus the number of 2-subsets, etc. This sum is therefore the total number of subsets and so (2.7) supplies another proof of Theorem I.2.1, which says that a set with n elements has 2^n subsets.

Example 2.2. The binomial theorem can be used to compute an approximate value of $(.99)^6$. For with $x = 1$ and $y = -.01$ in (2.5) or equivalently, with $t = -.01$ in (2.6), we obtain

$(.99)^6 = (1 - .01)^6$

$$= \binom{6}{0} + \binom{6}{1}(-.01) + \binom{6}{2}(-.01)^2 + \cdots + \binom{6}{6}(-.01)^6$$

$$= 1 - .06 + .0015 - .00002 + \cdots + .000000000001$$

$$= .941, \text{ to three decimal places.}$$

A convenient device for calculating and displaying the binomial coefficients is known as Pascal's triangle.* The first few rows are illustrated in Table 23. The row for $n = 0$ lists the one coefficient in the expansion of $(x + y)^0$, the row for $n = 1$ lists the two coefficients in the expansion of $(x + y)^1$, the row for $n = 2$ lists the three coefficients in the expansion of $(x + y)^2$, and so on. Since every set with n elements has exactly one null subset ($\emptyset$) and exactly one n-subset (namely, itself), we find 1's under the column headed $r = 0$ and also along the hypotenuse of the triangular array.

TABLE 23

n \ r	0	1	2	3	4	5	6
0	1						
1	1	1					
2	1	2	1				
3	1	3	3	1			
4	1	4	6	4	1		
5	1	5	10	10	5	1	
6	1	6	15	20	15	6	1

We observe that there is a simple relation among numbers in the triangle. For if we start at any number not on the hypotenuse and move to the right one number and then drop down to the row below we note that the sum of the first two numbers is precisely the number in the row below. For example, starting at the left in the row for $n = 4$ we obtain the numbers in the row for $n = 5$ as follows:

$$1 + 4 = 5, \quad 4 + 6 = 10, \quad 6 + 4 = 10, \quad 4 + 1 = 5.$$

The following result proves that our observation is generally true and can thus be used to extend the table one row at a time and thereby to compute binomial coefficients $\binom{n}{r}$ for larger and larger values of n.

Theorem 2.2. For any positive integers r and n with $r < n$,

---

* Pascal (1623–1662) was far from the first to form and study this triangular array of numbers, but somehow his name has become attached to it. For interesting historical notes, see C. B. Boyer, "Cardan and the Pascal Triangle," *American Mathematical Monthly*, vol. 57 (1950), 387–390.

(2.8) $$\binom{n}{r} = \binom{n-1}{r-1} + \binom{n-1}{r}.$$

Proof. Although it is possible to give a direct algebraic proof by writing the binomial coefficients in terms of factorials (see Problem 2.5c), we prefer the following argument. The $\binom{n}{r}$ r-subsets of a set with n elements can be divided into those that include a given element, and those that do not. The number of r-subsets including the given element is $\binom{n-1}{r-1}$, since fixing one element leaves us free to select $r-1$ others from the remaining $n-1$. The number of r-subsets that do not include the given element is $\binom{n-1}{r}$, since we are now choosing r from $n-1$ elements. Since we have now accounted for all r-subsets, formula (2.8) is proved.

Two other properties of the binomial coefficients are apparent from the Pascal triangle. The symmetry in each row of the triangle is due to the identity

(2.9) $$\binom{n}{r} = \binom{n}{n-r},$$

which expresses the fact that every selection of an r-subset is automatically a selection of its complementary $(n-r)$-subset, and vice versa. It follows that

$$\binom{n}{0} = \binom{n}{n}, \quad \binom{n}{1} = \binom{n}{n-1}, \quad \binom{n}{2} = \binom{n}{n-2}, \quad \text{etc.,}$$

and so in each row of the triangle, the first and last numbers are equal, as are the second and next to last, and so on.

We also observe that the binomial coefficients in any row increase as we move to the right and then eventually start to decrease. We leave for the problems the proof that this is generally true, and instead derive another important identity involving binomial coefficients.

As we shall see in the example that follows, it is convenient to extend the definition of the binomial coefficient $\binom{n}{r}$ to values of n and r when the symbol no longer has any combinatorial meaning. Indeed, it is possible and useful to have $\binom{n}{r}$ defined even when n is not an integer. (See Problem 2.9.) But here we make only the definition that for any positive integer n and r also an integer,

$$(2.10) \qquad \binom{n}{r} = 0 \quad \text{if either} \quad r > n \quad \text{or} \quad r < 0.$$

Since a set with n elements has no subsets with more than n elements and also does not have any subsets with a negative number of elements, Definition (2.10) is quite reasonable.

Example 2.3. We are given n defective and m acceptable items, $n + m$ items altogether, from a production line. They are mixed up and we are to draw a subset of r items from the lot. We count the number of possible r-subsets in two different ways, and thus derive a useful identity. First, there are clearly $\binom{n+m}{r}$ r-subsets that can be drawn from the $n + m$ items. To count again, note that the r-subsets can be classified according to the number of defective items they contain. We can select k defectives (and therefore $r - k$ acceptable items) in

$$\binom{n}{k}\binom{m}{r-k} \quad \text{ways.}$$

If we let k vary over all possible values and add these numbers, then we obtain *all* the r-subsets. Hence

$$\binom{n}{0}\binom{m}{r} + \binom{n}{1}\binom{m}{r-1} + \binom{n}{2}\binom{m}{r-2} + \cdots$$
$$+ \binom{n}{r}\binom{m}{0} = \binom{n+m}{r},$$

or

$$(2.11) \qquad \sum_{k=0}^{r} \binom{n}{k}\binom{m}{r-k} = \binom{n+m}{r}.$$

Note that some of the terms in the sum may be zero, since we cannot have more defective and acceptable items in the r-subset than there are defective and acceptable items in the whole lot. For example, if the number of acceptable items m happens to be equal to $r - 2$, then $\binom{m}{r}$ and $\binom{m}{r-1}$ both are zero according to (2.10). The value of the definition made in (2.10) lies in the fact that it allows us to write the sum in (2.11) without worrying about terms that should be omitted; instead of omitting them we made sure they would be zero.

Identity (2.11) will be used in Chapter 5 in connection with the so-called hypergeometric distribution in probability.

The method we used to prove Theorem 2.1 can be extended to prove the following result.

Theorem 2.3. (The multinomial theorem.) Let n be any positive integer and x_1, x_2, $\cdots$, x_k any k numbers. Then

$$(2.12) \qquad (x_1 + x_2 + \cdots + x_k)^n = \Sigma \binom{n}{n_1, n_2, \cdots, n_k} x_1^{n_1} x_2^{n_2} \cdots x_k^{n_k}$$

where the sum is taken over all nonnegative integers n_1, n_2, $\cdots$, n_k such that $n_1 + n_2 + \cdots + n_k = n$.

Proof. We again have n factors to multiply, but now each is the multinomial $(x_1 + x_2 + \cdots + x_k)$ instead of the binomial $(x + y)$ in (2.4). From each factor we must choose x_1, or x_2, $\cdots$, or x_k, multiply our n choices and add these products for all possible choices. For example, we get the product x_1^n by choosing x_1 from each factor, we get the product $x_1^{n-2} x_2 x_k$ by choosing x_1 from $n - 2$ of the factors, x_2 from one factor and x_k from another factor, etc.

For given nonnegative integers n_1, n_2, $\cdots$, n_k (whose sum is n) we get the product $x_1^{n_1} x_2^{n_2} \cdots x_k^{n_k}$ whenever we choose exactly

$$n_1 \ x_1\text{'s}, \quad n_2 \ x_2\text{'s}, \quad \cdots, \quad n_k \ x_k\text{'s}$$

from the n available factors. There are as many ways of making such a choice as there are ways of placing the n factors into k cells, the first cell containing the n_1 factors from which we choose x_1, the second cell containing the n_2 factors from which we choose x_2, etc. By (1.5) there are therefore

$$(2.13) \qquad \binom{n}{n_1, n_2, \cdots, n_k}$$

choices, each leading to the product $x_1^{n_1} x_2^{n_2} \cdots x_k^{n_k}$. Thus we have the multinomial theorem.

The numbers (2.13) are called *multinomial coefficients*. Since, as we noted in (1.10), a binomial coefficient is a special case of a multinomial coefficient, we see that the multinomial theorem becomes the binomial theorem if we put $k = 2$.

Example 2.4. To expand $(p + q + r)^3$ we write

$$(p + q + r)^3 = \Sigma \begin{pmatrix} 3 \\ n_1, n_2, n_3 \end{pmatrix} p^{n_1} q^{n_2} r^{n_3},$$

where the sum is taken over all nonnegative integers n_1, n_2, n_3 such that $n_1 + n_2 + n_3 = 3$. Hence

$$(p + q + r)^3 = \begin{pmatrix} 3 \\ 3, 0, 0 \end{pmatrix} p^3 + \begin{pmatrix} 3 \\ 0, 3, 0 \end{pmatrix} q^3 + \begin{pmatrix} 3 \\ 0, 0, 3 \end{pmatrix} r^3 + \begin{pmatrix} 3 \\ 2, 1, 0 \end{pmatrix} p^2 q$$

$$+ \begin{pmatrix} 3 \\ 1, 2, 0 \end{pmatrix} pq^2 + \begin{pmatrix} 3 \\ 2, 0, 1 \end{pmatrix} p^2 r + \begin{pmatrix} 3 \\ 1, 0, 2 \end{pmatrix} pr^2$$

$$+ \begin{pmatrix} 3 \\ 0, 2, 1 \end{pmatrix} q^2 r + \begin{pmatrix} 3 \\ 0, 1, 2 \end{pmatrix} qr^2 + \begin{pmatrix} 3 \\ 1, 1, 1 \end{pmatrix} pqr$$

$$= p^3 + q^3 + r^3 + 3p^2 q + 3pq^2 + 3p^2 r + 3pr^2$$
$$+ 3q^2 r + 3qr^2 + 6pqr.$$

We can give a probability interpretation to the terms in this sum by imagining that each person in a certain population is classified according to whether he answers "yes," "no," or "don't know," when asked a certain question. If we select one person at random from the population and record his answer, this trial is defined by the sample space $\{Y, N, DK\}$. We make the assignment of probabilities to simple events as follows:

$$P(\{Y\}) = p, \qquad P(\{N\}) = q, \qquad P(\{DK\}) = r,$$

where we assume $p \geq 0$, $q \geq 0$, $r \geq 0$, and $p + q + r = 1$. Now perform this trial three independent times, for instance, by choosing a random sample of three people with replacement from the population and asking each the question. Then the terms in the expansion of $(p + q + r)^3$ give the probabilities of all the possible combinations of answers. For example, the probability that all three people answer "yes" is p^3, the probability that exactly one person answers "yes" and two people answer "don't know" is $3pr^2$, etc.

PROBLEMS

2.1. Expand by the binomial theorem.

    (a) $(p + q)^5$ (b) $(1 - x)^4$ (c) $(a - 3b)^4$

2.2. What is the coefficient of $a^5 b^3$ in the expansion of $(2a - b)^8$?

2.3. By means of the binomial theorem, evaluate to three decimal places.

    (a) $(1.01)^7$ (b) $(1.02)^{10}$ (c) $(.98)^4$

2.4. Identify each expansion and thus evaluate each sum without computing individual terms in the sum.

(a) $\displaystyle\sum_{r=0}^{4}\binom{4}{r}2^{4-r}3^{r}$

(b) $\displaystyle\sum_{r=0}^{4}\binom{4}{r}2^{r}3^{4-r}$

(c) $\displaystyle\sum_{r=0}^{4}\binom{4}{r}2^{r}$

(d) $\displaystyle\sum_{r=0}^{4}(-1)^{r}\binom{4}{r}2^{r}$

(e) $\displaystyle\sum_{r=0}^{50}\binom{50}{r}\left(\frac{1}{3}\right)^{50-r}\left(\frac{2}{3}\right)^{r}$

(f) $\displaystyle\sum_{r=0}^{50}\binom{50}{r}\left(\frac{1}{2}\right)^{50}$

(g) $\displaystyle\sum_{r=1}^{5}\binom{4}{r-1}2^{5-r}3^{r-1}$

(h) $\displaystyle\sum_{r=0}^{6}\binom{4}{r-2}2^{6-r}3^{r-2}$

2.5. Write the binomial coefficients in terms of factorials and thus prove the following identities.

(a) $\displaystyle r\binom{n}{r}=n\binom{n-1}{r-1}$

(b) $\displaystyle n\binom{n}{r}=(r+1)\binom{n}{r+1}+r\binom{n}{r}=r\binom{n+1}{r+1}+\binom{n}{r+1}$

(c) $\displaystyle\binom{n}{r}=\binom{n-1}{r-1}+\binom{n-1}{r}$

2.6. Use the law of formation (2.8) to extend the Pascal triangle in Table 23 to $n=10$.

2.7. By writing the binomial coefficients in terms of factorials, derive the *recursion formula*

$$\binom{n}{r+1}=\frac{n-r}{r+1}\binom{n}{r}\qquad(r=0,1,2,\cdots,n-1).$$

This formula enables us to compute the numbers in any row of the Pascal triangle one by one, starting from $\binom{n}{0}=1$. Compute the binomial coefficients for $n=10$ this way.

2.8. Consider the binomial coefficients $\binom{n}{r}$ with n fixed and $r=0,1,2,$ $\cdots,n$. With this order, show that $\binom{n}{r}$ is greater than its predecessor if $r<\frac{1}{2}(n+1)$ and is smaller if $r>\frac{1}{2}(n+1)$. Show also that if n is an even integer, then there is one largest binomial coefficient, but if n is odd, then there are two equal binomial coefficients that are larger than all the others. (*Hint:* Consider the ratio of $\binom{n}{r}$ to $\binom{n}{r-1}$ and determine when this ratio is greater than. less than, or equal to 1.)

2.9. Using the notation introduced in Problem 1.4, define the generalized binomial coefficient $\binom{x}{r}$ for any number x and any positive integer r by the equation

$$\binom{x}{r} = \frac{(x)_r}{r!}.$$

(a) Show that this reduces to the familiar definition if x is a positive integer. (Consider the case $x < r$ as well as $x \geq r$.)

(b) Show that $\binom{-\frac{1}{2}}{n} = (-1)^n 2^{-2n} \binom{2n}{n}$.

2.10. Prove the following identities:

(a) $\binom{n}{0} - \binom{n}{1} + \binom{n}{2} - \cdots + (-1)^n \binom{n}{n} = 0.$

(b) $\binom{n}{0} + \binom{n}{2} + \binom{n}{4} + \cdots + \binom{n}{n} = 2^{n-1}$ if n is even.

(c) $\sum_{k=0}^{m} \binom{n-k}{n-m} \binom{n}{k} = 2^m \binom{n}{m}.$

(d) $\sum_{r=0}^{n} \binom{n}{r}^2 = \binom{2n}{n}.$

2.11. In the expansion of $(p + q + r + s)^{10}$, compute the coefficient of

(a) p^{10} (b) $p^5 q^5$ (c) $p^3 q^2 r s^4$.

2.12. Expand $(p + q + r)^4$ by the multinomial theorem and give a probability interpretation to each term in the expansion assuming p, q, and r are nonnegative numbers with sum 1.

SUPPLEMENTARY READING

In addition to the references listed at the end of Chapter 2 the following may also be consulted.

1. Borel, E. and A. Chéron, *Théorie Mathématique du Bridge à la Portée de Tous*, Gauthier-Villars, Paris, 1940.

2. Jacoby, O., *How to Figure the Odds*, Doubleday and Company, Inc., 1950.

3. Levinson, H. C., *The Science of Chance*, Rinehart and Company, 1950.

4. Riordan, J., *An Introduction to Combinatorial Analysis*, John Wiley and Sons, Inc., 1958.

5. Whitworth, W. A., *Choice and Chance*, Hafner Publishing Co., 1959 (reprint of 5th edition, 1901).

6. Whitworth, W. A., *DCC Exercises*, G. E. Stechert and Co., 1945.

Chapter 4

RANDOM VARIABLES

---

1. Random variables and probability functions

When we perform an experiment, we are often interested not in the particular outcome that occurs, but rather in some *number* associated with that outcome. For example, in the game of "craps" a player is interested not in the particular numbers on the two dice, but in their *sum;* in bridge, one often concentrates on the *number* of honor points in the hand rather than on the hand itself; in selecting a random sample of students from a certain college, we may want to compute the *proportion* of freshmen in the sample; in tossing a coin 50 times, we may be interested only in the *number* of heads obtained, and not in the particular sequence of heads and tails that constitutes the result of the 50 tosses; etc.

In all these examples, we have a rule which assigns to each outcome of the experiment a single real number. The mathematician says that a *function* is thereby defined. Indeed, a function is specified whenever we are given a set of elements (the *domain* of the function), together with a rule by which one and only one number is associated with each element of the domain. The number which the rule assigns to an element of the domain is called the *value* of the function for (or at) that element. The set of all values of a function is called the *range* of the function.

The reader is already familiar with the function concept. For example, the equation $y = x^2$ determines a function whose domain is the set of all real numbers and whose range is the set of nonnegative real numbers; i.e., to each real number x is associated the (necessarily nonnegative) real number $y = x^2$. As another familiar example, think of the function which assigns to each circle the number which is the circle's circumference. The domain of this function is the set of all circles. For any element (circle) in the domain, the value of this function is the number $2\pi r$, where r is the radius of the circle. For yet another example, consider the function whose domain is the set of all people to whom the federal tax laws apply, and which assigns to each such person the number which is his taxable income for a given year.

In probability theory, certain functions of special interest are given special names.

Definition 1.1. A function whose domain is a sample space and whose range is some set of real numbers is called a *random variable*. If the random variable is denoted by X and has the sample space $S = \{o_1, o_2, \cdots, o_n\}$ as domain, then we write $X(o_k)$ for the value of X at the element o_k. Thus $X(o_k)$ is the real number that the function rule assigns to the element o_k of S.

The reader may find it helpful to think of a function in terms of a machine. In Figure 17, we picture a machine for the random variable X. The possible inputs of the function-machine are the elements of the sample space S. Each such element o_k is "processed" by the machine, and what emerges is the output number $X(o_k)$. The set of all possible input elements is the sample space S, the domain of the function. The set of different output numbers is the range of the random variable X.

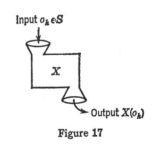

Figure 17

Let us now look at some examples of random variables.

Example 1.1. Let $S = \{1, 2, 3, 4, 5, 6\}$ and define X as follows:

$$X(1) = X(2) = X(3) = +1, \quad X(4) = X(5) = X(6) = -1.$$

Then X is a random variable whose domain is the sample space S and whose range is the set $\{1, -1\}$. X can be interpreted as the gain of a player in a game in which a die is rolled, the player winning $1 if the outcome is 1, 2, or 3 and losing $1 if the outcome is 4, 5, or 6.

Example 1.2. Two dice are rolled and we define the familiar sample space

$$S = \{(1, 1), (1, 2), \cdots, (6, 6)\}$$

containing 36 elements. Let X denote the random variable whose value for any element of S is the sum of the numbers on the two dice. Then the range of X is the set containing the 11 values of X:

$$2, 3, 4, 5, 6, 7, 8, 9, 10, 11, 12.$$

Each ordered pair of S has associated with it exactly one element of the range, as required by Definition 1.1. But, in general, the same value of X arises from many different outcomes. For example, $X(o_k) = 5$ if o_k is any one of the four elements of the event

$$\{(1, 4), (2, 3), (3, 2), (4, 1)\}.$$

A given input element in Figure 17 always leads to exactly one output number, but the same output number may be obtained from more than one input element.

Example 1.3. A coin is tossed, and then tossed again. We define the sample space

$$S = \{HH, HT, TH, TT\}.$$

If X is the random variable whose value for any element of S is the number of heads obtained, then

$$X(HH) = 2, \quad X(HT) = X(TH) = 1, \quad X(TT) = 0.$$

More than one random variable can be defined on the same sample space. For example, let Y denote the random variable whose value for any element of S is the number of heads minus the number of tails. Then

$$Y(HH) = 2, \quad Y(HT) = Y(TH) = 0, \quad Y(TT) = -2.$$

Naming a numerical-valued function defined on a sample space a random variable is singularly inappropriate. Like the alligator pear that is neither an alligator nor a pear and the biologist's white ant that is neither white nor an ant, the probabilist's random variable is neither random nor a variable. But this terminology has become standard by now, and we shall continue to use it, trusting that the reader will constantly keep in mind its true meaning, as given in Definition 1.1.

Suppose now that a sample space

$$S = \{o_1, o_2, \cdots, o_n\}$$

is given, and that some acceptable assignment of probabilities has been made to the simple events of S. Then if X is a random variable defined on S, we can ask for the probability that the value of X is some number, say x. The event that X has the value x is the subset of S containing those elements o_k for which $X(o_k) = x$. If we denote by $f(x)$ the probability of this event, then

(1.1) $$f(x) = P(\{o_k \in S \mid X(o_k) = x\}).$$

Because this notation is cumbersome, we shall write

(1.2) $$f(x) = P(X = x),$$

adopting the shorthand "$X = x$" to denote the event written out in (1.1).

Definition 1.2. The function f whose value for each real number x is given by (1.2), or equivalently by (1.1), is called the *probability function* of the random variable X.

In other words, the probability function of X has the set of all real numbers as its domain, and the function assigns to each real number x the probability that X has the value x.

Example 1.4. Continuing Example 1.1, if the die is fair, then

$$f(1) = P(X = 1) = \tfrac{1}{2}, \quad f(-1) = P(X = -1) = \tfrac{1}{2},$$

and $f(x) = 0$ if x is different from 1 or -1.

Example 1.5. If both dice of Example 1.2 are fair and the rolls are independent, so that each simple event of S has probability $\tfrac{1}{36}$, then we compute the value of the probability function at $x = 5$ as follows:

$$f(5) = P(X = 5) = P(\{(1, 4), (2, 3), (3, 2), (4, 1)\}) = \tfrac{4}{36}.$$

This is the probability that the sum of the numbers on the dice is 5. We can compute the probabilities $f(2), f(3), \cdots, f(12)$ in an analogous manner. These values are summarized in the following *probability table:*

x	2	3	4	5	6	7	8	9	10	11	12
$f(x)$	1/36	2/36	3/36	4/36	5/36	6/36	5/36	4/36	3/36	2/36	1/36

Let us agree, as here, to include in such probability tables only those numbers x for which $f(x) > 0$. Since we include *all* such numbers, the probabilities $f(x)$ in the table add to 1. From the probability table of a random variable X, we can tell at a glance not only the various values of X, but also the probability with which each value occurs. This information can also be presented graphically, as in,

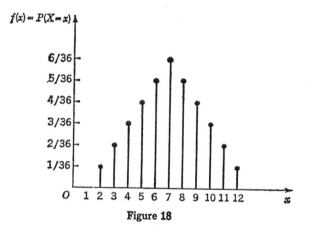

Figure 18

Figure 18, where the *probability chart* of the random variable X is drawn. In the probability chart, the various values of X are indicated on the horizontal x-axis, and the length of the vertical line drawn from the x-axis to the point with coordinates $(x, f(x))$ is the probability of the event that X has the value x.

We are often interested not in the probability that the value of a random variable X is a particular number, but rather in the probability that X has a value *less than or equal to* some number. For example, if you have ten items in inventory, you may want to know the probability that your inventory stocks will be sufficient to fill incoming orders. If X is the total number of items ordered, you are therefore interested in the probability of the event $X \leq 10$. As another example, if X denotes the percentage of votes cast in opposition to a school bond levy, then the school board is interested in the probability that the levy is approved, i.e., that X will be less than or equal to some critical number that separates victory from defeat.

In general, if X is defined on the sample space S, then the event that X is less than or equal to some number, say x, is the subset of S containing those elements o_k for which $X(o_k) \leq x$. If we denote by $F(x)$ the probability of this event (assuming an acceptable assignment of probabilities has been made to the simple events of S), then

(1.3) $F(x) = P(\{o_k \, \epsilon \, S \mid X(o_k) \leq x\}).$

In analogy with our agreement in (1.2), we adopt the shorthand "$X \leq x$" to denote the event written out in (1.3), and we then can write

(1.4) $F(x) = P(X \leq x).$

Definition 1.3. The function F whose value for each real number x is given by (1.4), or equivalently by (1.3), is called the *distribution function* of the random variable X.

In other words, the distribution function of X has the set of all real numbers as its domain, and the function assigns to each real number x the probability that X has a value less than or equal to (i.e., at most) the number x.

As our next example illustrates, it is an easy matter to calculate the values of F, the distribution function of a random variable X, when one knows f, the probability function of X. The distribution function can be presented in graphical or tabular form, as we also show.

Example 1.6. Let us continue with the dice experiment of Example 1.5. The event symbolized by $X \leq 1$ is the null event of the sample space S, since the sum of the numbers on the dice cannot be at most 1. Hence

$$F(1) = P(X \leq 1) = 0.$$

The event $X \leq 2$ is the subset $\{(1, 1)\}$, which is the same as the event $X = 2$. Thus,

$$F(2) = P(X \leq 2) = f(2) = \tfrac{1}{36}.$$

The event $X \leq 3$ is the subset $\{(1, 1), (1, 2), (2, 1)\}$, which is seen to be the union of the events $X = 2$ and $X = 3$. Hence,

$$F(3) = P(X \leq 3) = P(X = 2) + P(X = 3)$$
$$= f(2) + f(3)$$
$$= \tfrac{1}{36} + \tfrac{2}{36} = \tfrac{3}{36}.$$

Similarly, the event $X \leq 4$ is the union of the events $X = 2$, $X = 3$, and $X = 4$, so that

$$F(4) = P(X \leq 4) = P(X = 2) + P(X = 3) + P(X = 4)$$
$$= f(2) + f(3) + f(4)$$
$$= \tfrac{1}{36} + \tfrac{2}{36} + \tfrac{3}{36} = \tfrac{6}{36}.$$

Continuing in this way, we obtain the entries in the following *distribution table* for the random variable X:

(1.5)

x	2	3	4	5	6	7	8	9	10	11	12
$F(x)$	1/36	3/36	6/36	10/36	15/36	21/36	26/36	30/36	33/36	35/36	36/36

But remember that the domain of the distribution function F is the set of *all* real numbers. Hence, we must find the value $F(x)$ for *all* numbers x, not just those in the distribution table. For example, to find $F(2.6)$ we note that the event $X \leq 2.6$ is the subset $\{(1, 1)\}$, since the sum of the numbers on the dice is less than or equal to 2.6 if and only if the sum is exactly 2. Therefore,

$$F(2.6) = P(X \leq 2.6) = \tfrac{1}{36}.$$

In fact, $F(x) = \tfrac{1}{36}$ for all x in the interval $2 \leq x < 3$, since for any such x the event $X \leq x$ is the same subset, namely $\{(1, 1)\}$. Note that this interval contains $x = 2$, since $F(2) = \tfrac{1}{36}$, but does *not* contain $x = 3$, since $F(3) = \tfrac{3}{36}$. Thus, $F(x) = \tfrac{1}{36}$ for $x = 2.999\cdots$, no matter how many nines we write down, but at $x = 3$, the value of F jumps to $F(3) = \tfrac{3}{36}$. Similarly, we find $F(x) = \tfrac{3}{36}$ for all x in the interval $3 \leq x < 4$, but a jump occurs at $x = 4$, since $F(4) = \tfrac{6}{36}$; then $F(x) = \tfrac{6}{36}$ for all x in the interval $4 \leq x < 5$, but a jump occurs at $x = 5$, since $F(5) = \tfrac{10}{36}$; etc.

These facts are shown on the graph of the distribution function in Figure 19. The graph consists entirely of horizontal line segments. (A function having such a graph is appropriately called a *step function.*) We use a heavy dot in Figure 19 to indicate which of the two horizontal segments should be read at each jump (step) in the graph. Note that the magnitude of the jump at $x = 2$ is $f(2) = \tfrac{1}{36}$, the jump

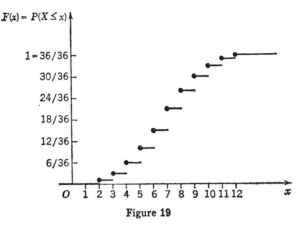

Figure 19

at $x = 3$ is $f(3) = \frac{2}{36}$, the jump at $x = 4$ is $f(4) = \frac{3}{36}$, etc. Finally, since the sum of the numbers on the dice is never less than 2 and always at most 12, we have $F(x) = 0$ if $x < 2$ and $F(x) = 1$ if $x \geq 12$.

If one knows the height of the graph of F at all points where jumps occur, then the entire graph of F is easily drawn. It is for this reason that we shall, as in (1.5), always list in the distribution table only those x-values at which jumps of F occur.

If we are given the graph of the distribution function F of a random variable X, then reading its height at any number x, we find $F(x)$, the probability that the value of X is less than or equal to x. Also, we can determine the places where jumps in the graph occur, as well as the magnitude of each jump, and so we can construct the probability function of X. Thus, we can obtain the probability function from the distribution function, or vice versa.

We have made our observations up to this point on the basis of some special examples, especially the two-dice example. We now turn to some general statements that apply to all probability and distribution functions of random variables defined on *finite* sample spaces.

Theorem 1.1. If a finite sample space S is given, if an acceptable assignment of probabilities is made to its simple events, and if a random variable X is defined with domain S, then f, the probability function of X, has the following properties:

(i) $f(x) \geq 0$ for all x, but there are at most a *finite* number, say N, of x-values for which $f(x) > 0$.

(ii) If $x_1, x_2, \cdots, x_N$ are *all* the x-values for which $f(x)$ is positive, i.e.,

(1.6) $$f(x_k) > 0 \qquad \text{for } k = 1, 2, \cdots, N,$$

then

(1.7) $$\sum_{k=1}^{N} f(x_k) = 1.$$

We leave the proof of this theorem for the problems.

The probability table of the random variable X thus has the following form:

(1.8)

x	x_1	x_2	$\cdots$	x_N
$f(x)$	$f(x_1)$	$f(x_2)$	$\cdots$	$f(x_N)$

Under these circumstances, it is customary to say that X is a random variable whose *possible values* are $x_1, x_2, \cdots, x_N$, and that the value x_k *occurs* with probability $f(x_k)$ for $k = 1, 2, \cdots, N$. This language, which we shall use from now on, should bring to the reader's mind the probability table in (1.8), whose entries satisfy (1.6) and (1.7).

Theorem 1.2. With the hypotheses of Theorem 1.1, the distribution function F of the random variable X has the following properties:

(i) There is a number, say m, such that $F(x) = 0$ if $x < m$; there is a number, say M, such that $F(x) = 1$ if $x \geq M$.

(ii) F is a *nondecreasing* function; i.e., $F(x) \geq F(y)$ if $x \geq y$.

(iii) F is a *step function* with a *finite* number of jumps or steps; i.e., the graph of F is made up of a finite number of horizontal line segments. The value of F at each step is given by the height of the *higher* of the two line segments forming that step. These steps occur at $x_1, x_2, \cdots, x_N$, and the magnitude of the jump at x_k is $f(x_k)$ for $k = 1, 2, \cdots, N$.

Proof. To prove (i), we let m be the smallest and M the largest of the possible values $x_1, x_2, \cdots, x_N$ of the random variable X. Then if $x < m$, the event $X \leq x$ is the null set $\emptyset$, and so

$$F(x) = P(X \leq x) = P(\emptyset) = 0.$$

On the other hand, if $x \geq M$, then the event $X \leq x$ is the entire sample space S, and so

$$F(x) = P(X \leq x) = P(S) = 1.$$

To prove (ii), note that if $x \geq y$, then the event $X \leq y$ is a subset of the event $X \leq x$. Hence, by Theorem II.4.2, we have

$$P(X \leq y) \leq P(X \leq x),$$

which was to be proved.

To prove (iii), suppose x_j and x_k are neighboring x-values, with $x_j < x_k$. Then the event $X \leq x$ is the same for all x in the interval $x_j \leq x < x_k$. Hence, $F(x)$ is the same number for all such x, and the graph of F is a horizontal line segment in this interval. At the *right*-hand endpoint of the interval, we find

$$P(X \leq x_k) = P(X < x_k) + P(X = x_k),$$

so that the jump at $x = x_k$ is

$$P(X \leq x_k) - P(X < x_k) = P(X = x_k) = f(x_k),$$

as claimed.

In Theorems 1.1 and 1.2, we stated our hypotheses very carefully in order to make clear that one must have a sample space, an acceptable assignment of probabilities to its simple events, and a random variable defined on the sample space, before talking about the probability function or the distribution function of the random variable. Nevertheless, in the probability literature (and starting in the next section in this book) one often sees definitions and theorems that begin with the words, "Let X be a random variable with probability function f," or "Let X be a random variable whose possible values x_1, x_2, $\cdots$, x_N occur with probabilities $f(x_1)$, $f(x_2)$, $\cdots$, $f(x_N)$, respectively," no mention being made of a sample space or assignment of probabilities to simple events. To understand why this is an acceptable state of affairs, we must first realize that the converse of Theorem 1.1 is true.

Theorem 1.3. Let a function f with properties (i) and (ii) in Theorem 1.1 be given. Then there is a finite sample space S, an acceptable assignment of probabilities to the simple events of S, and a random variable X whose domain is S, such that f is the probability function of X.

Proof. Define $S = \{x_1, x_2, \cdots, x_N\}$ and let $P(\{x_k\}) = f(x_k)$ for

$k = 1, 2, \cdots, N$. This is an acceptable assignment of probabilities, because we have assumed that (1.6) and (1.7) hold. Now define the random variable X as the identity function on S; i.e., to each element $x_k \in S$ we assign the number $X(x_k) = x_k$. Then

$$P(X = x_k) = P(\{x_k\}) = f(x_k) \qquad \text{for } k = 1, 2, \cdots, N$$

and

$$P(X = x) = P(\emptyset) = 0$$

if x is not one of the numbers $x_1, x_2, \cdots, x_N$. Hence f is the probability function of X, and the theorem is proved.

We also must realize that infinitely many different random variables can have the same probability function. (See Problem 1.13.) This possibility leads to the following oft-used terminology.

Definition 1.4. Two or more random variables are said to be *identically distributed* if and only if they have equal (i.e., identical) probability functions (and hence identical distribution functions).

Thus, whenever we want to make definitions or prove theorems that depend *only* on the probability function of a random variable, we are actually making definitions and proving theorems for any one of an infinite set of identically distributed random variables. The random variables in this set are all different, but this doesn't concern us if we are interested only in their probability functions. Hence, we do not mention the different sample spaces on which they are defined, knowing by virtue of Theorem 1.3 that their common probability function does indeed determine a sample space and an acceptable assignment of probabilities to its simple events. Moreover, the random variable defined on this sample space, as determined in the proof of Theorem 1.3, can serve as the prototype of all identically distributed random variables with the given probability function. This rather lengthy argument serves to explain why, in the following sections, there is so little mention of the underlying sample spaces and probability assignments to simple events, when we talk of random variables and their probability and distribution functions.

PROBLEMS

1.1. An experiment consists of three independent tosses of a fair coin. Let X be the random variable whose value for any outcome is the number of heads obtained.

(a) Find the probability function of X, and construct a probability table and a probability chart.

(b) Find the distribution function of X and draw its graph.

1.2. Repeat the preceding problem, but now (a) let X be the random variable whose value for any outcome is the number of heads minus the number of tails, (b) let X denote the gain of a player who wins $2 if the first head occurs at the first toss, wins $1 if the first head occurs at the second toss, loses $1 if the first head occurs at the third toss, and loses $2 if all three tosses are tails.

1.3. There are two defectives in a lot of eight articles. A sample of four articles is drawn at random (without replacement) from the lot. Let X denote the number of defectives in the sample.

(a) Determine the probability function of X and construct a probability table.

(b) Determine the distribution function of X and draw its graph.

1.4. The annual income of six people A, B, C, D, E, F is given in the following table. A committee of k people is selected from these six people,

Person	A	B	C	D	E	F
Income (in $1000's)	3	3	4	5	6	6

where $k = 1, 2, \cdots, 6$, and the random variable X_k is defined as the average income of the k committee members. For *each* value of k: (a) determine the probability function of X_k and construct the corresponding probability chart; (b) determine the distribution function of X_k and draw its graph. (c) Compare the six probability functions, noting especially how they change as k increases.

1.5. The random variable X has a probability function f of the following form, where k is some number:

$$f(x) = \begin{cases} k & \text{if } x = 0 \\ 2k & \text{if } x = 1 \\ 3k & \text{if } x = 2 \\ 0 & \text{otherwise.} \end{cases}$$

(a) Determine the value of k.

(b) Find $P(X < 2)$, $P(X \leq 2)$, $P(0 < X < 2)$.

(c) What is the smallest value of x for which $P(X \leq x) > .5$?

(d) Determine the distribution function of X.

1.6. Let X denote the number of hours you study during a randomly selected school day. Suppose the probability function of X has the following form, where k is some number:

$$f(x) = \begin{cases} .1 & \text{if } x = 0 \\ kx & \text{if } x = 1 \text{ or } 2 \\ k(5 - x) & \text{if } x = 3 \text{ or } 4 \\ 0 & \text{otherwise.} \end{cases}$$

(a) Find the value of k.

(b) Draw the probability chart.

(c) What is the probability that you study at least two hours? Exactly two hours? At most two hours?

(d) What number of hours is such that you study at least this number of hours with probability at least .70?

(e) Determine the distribution function of X.

(f) What is the conditional probability that you study three hours, given that you do study?

1.7. The distribution function of a random variable X is given as follows:

$$F(x) = \begin{cases} 0 & \text{if } x < -1 \\ \frac{1}{4} & \text{if } -1 \le x < 1 \\ \frac{1}{2} & \text{if } 1 \le x < 2 \\ \frac{2}{3} & \text{if } 2 \le x < 3 \\ 1 & \text{if } x \ge 3. \end{cases}$$

(a) Draw the graph of F.

(b) Find $P(X \le 1)$, $P(X = 1)$, $P(-1 < X \le 2)$, $P(-1 \le X < 2)$, $P(-1 \le X \le 2)$, $P(X < 3)$, $P(-2 < X \le 3.5)$, $P(1.5 < X < 2.7)$.

(c) Determine the probability function of X, and construct a probability table.

1.8. An urn contains three green and two red balls. Find the probability function and construct the probability table for each of the following random variables.

(a) The number of red balls in a random sample of three balls drawn with replacement.

(b) The number of red balls in a random sample of three balls drawn without replacement.

(c) The number of balls that are drawn (one by one, with replacement) in order to get a red ball.

(d) The number of balls that are drawn (one by one, without replacement) in order to get a red ball.

1.9. A bridge hand is dealt from a full deck. Let X denote the number of spades in the hand. Determine the probability function of the random variable X.

1.10. Let X be a random variable whose possible values $x_1, x_2, \cdots, x_N$ occur with probabilities $f(x_1), f(x_2), \cdots, f(x_N)$, respectively. If F is the distribution function of X, show that

$$F(x) = \sum_{x_k \le x} f(x_k),$$

where the sum is taken over all k-values for which $x_k \le x$. (Because values of F are obtained by successive additions of f-values, F is often called the *cumulative distribution function* of X.)

1.11. Let f and F be the probability and distribution function, respectively, of a random variable X. Show that for any numbers a and b ($a < b$),

(a) $P(a < X \le b) = F(b) - F(a)$
(b) $P(a \le X \le b) = F(b) - F(a) + f(a)$
(c) $P(a \le X < b) = F(b) - F(a) + f(a) - f(b)$
(d) $P(a < X < b) = F(b) - F(a) - f(b)$

1.12. (a) Let $x_1, x_2, \cdots, x_N$ be the possible values of a random variable X defined on a sample space S. Show that if no simple event of S is assigned zero probability, then

$$\{X = x_1, X = x_2, \cdots, X = x_N\}$$

is a partition of S.

(b) Prove Theorem 1.1.

1.13. A fair coin is tossed and you win \$2 if it falls heads, win \$1 if it falls tails. Call your gain X_1. A fair die is rolled and you win \$2 if a 1, 2, or 3 shows, win \$1 if a 4, 5, or 6 shows. Call your gain X_2.

(a) Show that X_1 and X_2 are *different* random variables, but have the same probability function. (*Note:* Two functions are equal if and only if they have the same domain and the same value for each element in their common domain.)

(b) Show that there are infinitely many different random variables that have the same probability function as X_1.

1.14. Let $S = \{o_1, o_2, \cdots, o_n\}$ and let E be any event of S. Define X_E, the *characteristic random variable* of event E, as follows:

$$X_E(o_k) = \begin{cases} 1 & \text{if } o_k \in E \\ 0 & \text{otherwise.} \end{cases}$$

In other words, X_E is equal to 1 if E occurs, and X_E is equal to 0 if E does not occur. Prove the following properties of characteristic random variables:

(a) $X_\emptyset$ is identically 0; i.e., $X_\emptyset(o_k) = 0$ for $k = 1, 2, \cdots, n$.
(b) X_S is identically 1; i.e., $X_S(o_k) = 1$ for $k = 1, 2, \cdots, n$.

(c) If $E = F$, then $X_E = X_F$, and conversely. (To say $X_E = X_F$ means $X_E(o_k) = X_F(o_k)$ for $k = 1, 2, \cdots, n$.)

(d) If $E \subseteq F$, then $X_E \leq X_F$, and conversely. (To say $X_E \leq X_F$ means $X_E(o_k) \leq X_F(o_k)$ for $k = 1, 2, \cdots, n$.)

(e) $X_E + X_{E'}$ is identically 1. (The value of $X_E + X_{E'}$ at o_k is defined to be $X_E(o_k) + X_{E'}(o_k)$.)

(f) $X_{E \cap F} = X_E X_F$. (The value of $X_E X_F$ at o_k is defined to be $X_E(o_k) X_F(o_k)$.)

(g) $X_{E \cup F} = X_E + X_F - X_{E \cap F}$.

2. The mean of a random variable

In many problems, the random variable under study has a rather complicated probability function. It is therefore desirable to be able to describe some features of the random variable by means of a few numbers that can be computed from its probability function. For some purposes, these numbers, rather than the entire function, are all that is needed. In this section, we concentrate on a number, called the *mean* of the random variable, that is a measure of location in the sense that it roughly locates a "middle" or "average" value of the random variable.

There are other often-used measures of location, in particular, the *median* and the *mode* of a random variable. But these are of lesser importance than the mean, and so we ask the interested reader to learn about them in the problems. (See Problems 2.21–2.22.)

Definition 2.1. Let X be a random variable whose possible values $x_1, x_2, \cdots, x_N$ occur with probabilities $f(x_1), f(x_2), \cdots, f(x_N)$, respectively. The *mean* of X, denoted by $E(X)$, is the number

$$(2.1) \qquad E(X) = \sum_{k=1}^{N} x_k f(x_k);$$

i.e., the mean of X is the *weighted average* of the possible values of X, each value being weighted by the probability with which it occurs.

Let us note that the concept of weighted average is a familiar one. When a student computes his average grade in a course in which his six grades are 75, 90, 75, 87, 75, and 90, he divides the sum of all his grades by the total number of grades:

$$\frac{75 + 90 + 75 + 87 + 75 + 90}{6} = \frac{492}{6} = 82.$$

But he can also write

$$\frac{75 + 90 + 75 + 87 + 75 + 90}{6} = \frac{75(3) + 87(1) + 90(2)}{6}$$

$$= 75\left(\frac{3}{6}\right) + 87\left(\frac{1}{6}\right) + 90\left(\frac{2}{6}\right),$$

from which we see that the average grade is the weighted average of the student's grades, each grade having as weight the proportion (or relative frequency) with which it occurs among all the grades.

The choice of the letter E to denote a mean is due to the fact that the concept of mean was first introduced with reference to games of chance, where the mean of the gain of a player is called his *mathematical expectation*. The mean is also called the *expected value* of X, but it is important to realize that this term is misleading, since the mean is not a value that we expect the random variable to assume. In fact, $E(X)$ can be different from all the possible values of the random variable X, as the following example shows. We should not be surprised by this, since it occurs in the illustration concerning grades, where the average grade was different from all the actual grades.

Example 2.1. Let X denote the number of points obtained in a throw of a fair die. Then the possible values of X are 1, 2, $\cdots$, 6, and each occurs with probability $\frac{1}{6}$. Hence, applying (2.1),

$$E(X) = 1(\tfrac{1}{6}) + 2(\tfrac{1}{6}) + 3(\tfrac{1}{6}) + 4(\tfrac{1}{6}) + 5(\tfrac{1}{6}) + 6(\tfrac{1}{6}) = \tfrac{7}{2}.$$

Example 2.2. Let X denote the sum of the numbers on two fair dice. In Example 1.5, we computed the probability function of the random variable X. Now we find

$$E(X) = 2(\tfrac{1}{36}) + 3(\tfrac{2}{36}) + 4(\tfrac{3}{36}) + 5(\tfrac{4}{36}) + 6(\tfrac{5}{36}) + 7(\tfrac{6}{36})$$
$$+ 8(\tfrac{5}{36}) + 9(\tfrac{4}{36}) + 10(\tfrac{3}{36}) + 11(\tfrac{2}{36}) + 12(\tfrac{1}{36}),$$

or

$$E(X) = 7.$$

Example 2.3. A florist stocks a perishable flower which costs him 50 cents and which he prices at \$1.50 on the first day it is in his shop. Any flowers not sold that first day are worthless and are thrown away. Let X be the random variable denoting the number of flowers that customers order on a randomly selected day. The florist has found that the probability function of X is given by the following probability table:

x	0	1	2	3
$f(x)$	.1	.4	.3	.2

How many flowers should the florist stock in order to maximize the mean (or expected value) of his net profit?

For $k = 0, 1, 2, 3$, let Y_k be the random variable denoting the florist's net profit when he stocks k flowers. We determine the probability function of each of these random variables, compute $E(Y_k)$ for each, and thus determine the value of k for which $E(Y_k)$ is largest.

If he stocks no flowers, then his profit Y_0 is equal to zero with probability 1, and so $E(Y_0) = 0$. If he stocks one flower, then he loses 50 cents if no flowers are ordered and makes a net profit of $150 - 50 = 100$ cents if at least one customer orders a flower. Hence the probability function of Y_1 is given by the table

y_1	-50	100
$P(Y_1 = y_1)$	.1	.9

and so

$$E(Y_1) = -50(.1) + 100(.9) = 85 \text{ cents.}$$

Let the reader check that the probability tables of Y_2 and Y_3 are given by

y_2	-100	50	200
$P(Y_2 = y_2)$	.1	.4	.5

y_3	-150	0	150	300
$P(Y_3 = y_3)$	.1	.4	.3	.2

from which we compute $E(Y_2) = 110$ cents and $E(Y_3) = 90$ cents. Thus the florist maximizes his mean net profit by stocking two flowers. (See Problem 2.7.)

If X is a random variable defined on a sample space S and x_k is a possible value of X, then it often happens that we are interested less in x_k itself than in some number determined by x_k, like $5x_k$, $3x_k - 2$, x_k^2, etc. (As a simple example, if X is the number of units demanded of a product that sells for \$7 and costs \$2 per unit, then the demand x_k determines the profit $5x_k$.) In such cases, we must first understand that a new random variable is determined. Then we can turn to methods of calculating the mean of this new random variable.

Definition 2.2. Let X be a random variable defined on the sample space S, and suppose g is a numerical-valued function whose domain includes the range of X. Then the *composite function* of g with X, denoted by $g(X)$, is defined as the function whose value for any element $o_k \in S$ is the real number $g(X(o_k))$.

Let us review this definition in terms of the function machines in Figure 20. We start with any element $o_k \in S$, and first obtain the

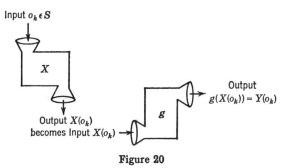

Figure 20

value $X(o_k)$. Now we have assumed in Definition 2.2 that any output number of the X-machine can serve as an input number of the g-machine. In particular, we can use $X(o_k)$ as input for the g-machine, and thereby obtain the value of g at $X(o_k)$. This final output number is therefore $g(X(o_k))$. The two machines taken together in the given order can be thought of as one composite machine that takes the input element $o_k \in S$ and produces the output number $g(X(o_k))$. The composite function $g(X)$ defined by this composite machine is therefore a random variable whose domain is S. If we let $Y = g(X)$, then the random variable Y assigns to each element $o_k \in S$ the number $Y(o_k) = g(X(o_k))$.

These ideas are illustrated in the next example.

Example 2.4. Suppose we toss two fair coins and take $S = \{HH, HT, TH, TT\}$ as sample space. Let X denote the number of heads obtained and let us take $g(x) = (x - 1)^2$, so that $Y = g(X) = (X - 1)^2$. In words, Y is the square of the deviation of the number of heads from 1. For each element $o_k \epsilon S$, we obtain the value of the random variable X and then find the corresponding value of the random variable Y:

o_k	$P(\{o_k\})$	$X(o_k)$	$Y(o_k)$
HH	1/4	2	$(2 - 1)^2 = 1$
HT	1/4	1	$(1 - 1)^2 = 0$
TH	1/4	1	$(1 - 1)^2 = 0$
TT	1/4	0	$(0 - 1)^2 = 1$

We thus obtain the probability functions of X and Y:

x	0	1	2
$P(X = x)$	1/4	1/2	1/4

y	0	1
$P(Y = y)$	1/2	1/2

And now we can compute the mean of Y by applying Definition 2.1. We find

$$(2.2) \qquad E(Y) = E[(X - 1)^2] = 0 \cdot \tfrac{1}{2} + 1 \cdot \tfrac{1}{2} = \tfrac{1}{2}.$$

Our next result shows that we can compute the mean of $Y = g(X)$ directly from the probability function of X *without* first finding the probability function of Y.

Theorem 2.1. Let X be a random variable whose possible values $x_1, x_2, \cdots, x_N$ occur with probabilities $f(x_1), f(x_2), \cdots, f(x_N)$, respectively. If $Y = g(X)$, then the mean of the random variable Y is given by

$$(2.3) \qquad E(Y) = E[g(X)] = \sum_{k=1}^{N} g(x_k)f(x_k).$$

Proof. Let the possible values of Y be $y_1, y_2, \cdots, y_M$. (We know that $M \leq N$, since it can happen that Y has the same value for two different values of X.) Then by the definition of $E(Y)$ we have

$$E(Y) = \sum_{j=1}^{M} y_j P(Y = y_j)$$

$$= \sum_{j=1}^{M} y_j P(g(X) = y_j)$$

$$= \sum_{j=1}^{M} y_j P(X = x, \text{ where } g(x) = y_j).$$

Now the probability of the event in this last expression is the sum of the probabilities of one or more (disjoint) events of the form $X = x_k$, where x_k is a possible value of X. And for each of these events we know that $y_j = g(x_k)$. As j varies from 1 to M, we include terms of the form $g(x_k)P(X = x_k)$ for each possible value x_k. Hence

$$E(Y) = \sum_{k=1}^{N} g(x_k)P(X = x_k),$$

which is precisely what we set out to prove.

Example 2.5. Let us illustrate the use of (2.3) by computing $E(Y)$ for the random variable $Y = (X - 1)^2$ in Example 2.4. We find

$$E(Y) = E[(X - 1)^2] = \sum_{k=1}^{3} (x_k - 1)^2 f(x_k)$$
$$= (0 - 1)^2 \tfrac{1}{4} + (1 - 1)^2 \tfrac{1}{2} + (2 - 1)^2 \tfrac{1}{4}$$
$$= \tfrac{1}{2},$$

as in (2.2).

Computing $E[g(X)]$ by means of Formula (2.3) is generally much easier than first determining the probability function of the random variable $g(X)$ and then computing its mean by using Definition 2.1. In our later work, for example, we shall use the formulas

$$(2.4) \qquad\qquad E(X^2) = \sum_{k=1}^{N} x_k^2 f(x_k),$$

$$(2.5) \qquad E[X - E(X)] = \sum_{k=1}^{N} [x_k - E(X)]f(x_k),$$

$$(2.6) \qquad E([X - E(X)]^2) = \sum_{k=1}^{N} [x_k - E(X)]^2 f(x_k),$$

which are obtained from (2.3) by putting $g(x)$ in turn equal to x^2, $x - E(X)$ and $[x - E(X)]^2$.

Example 2.6. If X denotes the number of points obtained in a roll of a fair die (see Example 2.1), then from (2.4) we find

$$E(X^2) = 1^2(\tfrac{1}{6}) + 2^2(\tfrac{1}{6}) + 3^2(\tfrac{1}{6}) + 4^2(\tfrac{1}{6}) + 5^2(\tfrac{1}{6}) + 6^2(\tfrac{1}{6}) = \tfrac{91}{6}.$$

Theorem 2.1 leads to a number of important results to which we now turn.

Theorem 2.2. If a and b are any numbers, then

(2.7) $$E(aX + b) = aE(X) + b.$$

Proof. We let $g(x) = ax + b$ in (2.3) and find

$$\begin{aligned}
E[g(X)] = E(aX + b) &= \sum_{k=1}^{N} (ax_k + b)f(x_k) \\
&= \sum_{k=1}^{N} ax_k f(x_k) + \sum_{k=1}^{N} bf(x_k) \\
&= a \sum_{k=1}^{N} x_k f(x_k) + b \sum_{k=1}^{N} f(x_k) \\
&= aE(X) + b,
\end{aligned}$$

the last equality following from the definition of $E(X)$, as given in (2.1), together with the fact, expressed in (1.7), that the sum of all the probabilities $f(x_k)$ is 1.

As special cases of (2.7), we have

$$E(X + b) = E(X) + b \quad \text{and} \quad E(aX) = aE(X).$$

In words, *adding a fixed amount to every value of a random variable changes the mean of the random variable by this same amount,* and *multiplying every value of a random variable by the same factor multiplies the mean by that factor.* It is comforting that our formulas yield these very appealing and reasonable results.

If we put $a = 1$ and $b = -E(X)$ in (2.7), we obtain

(2.8) $$E[X - E(X)] = E(X) - E(X) = 0.$$

Now $X - E(X)$ denotes the algebraic or signed deviation of X from its mean. This deviation is positive, zero, or negative depending upon whether the value of X is greater than, equal to, or less than the number $E(X)$. Thus, Formula (2.8) asserts that *the mean deviation of X from its mean is zero.*

It is possible to give a mechanical interpretation of some of our formulas. We think of N particles distributed along the x-axis at the points $x_1, x_2, \cdots, x_N$. The particle at point x_k has mass $f(x_k)$ for $k = 1, 2, \cdots, N$. Then (1.6) and (1.7) express the facts that each particle has positive mass and the total mass of all N particles is 1.

With this interpretation, the sum in (2.1) defines what the physicist calls the *center of gravity* of the system of N particles. Thus, the center of gravity is the weighted average of the x-values, each having weight equal to the mass concentrated at that x-value.

The number $x_k - E(X)$ is the *signed* distance of the particle at x_k from the center of gravity. If we imagine the x-axis as a lever suspended on a fulcrum placed at the center of gravity, then $x_k - E(X)$ is positive if x_k is to the right of the fulcrum and negative if x_k is to the left of the fulcrum. (See Figure 21.) The *moment* about this ful-

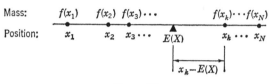

Figure 21

crum of the particle of mass $f(x_k)$ at x_k is the product of its mass and its signed distance from the fulcrum. (This signed distance is called the *moment arm* in mechanics.) The total moment (about the fulcrum) of the entire system of N particles is therefore precisely the sum in (2.5). When this total moment is zero, the lever is in equilibrium; i.e., it balances and does not turn about the fulcrum. Formula (2.8) is therefore merely the expression of the following property of the center of gravity: a distribution of mass particles is in equilibrium with respect to motion about a fulcrum placed at the center of gravity of the system. It is possible to show further that the center of gravity is the *only* location for a fulcrum if the lever is to be in equilibrium and not turn. (See Problem 2.14.)

In our concluding example, we find it convenient first to determine the distribution function of X, then the probability function, and finally $E(X)$. This example therefore serves as a quick review of some of the material presented up to now in this chapter.

Example 2.7. In a certain city, there are 25 officials with city-owned limousines carrying license plates numbered 1, 2, 3, $\cdots$, 25. In a 5-minute period in front of city hall we observe two official cars. Let us interpret this as a random sample of two cars drawn with replacement from the population of 25 cars. Let X denote the larger license plate number observed. (If the two numbers we observe are the same, then X is just the number observed.) We want to find the mean of the random variable X.

The possible values of X are clearly the integers 1, 2, $\cdots$, 25. The event $X \leq k$ occurs if and only if *both* license plate numbers are less than or equal to k. Hence for $k = 1, 2, \cdots, 25$,

$$F(k) = P(X \leq k) = \left(\frac{k}{25}\right)^2 = \frac{k^2}{625}.$$

But

$$f(k) = P(X = k) = P(X \leq k) - P(x \leq k - 1).$$

Therefore for $k = 1, 2, \cdots, 25$,

$$f(k) = \frac{k^2}{625} - \frac{(k-1)^2}{625} = \frac{2k-1}{625}.$$

Having found the probability function of X, we can use (2.1) to compute $E(X)$. The values $x_1, x_2, \cdots, x_N$ in (2.1) are now just the integers 1, 2, $\cdots$, 25. We thus find that

$$E(X) = \sum_{k=1}^{25} kf(k) = \sum_{k=1}^{25} k \cdot \frac{2k-1}{625}$$

$$= \frac{1}{625}\left[2\sum_{k=1}^{25} k^2 - \sum_{k=1}^{25} k\right].$$

But the sum of the first N positive integers and the sum of the squares of the first N positive integers are given by the formulas

$$(2.9) \quad \sum_{k=1}^{N} k = \frac{N(N+1)}{2}, \qquad \sum_{k=1}^{N} k^2 = \frac{N(N+1)(2N+1)}{6}.$$

It follows that

$$E(X) = \frac{1}{625}\left[\frac{2(25)(26)(51)}{6} - \frac{(25)(26)}{2}\right] = \frac{429}{25}.$$

Thus the mean of the larger of the two observed license plate numbers is 17.16.

A problem related to this one, but considerably more difficult, is the following. Suppose you do not know how many people are attending a convention, but you do know that as each person entered he was given an identification tag with a number on it. The tags are numbered serially from 1 to N, where N is the unknown number in attendance. You select a random sample of ten people, let us say, and observe that the largest number on their badges is 261. What estimate do you then make of the total attendance at the convention?

This is a problem in *statistical estimation* in which some characteristic of a population (the total number of people) is to be estimated on the basis of information (the largest of the ten selected badge numbers) obtained from a sample drawn from the population. In Example 2.7, we did a very simple problem of *sampling theory*, in which we answered a probability question about a sample on the assumption that we knew everything about the population from which the sample is drawn.

PROBLEMS

2.1. Let X denote the number of heads obtained in three independent tosses of a fair coin. (See Problem 1.1.)

(a) Find $E(X)$.

(b) Determine the probability function of the random variable $Y = X - E(X)$ and then verify (2.8) by computing $E(Y)$.

(c) Determine the probability function of the random variable $Z = [X - E(X)]^2$ and then calculate $E(Z)$. Check your result by also using Theorem 2.1 to compute $E(Z)$.

2.2. A coin (perhaps biased) is tossed. Let X denote the number of heads obtained. Determine the probability function of the random variable $Y = X(1 - X)$.

2.3. A thousand tickets are sold in a lottery in which there is one top prize of \$500, four prizes of \$100 each, and five prizes of \$10 each. A ticket costs \$1. If X is your net gain when you buy one ticket, find $E(X)$.

2.4. In roulette, the wheel has the 37 numbers 0, 1, 2, $\cdots$, 36 marked on equally spaced slots. A player bets \$1 on a given number. He receives \$36 from the croupier if the ball comes to rest in this slot; otherwise, he gets nothing. If X is the player's net gain, find $E(X)$.

2.5. Refer to Problem 1.3 and find the mean number of defectives in the sample.

2.6. Two defective tubes get mixed up with two good ones. You select and test one tube at a time until you have discovered both defectives. Let X be the number of tubes selected when the second defective is discovered. Determine the probability function of X and compute the mean of X. (Cf. Problem II.5.9.)

2.7. In Example 2.3, we assumed that there is no loss caused by inability to fill a customer's order. In actual practice, the florist might consider that turning a customer away for lack of stock is equivalent to sustaining a monetary loss, because customers may give their future business to another florist, good will is lost, etc. Suppose the florist counts each customer turned away for lack of stock as equivalent to a 50 cent loss. (a) Show that he should still stock two flowers. (b) How large must be the equivalent monetary loss of each customer turned away for lack of stock before the florist maximizes his mean net profit by stocking three flowers?

2.8. Player A pays B $1 and two fair dice are rolled. A receives $2 from B if one six appears, $4 if two sixes appear, and he gets nothing if no six appears. Let X denote player A's net gain. (a) Find $E(X)$. (b) What must A pay B as entrance fee (instead of $1) in order to have $E(X) = 0$?

2.9. Player A bets $1 against B's b that if two cards are dealt from a standard deck, both cards will be of the same color. If X is player A's net gain, what value of b is required to make $E(X) = 0$? With the value of b so determined, what is $E(Y)$ if Y is player B's net gain?

2.10. Compute the means of the random variables defined in Problem 1.8.

2.11. Suppose you have convinced a friend to play the following game with you. A fair coin is to be tossed until the first head appears, but the game is over if no head appears after 20 tosses. Your friend agrees to pay you $2 if a head turns up on the first toss, 2^2 (= $4) if the first head comes up on the second toss, $\cdots$, 2^{20} (= $1,048,576) if the first head comes up on the twentieth toss. You receive nothing if the 20 tosses yield no head. What entrance fee should you pay your friend before the game to make your net gain have mean zero?

2.12. A store sells an item which yields a profit of $3 per item. If the item is out of stock, customers buy elsewhere. At the end of one day, the store manager notices that there are only five items left in stock. The number of items demanded by customers each day is a random variable D such that $E(D) = 12$. Assuming additional stock is unavailable, let X denote the profit lost due to the manager's failure to reorder. Find $E(X)$. What theorem have you used?

2.13. Let X denote the sum of the numbers obtained when two fair dice are rolled. Find $E(X^2)$. Is $E(X^2) = [E(X)]^2$? (Refer to Example 2.2.)

2.14. Show that

$$\text{if} \quad \sum_{k=1}^{N} (x_k - c)f(x_k) = 0, \quad \text{then} \quad c = E(X).$$

2.15. In the carnival game known as *chuck-a-luck*, a player pays an amount e, his entrance fee for playing the game. He selects one number from the six numbers 1, 2, $\cdots$, 6 and then rolls three dice. If all three dice show the number the player selected, the player is paid four times his entrance fee; if two of the dice show the number, the player is paid three times his entrance fee; and if only one of the dice shows the number, the player receives an amount equal to twice his entrance fee. If his number does not show up, then he receives nothing. Let X denote the player's net gain in a single play of this game. Assuming the dice are fair: (a) determine the probability function of the random variable X; (b) compute $E(X)$ and thus show, in particular, that if the entrance fee is \$1, then the player sustains a mean loss per game of about 8 cents.

2.16. After working together on many jobs, four people A, B, C, and D are each asked to write on a slip of paper the name of that person (from among his three co-workers) who is most cooperative. Let X denote the number of people who are considered most cooperative by *none* of their co-workers. Assuming that each person selects one of his co-workers at random and writes his name on the slip of paper, find $E(X)$.

2.17. A drunk reaches home and wants to open his front door. He has five keys on his key chain and tries them one at a time and at random. He is alert enough to eliminate unsuccessful keys from subsequent selections. Let X denote the number of keys he tries in order to find the one that opens his door. Find $E(X)$.

2.18. Refer to Problem 1.4 and show that $E(X_k)$ has the same value for $k = 1, 2, \cdots, 6$.

2.19. An urn has ten balls, numbered from 1 to 10. You are offered the following options:

(1) Pay \$1, draw a ball from the urn, and be paid a number of dollars equal to the number on the ball.
(2) Pay \$1, draw a ball from the urn. If the number on the ball is greater than 5, then be paid a number of dollars equal to the number on the ball. If the number on the ball is 5 or less, then put the ball back in the urn, pay \$3, draw another ball from the urn and be paid a number of dollars equal to the number on the ball.

Let X_1 and X_2 be your net profit when you accept options 1 and 2, respectively. (a) Determine the probability functions of X_1 and X_2. (b) If you want to maximize your mean net profit, which option do you accept?

2.20. (a) One number is selected at random from the first ten positive integers. Let X denote the number obtained. Find $E(X)$.
(b) Two numbers are selected at random (with replacement) from the first ten positive integers. Let X denote the larger of the two numbers obtained. Find $E(X)$.

(c) Three numbers are selected at random (with replacement) from the first ten positive integers. Let X denote the largest of the three numbers obtained. Find $E(X)$.

(d) Redo Parts (b) and (c), assuming the numbers are selected *without* replacement.

2.21. Let X be a random variable with distribution function F. As we know, the graph of F is a step function. Imagine vertical line segments drawn connecting the lower and upper pieces of the graph at $x_1, x_2, \cdots, x_N$ (where jumps occur) and call the new graph the *extended* graph of F. Select any probability p on the vertical axis and consider the horizontal line at this height. The x-coordinate of *any* point where this horizontal line intersects the extended graph of F is called *a* $100p$th *percentile* of the random variable X. A 25th percentile is called a *lower quartile;* a 75th percentile is an *upper quartile;* a 50th percentile is a *median* of X.

(a) In terms of the construction just described, state when there is a unique $100p$-th percentile and when there are more than one.

(b) Show that a median of X can equivalently be defined as any number m such that $P(X \leq m) \geq \frac{1}{2}$ and $P(X \geq m) \geq \frac{1}{2}$. Formulate a corresponding definition for a $100p$-th percentile of X.

(c) Let X denote the sum of the numbers on two fair dice. Show that the median of X is 7, the lower quartile is 5, and the upper quartile is 9. (Cf. Example 2.2.)

(d) Consider the random variable X defined in Example 2.3. Show that $E(X) = 1.6$. Also show that any number between 1 and 2 inclusive is a median of X.

2.22. A possible value of X that occurs with a probability at least as large as the probability of any other value of X is said to be a *mode* (or modal value) of the random variable X.

(a) Let X be the sum of the numbers on two fair dice. Show that the mode of X is 7 so that this random variable happens to have its mean, median, and mode all equal.

(b) A and B match pennies four times. On each match A wins one penny with probability $\frac{1}{2}$ and loses one penny with probability $\frac{1}{2}$. Let X denote the number of times during the course of the game that A is ahead. Find the mean, median, and mode of the random variable X.

2.23. Suppose the probability function f is symmetrical about the line $x = a$, i.e., $f(a + x) = f(a - x)$ for all x. Show that $E(X) = a$.

3. The variance and standard deviation of a random variable

The mean of a random variable X is an "average" value of X; it gives us no information about the variability of the values of X. For many purposes, we also require a measure of this variability, of the "spread" or "dispersion" of the values of the random variable. This requirement is especially apparent as soon as one realizes that random variables with different probability functions can have *equal* means. For example, we have tabulated below the probability functions of four different random variables.

(3.1) X_1:

x	1	2	3	4
$f_1(x)$	1/8	2/8	3/8	2/8

$E(X_1) = 2.75$

(3.2) X_2:

x	-1	0	4	5	6
$f_2(x)$	1/8	2/8	3/8	1/8	1/8

$E(X_2) = 2.75$

(3.3) X_3:

x	4	5	6	7
$f_3(x)$	1/8	2/8	3/8	2/8

$E(X_3) = 5.75$

(3.4) X_4:

x	2	4	6	8
$f_4(x)$	1/8	2/8	3/8	2/8

$E(X_4) = 5.50$

The corresponding probability charts are drawn in Figure 22.

The reader can check that X_1 and X_2 have equal means. To distinguish X_1 from X_2 requires that we have a measure of the extent to which the values of the random variables spread out along the horizontal axis. We would certainly expect of such a measure that it would be larger for X_2 than for X_1, reflecting the fact that graph (b) is more spread out than graph (a), in Figure 22.

The random variable X_3 is obtained by adding 3 to each value of X_1; i.e., $X_3 = X_1 + 3$. As we showed in the preceding section, the mean is thereby increased by 3. A glance at graphs (a) and (c) in Figure 22 shows they are identical, except that graph (c) is 3 units further along the x-axis. But the graphs show the same variability in the values of X_1 and X_3, and we therefore expect our measure of dispersion to be equal for X_1 and X_3.

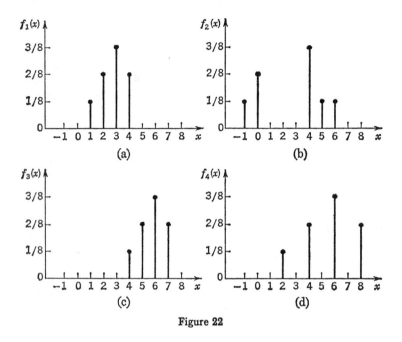

Figure 22

On the other hand, X_4 is obtained by multiplying each value of X_1 by 2; i.e., $X_4 = 2X_1$. The mean is thereby also doubled, but now graph (d) is more spread out on the axis than graph (a) in Figure 22. We therefore expect our measure of dispersion to be larger for X_4 than for X_1. Graphs (b) and (d) are harder to compare by eye, and it is clear that we must now leave these special examples and somehow obtain a numerical measure of dispersion that will apply to any random variable.

A first attempt to formulate a precise definition of the spread of the values of a random variable might proceed as follows. Choose some central or average value of X, say $E(X)$. For each possible value x_k of the random variable X, the number $x_k - E(X)$ measures the deviation of x_k from $E(X)$. Compute this deviation for $k = 1, 2, \cdots, N$. Finally, form the weighted average of these deviations, using as weight for the kth deviation the probability $f(x_k)$ with which the value x_k (and hence the deviation $x_k - E(X)$) occurs. We are thus led to the number

(3.5) $$\sum_{k=1}^{N} [x_k - E(X)]f(x_k) = E[X - E(X)],$$

which, disappointingly, is useless as a measure of spread, since we showed in (2.8) that it is zero for *all* random variables.

A second attempt would follow the realization that the sum in (3.5) is the weighted average of the *algebraic* or *signed* deviations $x_k - E(X)$ and that, after being properly weighted, these deviations, some positive and some negative, add to zero. When measuring the spread of the values of a random variable, we should be concerned with the *magnitude* of $x_k - E(X)$, but *not* with its sign. In other words, we care only about how far x_k is from the mean, not about whether it is less than or greater than the mean. Although not the only way to accomplish this (see Problem 3.13), the most mathematically tractable way is to *square* each deviation and then compute the weighted average of these squared deviations. We are thus led to the following definition. (To simplify notation, we here introduce the symbol μ_X for the mean of the random variable X. From now on, we use μ_X and $E(X)$ interchangeably for the mean of X.)

Definition 3.1. Let X be a random variable whose possible values $x_1, x_2, \cdots, x_N$ occur with probabilities $f(x_1), f(x_2), \cdots, f(x_N)$, respectively. Let $\mu_X = E(X)$ be the mean of X. The *variance* of X, denoted by $\mathrm{Var}(X)$ or σ_X^2, is defined as the number

(3.6) $$\sigma_X^2 = \mathrm{Var}(X) = E[(X - \mu_X)^2]$$

or equivalently, by (2.6),

(3.7) $$\sigma_X^2 = \mathrm{Var}(X) = \sum_{k=1}^{N} (x_k - \mu_X)^2 f(x_k).$$

The nonnegative number

(3.8) $$\sigma_X = \sqrt{\mathrm{Var}(X)}$$

is called the *standard deviation* of the random variable X.

Let us use this definition to compute the variances of the random variables whose probability functions are given in (3.1)–(3.4). The computation of $\mu_1 = E(X_1)$ and $\sigma_1^2 = \mathrm{Var}(X_1)$ is summarized in Table 24. (Note that we avoid subscripts on subscripts by writing μ_1 and σ_1 in place of μ_{X_1} and σ_{X_1}.)

TABLE 24

x_k	$f_1(x_k)$	$x_k f_1(x_k)$	$x_k - \mu_1$	$(x_k - \mu_1)^2$	$(x_k - \mu_1)^2 f_1(x_k)$
1	1/8	1/8	$-7/4$	49/16	49/128
2	2/8	4/8	$-3/4$	9/16	18/128
3	3/8	9/8	1/4	1/16	3/128
4	2/8	8/8	5/4	25/16	50/128
Sums:	1	22/8			120/128

$$\mu_1 = E(X_1) = \sum_{k=1}^{4} x_k f_1(x_k) = \frac{22}{8} = 2.75$$

$$\sigma_1^2 = \text{Var}(X_1) = \sum_{k=1}^{4} (x_k - \mu_1)^2 f_1(x_k) = \frac{120}{128} = .9375$$

$$\sigma_1 = \sqrt{\text{Var}(X_1)} = .97, \text{ approx.}$$

Proceeding in this way, we compute the following values:

$$\sigma_1^2 = \text{Var}(X_1) = .9375 \qquad \sigma_1 = .97$$
$$\sigma_2^2 = \text{Var}(X_2) = 6.1875 \qquad \sigma_2 = 2.49$$
$$\sigma_3^2 = \text{Var}(X_3) = .9375 \qquad \sigma_3 = .97$$
$$\sigma_4^2 = \text{Var}(X_4) = 3.75 \qquad \sigma_4 = 1.94$$

If we refer back to our discussion of the spread exhibited by the probability charts in Figure 22, we can compare our intuitive observations with the numbers just computed. If we take the variance as a measure of spread, then X_2 shows the largest spread, X_4 spreads less than X_2 but more than X_1, and X_1 and X_3 show the same spread. We also observe the same relative magnitudes among the standard deviations. In addition, we recall that $X_4 = 2X_1$ and then note with interest the fac that $\sigma_4 = 2\sigma_1$. This is a special case of a general result soon to be proved. As these examples seem to indicate, the variance or the standard deviation can serve as our measure of the spread or dispersion of the values of a random variable. The usefulness of these concepts in the general theory will become apparent as our work develops.

One difficulty with the variance is that it does not measure dispersion in the same units as the values of X. Thus, if X has dollar values, then $E(X)$ is a certain number of dollars but, since the variance is the mean *square* deviation, $\text{Var}(X)$ is measured in dollars squared. It is in order to have a measure of dispersion in the same units as the values of X that we define the standard deviation as the square root of the variance.

We turn now to some general results concerning the variance of a random variable. Each term in the sum (3.7) that defines $\text{Var}(X)$ is nonnegative, and so the entire sum is either zero or a positive number. The sum is zero if and only if each term in the sum is zero. Since $f(x_k) > 0$, this means $x_k = \mu_X$ for all k. We have therefore proved the following result.

Theorem 3.1. For any random variable X, we have

$$(3.9) \qquad \text{Var}(X) \geq 0,$$

the equality holding if and only if there is only one possible value of X, this value therefore occurring with probability 1.

Although the calculations summarized in Table 24 are not difficult, it would be gratifying to have a simpler way of computing the variance of a random variable than by the use of the defining equation (3.7). Our next result gives us the required formula.

Theorem 3.2. The variance of X is obtained by subtracting the square of the mean of X from the mean of X^2. In symbols,

$$(3.10) \qquad \text{Var}(X) = E(X^2) - \mu_X^2.$$

Proof. We expand the summand in (3.7) and obtain

$$\text{Var}(X) = \sum_{k=1}^{N} (x_k^2 - 2\mu_X x_k + \mu_X^2) f(x_k)$$
$$= \sum_{k=1}^{N} x_k^2 f(x_k) - 2\mu_X \sum_{k=1}^{N} x_k f(x_k) + \mu_X^2 \sum_{k=1}^{N} f(x_k).$$

The first sum is $E(X^2)$ by (2.4), the second sum defines μ_X, and the probabilities in the third sum add to 1. Hence

$$\text{Var}(X) = E(X^2) - 2\mu_X^2 + \mu_X^2,$$

from which (3.10) follows immediately.

It is important to distinguish clearly between the "mean square" $E(X^2)$ and the "square mean" $\mu_X^2 = [E(X)]^2$ in Formula (3.10). This formula is usually used to compute $\text{Var}(X)$. In Table 25, we summarize the calculations involved in finding the variance of the random variable X_1 whose probability function is given in (3.1). A comparison with Table 24 will show how much simpler it is to use (3.10) rather than (3.7) to compute the variance.

TABLE 25

x_k	$f_1(x_k)$	$x_k f_1(x_k)$	$x_k^2 f_1(x_k)$
1	1/8	1/8	1/8
2	2/8	4/8	8/8
3	3/8	9/8	27/8
4	2/8	8/8	32/8
Sums:	1	22/8	68/8

$$\mu_1 = E(X_1) = \sum_{k=1}^{4} x_k f_1(x_k) = \frac{22}{8} = 2.75$$

$$E(X_1^2) = \sum_{k=1}^{4} x_k^2 f_1(x_k) = \frac{68}{8} = 8.5$$

$$\sigma_1^2 = \mathrm{Var}(X_1) = E(X_1^2) - \mu_1^2 = 8.5 - (2.75)^2 = .9375$$

$$\sigma_1 = \sqrt{\mathrm{Var}(X_1)} = .97, \text{ approx.}$$

Example 3.1. If X denotes the number of points obtained in a roll of a fair die, then we computed $E(X) = \frac{7}{2}$ in Example 2.1 and $E(X^2) = \frac{91}{6}$ in Example 2.6. Applying (3.10) we find

$$\mathrm{Var}(X) = \frac{91}{6} - (\frac{7}{2})^2 = \frac{35}{12} \quad \text{and} \quad \sigma_X = \sqrt{\frac{35}{12}} = 1.7, \text{ approx.}$$

In Theorem 2.2, we studied the effect on the mean of changing each value of a random variable by (1) adding or subtracting a fixed number, and (2) multiplying or dividing by a fixed number. Changes of the first kind are known as changes in the *location of the origin* on the horizontal axis of the probability chart of the random variable; changes of the second kind are known as changes in the *scale* on this axis. For example, in Figure 22 the graph of X_3 in (c) can be obtained from the graph of X_1 in (a) by a change in location of the origin: if we shift the number 0 (and all other numbers) three units to the left, then with this relabeling of the axis graph (a) becomes graph (c). But the graph of X_4 in (d) is obtained from the graph of X_1 in (a) by a change in scale: if we make each unit on the axis in (a) two units, then with this relabeling graph (a) becomes graph (d). It is as if the axis in (a) measured the values of X_1 in units of quarts, let us say, whereas the axis in (d) measured the same variable in units of pints. In the following theorem, we study the effect on the variance and the standard deviation of changes in location of the origin and changes in scale.

Theorem 3.3. If a and b are any numbers, then

$$(3.11) \qquad \text{Var}(aX + b) = a^2 \, \text{Var}(X),$$
$$(3.12) \qquad \sigma_{aX+b} = |a| \, \sigma_X.$$

Proof. If we apply the defining equation (3.6) to the random variable $aX + b$, then we find

$$\text{Var}(aX + b) = E([aX + b - E(aX + b)]^2).$$

By (2.7), this simplifies as follows:

$$\begin{aligned}
\text{Var}(aX + b) &= E([aX + b - aE(X) - b]^2) \\
&= E(a^2[X - E(X)]^2) \\
&= a^2 E([X - E(X)]^2) \\
&= a^2 \, \text{Var}(X).
\end{aligned}$$

Hence

$$\begin{aligned}
\sigma_{aX+b} &= \sqrt{\text{Var}(aX + b)} \\
&= \sqrt{a^2} \sqrt{\text{Var}(X)} \\
&= |a| \, \sigma_X,
\end{aligned}$$

where $|a|$ denotes the *absolute value* of the number a. (Note that when we write the square root sign, then by definition we mean the *nonnegative* square root. Hence it is *not* correct to replace $\sqrt{a^2}$ by a. Instead, $\sqrt{a^2} = a$ if $a \geq 0$ and $\sqrt{a^2} = -a$ if $a < 0$; i.e., $\sqrt{a^2} = |a|$.)

From (3.11) and (3.12) we conclude that

$$\text{Var}(X + b) = \text{Var}(X), \qquad \sigma_{X+b} = \sigma_X,$$

and

$$\text{Var}(aX) = a^2 \, \text{Var}(X), \qquad \sigma_{aX} = |a| \, \sigma_X.$$

In words, *adding a fixed amount to every value of a random variable has no effect on the variance and the standard deviation of the random variable, but multiplying each value of a random variable by the same factor a multiplies the variance by a^2 and the standard deviation by $|a|$.*

Because of its importance in our later work, we state the following special case of Theorem 3.3. The proof is left for the problems.

Theorem 3.4. Let X be any random variable with mean μ_X and standard deviation $\sigma_X > 0$. Let the random variable $X*$ be defined as follows:

(3.13) $$X^* = \frac{X - \mu_X}{\sigma_X}.$$

(X^* is called the *standardized* random variable corresponding to X.)
Then

(3.14) $E(X^*) = 0$ and $\text{Var}(X^*) = 1;$

i.e., the standardized random variable has mean 0 and standard deviation 1.

Example 3.2. In a manufactured lot, there is a proportion p of
defective items. An item is chosen at random from the lot. Let X
have the value 1 if the selected item is defective, and 0 otherwise.
Thus the possible values of X are 1 and 0, and these occur with probability p and $q = 1 - p$, respectively. Hence

$$\mu_X = 1 \cdot p + 0 \cdot q = p,$$
$$E(X^2) = 1^2 \cdot p + 0^2 \cdot q = p,$$
$$\sigma_X^2 = E(X^2) - \mu_X^2 = p - p^2 = p(1 - p) = pq,$$

and the standardized random variable corresponding to X is

$$X^* = \frac{X - p}{\sqrt{pq}}.$$

Example 3.3. To each value of a random variable X there corresponds a value of the corresponding standardized variable X^*, and
vice versa. If the value of X is a test score, then X^* is the corresponding *standard score*. To interpret the standard score, we solve (3.13)
for X and obtain

$$X = \mu_X + X^*\sigma_X.$$

Thus, if we are told that the value of the standard score X^* is some
number, say z, then the corresponding value of the actual score X is
z standard deviations removed from the mean, being above the mean
if $z > 0$ and below if $z < 0$. A standard score of $+2$ means an actual
score of 2 standard deviations above the mean score, etc.

The following theorem, due to the Russian mathematician P. L.
Chebyshev (1821–1894), gives us further insight into the significance
of the standard deviation as a measure of the dispersion of the values
of a random variable about the mean.

Theorem 3.5. Let X be a random variable with mean μ_X and
standard deviation $\sigma_X > 0$. Let c be any positive number. Then the

probability that a value of X occurs that differs from μ_X by more than c is less than σ_X^2/c^2. In symbols,

$$(3.15) \qquad P(|X - \mu_X| > c) < \frac{\sigma_X^2}{c^2}.$$

Proof. We start with Formula (3.7) for the variance of X:

$$\sigma_X^2 = \sum_{k=1}^{N} (x_k - \mu_X)^2 f(x_k).$$

Since each term of this sum is nonnegative, omitting some terms cannot increase the value of the sum. Therefore, if we delete all terms (if any) for which $|x_k - \mu_X| \leq c$, we obtain

$$\sigma_X^2 \geq \sum_k{}^* (x_k - \mu_X)^2 f(x_k)$$

where the asterisk indicates that the summation extends only over those k for which $|x_k - \mu_X| > c$. It follows that we further decrease this sum if we replace each $|x_k - \mu_X|$ by c; i.e.,

$$\sigma_X^2 > \sum_k{}^* c^2 f(x_k) = c^2 \sum_k{}^* f(x_k).$$

But

$$\sum_k{}^* f(x_k) = \sum_k{}^* P(X = x_k) = P(|X - \mu_X| > c).$$

Hence

$$\sigma_X^2 > c^2 P(|X - \mu_X| > c)$$

and the result follows by dividing both sides by c^2.

From Formula (3.15) we see that with c fixed, the smaller the variance of X, the lower the probability that a value of X occurs that deviates from μ_X by more than c. Thus the variance in this sense controls the spread or dispersion of the values of the random variable X about the mean. To be somewhat more precise, it is convenient to obtain an alternate form of (3.15) by substituting $z\sigma_X$ for c. One thus obtains

$$(3.16) \qquad P(|X - \mu_X| > z\sigma_X) < \frac{1}{z^2},$$

or equivalently,

$$(3.17) \qquad P(|X - \mu_X| \leq z\sigma_X) > 1 - \frac{1}{z^2}.$$

Formula (3.17) can be written more succinctly if we introduce the standardized random variable X^* defined in Theorem 3.4. We obtain

(3.18) $P(|X^*| \le z) > 1 - \dfrac{1}{z^2}.$

Formulas (3.15)–(3.18) are alternate forms of *Chebyshev's Inequality*.

If $0 < z < 1$, then the inequality does not yield any useful information. For then $1/z^2 > 1$, and (3.16) merely asserts the obvious fact that a probability is less than a number greater than 1.

But if $z > 1$, then Chebyshev's inequality gives us some information about the probability function of X. For example, if we put $z = 2$ in (3.17), then $P(|X - \mu_X| \le 2\mu_X) > \frac{3}{4}$. In words, the event that a random variable assumes a value that is within two standard deviations of its mean has probability greater than $\frac{3}{4}$. Put differently, a total probability of more than $\frac{3}{4}$ is accounted for by values of X in the interval $[\mu_X - 2\sigma_X, \mu_X + 2\sigma_X]$. If $z = 3$, then we similarly conclude that a total probability of more than $\frac{8}{9}$ is accounted for by values of X in the interval $[\mu_X - 3\sigma_X, \mu_X + 3\sigma_X]$.

By using (3.18), these facts can equally well be expressed in terms of the standardized random variable X^*. For example,

$$P(-2 \le X^* \le 2) > \tfrac{3}{4}, \quad P(-3 \le X^* \le 3) > \tfrac{8}{9}, \quad \text{etc.}$$

Either way, we see how the spread or dispersion of the values of the random variable X about the mean μ_X is controlled by the standard deviation σ_X.

Theorem 3.5 is extraordinarily general; the probability statements given by Chebyshev's inequality apply to *any* random variable. One pays a price for such generality, since one cannot expect an inequality that applies to all random variables to be especially sharp and definitive when applied to some specific random variable. (See Problem 3.17.) Nevertheless, Chebyshev's theorem is an important analytic tool in the theory of probability. We shall have occasion to use it in a later section when we prove the so-called law of large numbers.

PROBLEMS

3.1. A questionnaire sent to four families yields the following information.

	"Own TV set?"	Total Income	Number of Children
A	yes	$10,000	2
B	yes	5,000	3
C	yes	8,000	0
D	no	5,000	2

One of these families is chosen at random. Let X have the value 1 if the family owns a television set and the value 0 otherwise, let Y be the income of the family, and let Z be the number of children in the family. Find the mean, variance, and standard deviation of each of these random variables.

3.2. Suppose 70 percent of the voters favor a certain proposal, 30 percent being opposed. A voter is selected at random and we let $X = 0$ if he is opposed, $X = 1$ if he is in favor. Find $E(X)$ and $\text{Var}(X)$.

3.3. Let X denote the sum of the numbers obtained when two fair dice are rolled. Find the variance and standard deviation of X. (Cf. Problem 2.13.)

3.4. (a) Consider the random variable X defined in Example 3.2. Show that $\text{Var}(X) \leq \frac{1}{4}$. For what value of p is $\text{Var}(X) = \frac{1}{4}$? (b) Generalize (a) by showing that if X is *any* random variable such that $E(X^2) = E(X)$, then $\text{Var}(X) \leq \frac{1}{4}$.

3.5. (a) Let X_k be the number of heads obtained when a fair coin is tossed k independent times. For $k = 1, 2, 3, 4$, calculate the variance and standard deviation of X_k.

(b) Redo part (a), but now assume the coin is biased so that the probability is p $(0 \leq p \leq 1)$ that it falls heads on any toss.

3.6. In Problem 2.18, six random variables were shown to have the same mean. Find the variance of each random variable. Interpreting variance as a measure of spread, does it seem reasonable that $\text{Var}(X_k)$ should decrease as k increases?

3.7. The mean and variance of X are 50 and 4, respectively. Evaluate (a) the mean of X^2, (b) the variance of $2X + 3$, (c) the standard deviation of $2X + 3$, (d) the variance of $-X$, (e) the standard deviation of $-X$.

3.8. Prove the following result for any number a:

$$E([X - a]^2) = \text{Var}(X) + (E(X) - a)^2.$$

Use this formula to show that $E([X - a]^2)$ is minimized when $a = E(X)$; i.e., the mean of the squared deviations of the values of a random variable is as small as possible when the deviations are computed from the mean of the random variable. (*Note.* We remarked in Section 2 that if $f(x_k)$ is thought of as the mass of a particle at the point x_k on the x-axis, then $E(X)$ as given by Formula (2.1) is the center of gravity of the system of masses at $x_1, x_2, \cdots, x_N$. With this same interpretation, $\text{Var}(X)$ as given by Formula (3.7) becomes what the physicist calls the *moment of inertia* of the mass system with respect to an axis through the center of gravity and perpendicular to the x-axis. The reader can verify that the equation established here is the mathematical formulation of the following *parallel axis theorem:* The moment of inertia of a mass system about *any* given axis is the moment of inertia of the system about a parallel axis through the center of gravity, plus the moment of inertia about the given axis if all the mass were concentrated at the center of gravity.)

3.9. The random variable X is given and we define a new random variable $Y = g(X)$, as in Definition 2.2. If $g(x) = a + bx + cx^2$, show that
$$E(Y) = a + bE(X) + c[E(X)]^2 + c\,\mathrm{Var}(X).$$

3.10. An urn contains six balls. Three have 1's on them, one has a 2, and two have 3's. One ball is drawn from the urn and then, without replacing the first, another is drawn. Let X_1 be the number on the first ball and X_2 the number on the second ball. Find the standard deviations of X_1 and X_2.

3.11. A subject is shown a deck of three cards numbered 1, 2, and 3. The cards are shuffled and placed face down on the table. The subject is asked to call the order of the cards. Let X denote the number of correct calls made by the subject. Consider the following possible ways that a subject might *guess:*

(1) He chooses one card and calls it three straight times.

(2) He makes three independent guesses. For example, he can roll a fair die and guess the first card is 1 if a 1 or a 2 comes up, guess the first card is 2 if a 3 or 4 comes up, and guess the first card is 3 if a 5 or 6 comes up. The die is then thrown twice more to determine the subject's second and third calls.

(3) He chooses at random one permutation from among all the permutations of the numbers 1, 2, 3 and calls his guesses in the order specified by the selected permutation. (Cf. Problem II.3.9.)

For *each* of these methods of guessing, find (a) the probability function of the random variable X, (b) the mean of X, and (c) the standard deviation of X.

3.12. Let X denote the number obtained when one number is selected at random from the numbers 1, 2, 3, $\cdots$, N. Show that
$$E(X) = \frac{N+1}{2}, \qquad \mathrm{Var}(X) = \frac{N^2-1}{12}.$$

3.13. Another measure of the spread* of the values of a random variable is the *mean absolute deviation* defined as the number
$$\sum_{k=1}^{N} |x_k - \mu_X|\, f(x_k).$$

* For an interesting discussion of measures of variability and their use to measure risk in a portfolio of securities, see H. Markowitz, *Portfolio Selection.* John Wiley and Sons, Inc., 1959, pp. 286–297.

Compute the mean absolute deviation of the random variables whose probability functions are given in (3.1)–(3.4) and compare with their standard deviations.

3.14. Prove Theorem 3.4.

3.15. A random variable X has mean 100 and standard deviation 10. X^* is the standardized random variable corresponding to X. (a) What value of X^* corresponds to each of the following values of X: 85, 100, 103? (b) What value of X corresponds to each of the following values of X^*: $-2, -1, -0.4, 1.3$?

3.16. In the proof of Theorem 3.5, where was the hypothesis $\sigma_X > 0$ used? Is (3.15) true if $\sigma_X = 0$?

3.17. For each of the following random variables, calculate
$$P(|X - \mu_X| \leq z\sigma_X)$$
for $z = 1.5$ and $z = 2$, and compare these probabilities with the corresponding estimates given by Chebyshev's inequality.

(a) X, the number of points obtained in a roll of a fair die.

(b) X, the sum of the number of points on two fair dice.

(c) X, the number of heads obtained when four fair coins are tossed.

3.18. You are told that no possible value of a random variable X is more than one standard deviation from the mean; i.e., all possible values are in the interval $[\mu_X - \sigma_X, \mu_X + \sigma_X]$. Show that X either has only one possible value, this value therefore occurring with probability 1, or X has two possible values, each occurring with probability $\frac{1}{2}$.

3.19. Let X be any random variable and consider the statement
$$P(-z \leq X^* \leq z) > p.$$
For each of the following values of p find the smallest value of z (according to Chebyshev's inequality) that makes the statement true: $p = 0.5$, $p = 0.9$, $p = 0.95$, $p = 0.99$.

3.20. The random variable X is given and the new random variable $Y = g(X)$ is defined. Suppose that the possible values of Y are all nonnegative and that not all are zero. For any positive number, say c^2, prove that
$$P(Y > c^2) < \frac{E(Y)}{c^2}.$$
[*Note.* This formula generalizes Chebyshev's theorem, for we obtain (3.15) as a special case if we put $g(x) = (x - \mu_X)^2$.]

4. Joint probability functions; independent random variables

When an experiment is performed, we are often interested in more than one characteristic of the resulting outcome. If 13 cards are dealt from a full deck, we might be interested in the number of spades in the hand *and* the number of aces; if a person is selected from a certain population, we might want to record his height *and* weight, his IQ test score *and* the average number of hours he watches television, etc. In such cases, we are interested not only in studying each characteristic separately, but also in determining interrelationships that exist among the characteristics.

In mathematical terms, we are given a sample space S and n random variables defined on S, where n is an integer greater than or equal to 2. In this section, we study the bivariate case ($n = 2$), concluding with some remarks on the more general multivariate case ($n \geq 2$). We begin with an example that serves to prepare the way for the formal development that follows.

Example 4.1. A fair coin is tossed three independent times. We choose the familiar set

$$S = \{\text{HHH, HHT, HTH, THH, HTT, THT, TTH, TTT}\}$$

as sample space and assign probability $\frac{1}{8}$ to each simple event. We define the following random variables:

$$X = \begin{cases} 0 & \text{if the first toss is a tail,} \\ 1 & \text{if the first toss is a head,} \end{cases}$$

$Y =$ the total number of heads,

$Z =$ the absolute value of the difference between the number of heads and tails.

(Note that when we define random variables in this way, the equality sign is used as shorthand for "is the random variable whose value for any outcome (element of S) is". The distinction between the random variable and the value of the random variable should be kept clearly in mind even when, as here, the customary notation is somewhat misleading.)

We list in Table 26 the values of these three random variables for each element of the sample space S. Consider first the pair X, Y. We want to determine not only the possible pairs of values of X and Y, but also the probability with which each such pair occurs. To say, for example, that X has the value 0 and Y the value 1 is to say that the event $\{\text{THT, TTH}\}$ occurs. The probability of this event is therefore $\frac{2}{8}$ or $\frac{1}{4}$. We write

$$P(X = 0, Y = 1) = \tfrac{1}{4},$$

TABLE 26

Element of S	Value of X	Value of Y	Value of Z
HHH	1	3	3
HHT	1	2	1
HTH	1	2	1
THH	0	2	1
HTT	1	1	1
THT	0	1	1
TTH	0	1	1
TTT	0	0	3

adopting the usual convention in which a comma is used in place of $\cap$ to denote the *intersection* of the two events $X = 0$ and $Y = 1$. We similarly find

$$P(X = 0, Y = 0) = P(\{TTT\}) = \tfrac{1}{8},$$
$$P(X = 1, Y = 0) = P(\emptyset) = 0, \quad \text{etc.}$$

In this way we obtain the probabilities of all possible pairs of values of X and Y. These probabilities are conveniently arranged in Table 27, the so-called *joint probability table* of X and Y.

TABLE 27

x \ y	0	1	2	3	$P(X = x)$
0	1/8	1/4	1/8	0	1/2
1	0	1/8	1/4	1/8	1/2
$P(Y = y)$	1/8	3/8	3/8	1/8	1

We can also represent these results graphically as in Figure 23. In Figure 23(a) a heavy dot is located at each point (x, y) for which $P(X = x, Y = y)$ is positive, and this probability appears next to the dot. In Figure 23(b) we draw a three dimensional chart in which $P(X = x, Y = y)$ is the height of a vertical line drawn above the point (x, y) in the horizontal x-y plane.

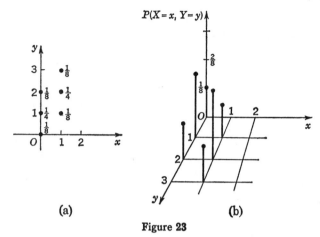

(a)   (b)

Figure 23

The event $Y = 0$ is the union of the mutually exclusive events $(X = 0, Y = 0)$ and $(X = 1, Y = 0)$. Hence

$$P(Y = 0) = P(X = 0, Y = 0) + P(X = 1, Y = 0)$$
$$= \tfrac{1}{8} + 0 = \tfrac{1}{8}.$$

In Table 27, this probability is obtained as the sum of the entries in the column headed $y = 0$. By adding the entries in the other columns, we similarly find

$$P(Y = 1) = \tfrac{3}{8}, \quad P(Y = 2) = \tfrac{3}{8}, \quad P(Y = 3) = \tfrac{1}{8}.$$

In this way we obtain the probability function of the random variable Y from the joint probability table of X and Y. Since values of this probability function are written in the lower margin of the joint table, the function is commonly called the *marginal* probability function of Y, in spite of the fact that the adjective "marginal" is redundant. By adding across the rows in the joint table, one similarly obtains the (marginal) probability function of X.

We add one final observation about the random variables X and Y. It is clear from the meaning of X and Y that knowing the value of X changes the probability that a given value of Y occurs. For example, $P(Y = 2) = \tfrac{3}{8}$. But if we are told that the value of X is 1, then the conditional probability of the event $Y = 2$ becomes $\tfrac{1}{2}$. For, by the definition of conditional probability,

$$P(Y = 2 | X = 1) = \frac{P(X = 1, Y = 2)}{P(X = 1)} = \frac{\tfrac{1}{4}}{\tfrac{1}{2}} = \frac{1}{2}.$$

As we expect, the events $X = 1$ and $Y = 2$ are *not* independent: knowing that the first toss results in a head *increases* the probability of obtaining exactly two heads in the three tosses.

What we have done for the pair X, Y can also be done for X, Z. We give the results only, asking the reader to check our calculations. The joint probability table of X and Z is given in Table 28. We have, as before, written in the margins the row-sums and the column-

TABLE 28

z x	1	3	$P(X = x)$
0	3/8	1/8	1/2
1	3/8	1/8	1/2
$P(Z = z)$	3/4	1/4	1

sums which determine the (marginal) probability functions of X and Z, respectively. In Figure 24, we graph these results as we did for X, Y in Figure 23.

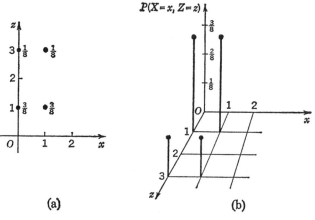

(a)   (b)

Figure 24

Finally, let us observe that the events $X = 0$ and $Z = 1$ are independent, since we find $P(X = 0, Z = 1) = \frac{3}{8}$, and since this is equal to the product of $P(X = 0) = \frac{1}{2}$ and $P(Z = 1) = \frac{3}{4}$. This is reflected in Table 28 by the fact that the probability appearing in the cell determined by the row labeled $x = 0$ and the column labeled $z = 1$ is the product of the marginal totals for that row and column. Indeed, this multiplication property holds for each of the four entries in the joint probability table of X and Z. A comparison with Table 27 will show that the entries in the cells of the joint probability table of X and Y are *not* products of the corresponding marginal probabilities. Thus the random variables X and Y have a relationship to each other that is different from that shown by X and Z. According to the definitions to be given below, we say that the random variables X and Y are *dependent*, but that X and Z are *independent*.

With this particular example understood, we can now proceed to discuss the general case of any two random variables defined on the same sample space.

Definition 4.1. Let a sample space $S = \{o_1, o_2, \cdots, o_n\}$ be given together with an acceptable assignment of probabilities to its simple events. Let X and Y be random variables defined on S. Then the function h whose value at the point (x, y) is given by

$$(4.1) \qquad \begin{aligned} h(x, y) &= P(X = x, Y = y) \\ &= P(\{o_i \in S \mid X(o_i) = x \text{ and } Y(o_i) = y\}) \end{aligned}$$

is called the *joint probability function* of the random variables X and Y. (The domain of the function h is the set of *all* ordered pairs of real numbers, although h has nonzero values for only a finite number of such pairs.)

Let us suppose that X has possible values $x_1, x_2, \cdots, x_M$ and probability function f. Then

$$(4.2) \qquad f(x_j) = P(X = x_j) > 0, \qquad \sum_{j=1}^{M} f(x_j) = 1.$$

Similarly, if Y has possible values $y_1, y_2, \cdots, y_N$ and probability function g, then

$$(4.3) \qquad g(y_k) = P(Y = y_k) > 0, \qquad \sum_{k=1}^{N} g(y_k) = 1.$$

With this notation, the *joint probability table* of X and Y is defined as the double-entry array in Table 29. The probabilities listed in this table have the following properties:

$$(4.4) \qquad h(x_j, y_k) \geq 0 \qquad \text{for } j = 1, 2, \cdots, M; k = 1, 2, \cdots, N.$$

(4.5) $\quad \sum_{\text{all } j,k} h(x_j, y_k) = 1.$

(4.6) $\quad \sum_{k=1}^{N} h(x_j, y_k) = f(x_j) \qquad$ for $j = 1, 2, \cdots, M.$

(4.7) $\quad \sum_{j=1}^{M} h(x_j, y_k) = g(y_k) \qquad$ for $k = 1, 2, \cdots, N.$

The inequality in (4.4) expresses the obvious fact that the probability of the joint occurrence of the events $X = x_j$ and $Y = y_k$ is nonnegative. We note, however, that although we have assumed in (4.2) and (4.3) that the events $X = x_j$ and $Y = y_k$ occur with *positive* probability, we must allow in (4.4) for the possibility that the intersection of these events is the empty set, and thus has probability zero.

TABLE 29

x \ y	y_1	y_2	$\cdots$	y_k	$\cdots$	y_N	$P(X = x)$
x_1	$h(x_1,y_1)$	$h(x_1,y_2)$	$\cdots$	$h(x_1,y_k)$	$\cdots$	$h(x_1,y_N)$	$f(x_1)$
x_2	$h(x_2,y_1)$	$h(x_2,y_2)$	$\cdots$	$h(x_2,y_k)$	$\cdots$	$h(x_2,y_N)$	$f(x_2)$
·	·	·		·		·	·
·	·	·		·		·	·
·	·	·		·		·	·
x_j	$h(x_j,y_1)$	$h(x_j,y_2)$	$\cdots$	$h(x_j,y_k)$	$\cdots$	$h(x_j,y_N)$	$f(x_j)$
·	·	·		·		·	·
·	·	·		·		·	·
·	·	·		·		·	·
x_M	$h(x_M,y_1)$	$h(x_M,y_2)$	$\cdots$	$h(x_M,y_k)$	$\cdots$	$h(x_M,y_N)$	$f(x_M)$
$P(Y = y)$	$g(y_1)$	$g(y_2)$	$\cdots$	$g(y_k)$	$\cdots$	$g(y_N)$	1

In (4.5), we merely observe that the sum of all MN probabilities in the joint probability table (not including the entries listed in the margins) is 1. This sum can be calculated in any number of ways, but two methods are especially noteworthy. We can sum each row first, then add the row-sums; i.e.,

(4.8) $\quad \sum_{\text{all } j,k} h(x_j, y_k) = \sum_{j=1}^{M} \sum_{k=1}^{N} h(x_j, y_k) = \sum_{j=1}^{M} f(x_j) = 1;$

or we can sum each column first, then add the column-sums; i.e.,

(4.9) $\displaystyle\sum_{\text{all } j,k} h(x_j, y_k) = \sum_{k=1}^{N} \sum_{j=1}^{M} h(x_j, y_k) = \sum_{k=1}^{N} g(y_k) = 1.$

In the first method, the row-sums are the probabilities with which the possible values of X occur. This fact is recorded in (4.6) and follows from the observation that the event $X = x_j$ occurs whenever one of the joint events $(X = x_j, Y = y)$ occurs for some value y of the random variable Y. For different values of y, these joint events are clearly mutually exclusive, and so

$$P(X = x_j) = \sum_y P(X = x_j, Y = y).$$

This equality is equivalent to (4.6), since only *possible* values of Y can contribute to the sum. We similarly can show that the sum of the entries in any column of the joint table is the probability with which the value of Y determining that column occurs. This fact is recorded in (4.7).

Thus we see that from the joint probability table we can recover the probability functions of the random variables X and Y by adding rows and columns. Since the resulting probabilities $f(x_j)$ for $j = 1, 2, \cdots, M$ and $g(y_k)$ for $k = 1, 2, \cdots, N$ are written in the margins of the table, these probabilities are known as *marginal probabilities*, and f and g are referred to as the *marginal probability functions* of X and Y, respectively. In both cases we note that the adjective "marginal" is technically redundant.

Let us now turn to the task of defining the important concept of independent random variables, to which we alluded at the end of our discussion in Example 4.1. (We know the meaning of independent *events* and independent *trials* from Chapter 2.) The following definition seems reasonable, in view of our observations in Example 4.1.

Definition 4.2. Two random variables X and Y defined on the same sample space S are said to be *independent* if and only if

(4.10) $P(X = x_j, Y = y_k) = P(X = x_j)P(Y = y_k)$

for $j = 1, 2, \cdots, M$ and $k = 1, 2, \cdots, N$. In other words, "X and Y are *independent random variables*" means that the events $X = x_j$ and $Y = y_k$ are *independent events* for *all* pairs of possible values x_j and y_k. Random variables that are not independent are said to be *dependent*.

Equivalently, we see that X and Y are independent random variables if and only if

(4.11) $$h(x_j, y_k) = f(x_j)g(y_k),$$

i.e., if and only if the joint probability table assumes the form of a *multiplication table* in which $h(x_j, y_k)$, the entry in any row and column, is the product of $f(x_j)$, the probability in the row margin, and $g(y_k)$, the probability in the column margin.

With this definition before us, a quick glance at Tables 27 and 28 shows that in Example 4.1, as we anticipated, X and Y are dependent random variables, but X and Z are independent.

Example 4.2. An urn contains three red and two green balls. A random sample of two balls is drawn (a) with replacement, and (b) without replacement. In either case, we define

$$X = \begin{cases} 0 & \text{if the first ball is green} \\ 1 & \text{if the first ball is red,} \end{cases}$$

$$Y = \begin{cases} 0 & \text{if the second ball is green} \\ 1 & \text{if the second ball is red.} \end{cases}$$

We find the two joint probability tables given in Table 30. Note that,

TABLE 30

x \ y	0	1	$P(X=x)$
0	$\frac{2}{5} \cdot \frac{2}{5}$	$\frac{2}{5} \cdot \frac{3}{5}$	$\frac{2}{5}$
1	$\frac{3}{5} \cdot \frac{2}{5}$	$\frac{3}{5} \cdot \frac{3}{5}$	$\frac{3}{5}$
$P(Y=y)$	$\frac{2}{5}$	$\frac{3}{5}$	1

x \ y	0	1	$P(X=x)$
0	$\frac{2}{5} \cdot \frac{1}{4}$	$\frac{2}{5} \cdot \frac{3}{4}$	$\frac{2}{5}$
1	$\frac{3}{5} \cdot \frac{2}{4}$	$\frac{3}{5} \cdot \frac{2}{4}$	$\frac{3}{5}$
$P(Y=y)$	$\frac{2}{5}$	$\frac{3}{5}$	1

(a) with replacement   (b) without replacement

in (a), X and Y are identically distributed and are independent random variables. In (b), X and Y are also identically distributed, but now they are dependent random variables. Although it is always possible to derive the probability functions of X and Y from their joint probability function, as this example demonstrates it is generally impossible to reconstruct the joint probability table if only the marginal probabilities of X and Y are known.

Further insight into the reasonableness of our definition of independence of random variables is obtained by looking at conditional probabilities. Suppose we are interested in the event that X has the value x_j, given that Y has the value y_k. We find directly from the definition of conditional probability that

$$(4.12) \qquad P(X = x_j | Y = y_k) = \frac{P(X = x_j, Y = y_k)}{P(Y = y_k)} = \frac{h(x_j, y_k)}{g(y_k)}.$$

(Recall that in (4.3) we have assumed $g(y_k) \neq 0$.) If we write

$$(4.13) \qquad\qquad f(x_j \mid y_k) = P(X = x_j | Y = y_k),$$

then for *fixed* k, we have a function defined with domain the set of possible values of the random variable X. To distinguish clearly the function from its value, we shall write $f(\cdot \mid y_k)$ for the function and $f(x_j \mid y_k)$ for the value of the function at $x = x_j$. We have N such functions, one for each possible value y_k of Y. In terms of a function-machine, the inputs of the $f(\cdot \mid y_k)$-machine are the possible values of the random variable X. If x_j is the input, then the corresponding output number is the conditional probability $f(x_j \mid y_k)$ given in (4.13).

We now show that each of the functions $f(\cdot \mid y_k)$ is a probability function. By Theorem 1.3, it suffices to show that the values of the function are nonnegative and that these values add to 1. Since $f(x_j \mid y_k)$ is defined in (4.13) as a probability, it is clearly nonnegative. Furthermore,

$$(4.14) \qquad \sum_{j=1}^{M} f(x_j \mid y_k) = \frac{1}{g(y_k)} \sum_{j=1}^{M} h(x_j, y_k) = \frac{g(y_k)}{g(y_k)} = 1.$$

Hence $f(\cdot \mid y_k)$ is a probability function. It is important enough to deserve a special name.

Definition 4.3. Let y_k be any possible value of Y. The function $f(\cdot \mid y_k)$ whose domain is the set of possible values of X and whose value $f(x_j \mid y_k)$ is given by (4.13), or by (4.12), is called the *conditional probability function of X, given $Y = y_k$*. The *conditional probability function of Y, given $X = x_j$*, is similarly defined as the function $g(\cdot \mid x_j)$ whose value at y_k is given by

$$(4.15) \qquad\qquad g(y_k \mid x_j) = P(Y = y_k | X = x_j) = \frac{h(x_j, y_k)}{f(x_j)}.$$

Example 4.3. Refer to Example 4.2 and suppose that Y has the value 1, i.e., that the second ball drawn is red. We want to determine the conditional probability function of X when the balls are drawn (a) with replacement, and (b) without replacement. We must therefore calculate $f(0 \mid 1)$ and $f(1 \mid 1)$ in both cases. The probabilities we need in order to use Formula (4.12) can be read directly from Table 30, and we obtain the following results.

(a) With replacement:

$$f(0 \mid 1) = \frac{h(0, 1)}{g(1)} = \frac{\frac{6}{25}}{\frac{3}{5}} = \frac{2}{5}$$

$$f(1 \mid 1) = \frac{h(1, 1)}{g(1)} = \frac{\frac{9}{25}}{\frac{3}{5}} = \frac{3}{5}.$$

(b) Without replacement:

$$f(0 \mid 1) = \frac{h(0, 1)}{g(1)} = \frac{\frac{3}{10}}{\frac{3}{5}} = \frac{1}{2}$$

$$f(1 \mid 1) = \frac{h(1, 1)}{g(1)} = \frac{\frac{3}{10}}{\frac{3}{5}} = \frac{1}{2}.$$

Observe that in case (a), where X and Y are independent, the conditional probability function of X given $Y = 1$ has the same values as the probability function of X; i.e., $f(0 \mid 1) = f(0)$ and $f(1 \mid 1) = f(1)$. But in case (b), where X and Y are dependent, knowing that the second ball drawn is red changes the probabilities of drawing a red or green ball on the first draw. For example, $f(1) = \frac{3}{5}$ is the probability that the first ball is red in the absence of any information. When we are told that the second ball drawn is red, the conditional probability that the first ball is red decreases to $f(1 \mid 1) = \frac{1}{2}$.

The following result reformulates the definition of independence of random variables in terms of the conditional probability function. We leave the proof for the problems.

Theorem 4.1. The random variables X and Y are independent if and only if, for every possible value y_k of Y, the conditional probability function of X given $Y = y_k$ and the (marginal) probability function of X have equal values for each possible value of X; i.e., if and only if

(4.16) $\qquad f(x_j \mid y_k) = f(x_j) \qquad$ for $j = 1, 2, \cdots, M; k = 1, 2, \cdots, N.$

This result shows that X and Y are independent whenever knowing the value of Y does not change the probability with which X has any of its values, or equivalently (see Problem 4.14), whenever knowing the value of X does not change the probability with which Y has any of its values.

We shall return to conditional probability functions in a later section, but now we take up two results that are strongly suggested by our intuition. The first of these asserts that if X and Y are independent, then so are $u(X)$ and $v(Y)$ for any functions u and v. For example, if X and Y are independent, then with $u(x) = x - \mu_X$ and $v(y) = y - \mu_Y$ this theorem will permit us to conclude that $X - \mu_X$ and $Y - \mu_Y$ are independent; with $u(x) = x^2$ and $v(y) = y^2$, that X^2 and Y^2 are independent; etc.

Theorem 4.2. Let X and Y be independent random variables. Let u and v be functions for which $u(X)$ and $v(Y)$ are defined in the sense of Definition 2.2. Then $u(X)$ and $v(Y)$ are also independent random variables.

Proof. According to Definition 4.2, to prove $u(X)$ and $v(Y)$ are independent it suffices to prove that

$$P(u(X) = x, v(Y) = y) = P(u(X) = x)P(v(Y) = y)$$

for every pair of numbers x and y. Now

$$P(u(X) = x, v(Y) = y) = \sum_{j,k}{}^* P(X = x_j, Y = y_k),$$

where the asterisk indicates that the sum is to be taken over only those values of j and k for which $u(x_j) = x$ and $v(y_k) = y$. Since X and Y are independent by hypothesis, we can apply (4.10) to obtain

$$\begin{aligned}
P(u(X) = x, v(Y) = y) &= \sum_{j,k}{}^* P(X = x_j)P(Y = y_k) \\
&= \sum_{j}{}^* P(X = x_j) \sum_{k}{}^* P(Y = y_k) \\
&= P(u(X) = x)P(v(Y) = y),
\end{aligned}$$

and the proof is complete.

Our next result concerns two random variables X and Y such that the value of X is determined by the first trial and the value of Y is determined by the second trial of a two-trial experiment. If the trials are independent (as defined in Section II.9), then we would be dissatisfied with our theory if it did not enable us to prove that X and Y are independent random variables. For example, let X and Y denote respectively the sum obtained in the first and second rolls of a pair of dice. If the two rolls (trials) are independent, then we expect that X and Y are independent random variables. We leave for the reader the task of showing that the following result is an immediate consequence of Theorem II.9.1.

Theorem 4.3. Let an experiment consist of two independent trials. If the value of a random variable X is determined by the first trial and the value of a random variable Y is determined by the second trial, then X and Y are independent.

We conclude this section by recording for later use the extension of some of our results to the case where more than two random variables are defined on the same sample space. First we make the natural extension of Definition 4.2.

Definition 4.4. Let n be any positive integer greater than 1. The random variables V_1, V_2, $\cdots$, V_n defined on a sample space S are said to be *independent* if and only if

$$(4.17) \qquad P(V_1 = v_1, V_2 = v_2, \cdots, V_n = v_n)$$
$$= P(V_1 = v_1)P(V_2 = v_2) \cdots P(V_n = v_n)$$

for *all* combinations of possible values v_1 of V_1, v_2 of V_2, $\cdots$, v_n of V_n. In other words, "V_1, V_2, $\cdots$, V_n are independent random variables" means that $V_1 = v_1$, $V_2 = v_2$, $\cdots$, $V_n = v_n$ are independent events (in the sense of Definition II.8.3) for all possible values v_1, v_2, $\cdots$, v_n.

Corresponding to Theorems 4.2 and 4.3 we have the following results whose proofs we leave for the reader.

Theorem 4.4. Let V_1, V_2, $\cdots$, V_n be independent random variables. Let u_1, u_2, $\cdots$, u_n be functions for which $u_1(V_1)$, $u_2(V_2)$, $\cdots$, $u_n(V_n)$ are defined in the sense of Definition 2.2. Then $u_1(V_1)$, $u_2(V_2)$, $\cdots$, $u_n(V_n)$ are also independent random variables.

Theorem 4.5. Let an experiment consist of n independent trials. If the value of random variable V_j is determined by the jth trial for $j = 1, 2, \cdots, n$, then the random variables V_1, V_2, $\cdots$, V_n are independent.

We continue our study of joint probability functions in the next section.

PROBLEMS

4.1. Let Y and Z be the random variables defined in Example 4.1. Construct the joint probability table of Y and Z, sketch the corresponding three-dimensional probability chart, and determine whether or not Y and Z are independent.

4.2. Modify Example 4.1 by redefining Z as the *algebraic* difference of the number of tails from the number of heads. Construct the joint probability table of X, Z and sketch the corresponding three-dimensional probability chart. Are X and Z independent?

4.3. Suppose three indistinguishable objects are distributed at random into three numbered cells. Let X be the number of empty cells and Y the number of objects in the first cell. Construct the joint probability table of X and Y. Are X and Y independent?

4.4. Let X be the larger of the two numbers and Y be the sum of the numbers showing when two fair dice are rolled. Construct the joint probability table of X and Y. Are X and Y independent random variables?

4.5. Let X denote the number of spades and Y the number of hearts in a bridge hand. Write a formula for

$$h(x, y) = P(X = x, Y = y)$$

and prove that X and Y are dependent.

4.6. Three cards are drawn from the 12 face cards of an ordinary deck. Let X be the number of red jacks and Y be the number of red queens. Construct the joint probability table of X and Y, sketch the corresponding three-dimensional probability chart, and show that X and Y are identically distributed but dependent random variables.

4.7. A fair coin is tossed four independent times. Let X be the number of heads obtained in the first two tosses and Y the number of heads obtained in the last two tosses. Construct the joint probability table of X and Y and sketch the corresponding three-dimensional probability chart. Show that X and Y are independent by using Definition 4.2 and also by invoking Theorem 4.3.

4.8. The joint probability function of X and Y is given by

$$h(x, y) = \tfrac{1}{32}(x^2 + y^2) \qquad \text{for } x = 0, 1, 2, 3 \text{ and } y = 0, 1.$$

(a) Show that the marginal probability function of X is given b

$$f(x) = \tfrac{1}{32}(2x^2 + 1) \qquad \text{for } x = 0, 1, 2, 3.$$

(b) Show that the marginal probability function of Y is given by

$$g(y) = \tfrac{1}{16}(2y^2 + 7) \qquad \text{for } y = 0, 1.$$

(c) Show that the conditional probability function of X given $Y = y$ is given by

$$f(x \mid y) = \frac{1}{2} \frac{x^2 + y^2}{2y^2 + 7} \qquad \text{for } x = 0, 1, 2, 3 \text{ and } y = 0, 1.$$

(d) Show that the conditional probability function of Y given $X = x$ is given by

$$g(y \mid x) = \frac{x^2 + y^2}{2x^2 + 1} \qquad \text{for } x = 0, 1, 2, 3 \text{ and } y = 0, 1.$$

4.9. We select one of the integers 1, 2, 3, 4, 5. After discarding all integers (if any) less than the selected integer, we draw one of the remaining integers. (For example, if we select 3 first, then the second draw is made from the integers 3, 4, 5.) Let X and Y denote the numbers obtained on the first and second draws, respectively.

(a) Construct the joint probability table of X and Y.
(b) Determine the marginal probability functions of X and of Y.
(c) Determine the conditional probability function of Y given $X = 3$.
(d) Determine the conditional probability function of X given $Y = 3$.
(e) Find $P(X + Y > 7)$ and $P(Y - X > 0)$.

4.10. Suppose X has only one possible value, this value therefore occurring with probability 1. Show that X and Y are independent for any random variable Y.

4.11. For the random variables X and Z defined in Example 4.1 of the text, determine the value of $P(X \leq x, Z \leq z)$ for all real numbers x and z. (*Hint.* Refer to Figure 24(a) and divide the x-z plane into a number of regions such that $P(X \leq x, Z \leq z)$ has the same value at all points of any one region.)

4.12. The *joint distribution function* of X and Y is defined for all real numbers x and y by the equation

$$H(x, y) = P(X \leq x, Y \leq y).$$

(a) If h is the joint probability function of X and Y, show that

$$H(x, y) = \sum_{x_j \leq x, y_k \leq y} h(x_j, y_k)$$

where the sum is taken over all j- and k-values for which $x_j \leq x$ and $y_k \leq y$. (This is the bivariate analogue of the formula in Problem 1.10.)

(b) Let a, b, c, d be any numbers with $a < b$ and $c < d$. Show that (cf. Problem 1.11)

$$P(a < X \leq b, c < Y \leq d) \\ = H(b, d) - H(a, d) - H(b, c) + H(a, c).$$

(c) Show that for fixed x, $H(x, y)$ is a nondecreasing function of y, and that for fixed y, $H(x, y)$ is a nondecreasing function of x.

(d) Show that numbers r and R exist so that $H(x, y) = 0$ if $x \leq r$ and $y \leq r$, whereas $H(x, y) = 1$ if $x \geq R$ and $y \geq R$.

(e) Above each point (x, y) in the x-y plane, imagine a point drawn at the height $H(x, y)$. The set of all points drawn in this way is a *surface* which is the three-dimensional graph of the joint distribution function H. Describe the kind of surface one obtains.

4.13. Let X and Y be independent random variables. Choose any two rows of their joint probability table. Show that there is a number (which will depend on the rows you choose) such that the probabilities in one row are obtained by multiplying the corresponding probabilities in the other row by this number.

4.14. Let $g(y_k \mid x_j)$ denote the conditional probability of $Y = y_k$ given $X = x_j$. Show that if $f(x_j \mid y_k) = f(x_j)$ for all possible values x_j of X and y_k of Y, then for all these values also $g(y_k \mid x_j) = g(y_k)$.

4.15. (a) Prove Theorem 4.1.
   (b) Prove Theorem 4.3.
   (c) Show that the converse of Theorem 4.2 is false by giving an example of two *dependent* random variables X and Y for which X^2 and Y^2 are *independent*.

4.16. (a) Show from Definition 4.4 that if $V_1, V_2, \cdots, V_n$ are independent, then any smaller number of random variables taken from these n are also independent.
   (b) Let an experiment consist of n independent trials. For any positive integer $k < n$, we can think of this experiment as made up of two supertrials, the first k trials being the first supertrial and the last $n - k$ trials being the second supertrial. Show that these supertrials are independent, and hence conclude from Theorem 4.3 that if X is a random variable determined by the first k trials and Y is a random variable determined by the last $n - k$ trials, then X and Y are independent.

5. Mean and variance of sums of random variables; the sample mean

We shall see in this section that if two random variables X and Y are defined on a sample space S, then there are automatically many other random variables also defined on S. In particular, the sum $X + Y$ and the product XY turn out to be especially important. We also extend our results to the case where more than two random variables are defined on S, and are then able to prove some theorems of great interest in the branch of statistics known as sampling theory.

Our first task is to develop the bivariate analogue of Theorem 2 1 as an aid to computing means of random variables that are functions of X and Y. Suppose z is a numerical-valued function whose domain is a set of ordered pairs of real numbers and let $z(x, y)$ denote the value of z at the ordered pair or point (x, y). If the domain of z includes all the ordered pairs of values of X and Y, then for each element $o_i \in S$ we can first find the corresponding values $X(o_i)$ and $Y(o_i)$, and then evaluate z at the point $(X(o_i), Y(o_i))$. See Figure 25. In

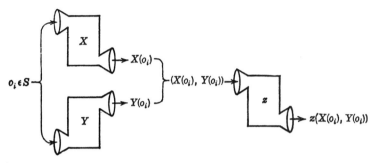

Figure 25

this way, to $o_i \in S$ (the input) we make correspond the (output) real number $z(X(o_i), Y(o_i))$, and thus we have a new *random variable* defined on S. This random variable is denoted by $z(X, Y)$. For example, if $z(x, y) = x + y$, then

$$z(X, Y) = X + Y$$

is the sum of the random variables X and Y; if

$$z(x, y) = (x - \mu_X)(y - \mu_Y),$$

then

$$z(X, Y) = (X - \mu_X)(Y - \mu_Y)$$

is the product of the deviations of X and Y from their respective means; etc. In the example that follows, we illustrate how to determine the probability function of the random variable $z(X, Y)$ from the joint probability table of X and Y. The mean of $z(X, Y)$ can then easily be computed.

Example 5.1. Consider the random variables X and Y of Example 4.1. The possible values of X and Y, together with their joint probabilities, are given in Table 27. Let $z(x, y) = x + y$ so that $U = z(X, Y) = X + Y$. From the joint probability table, we can determine the possible values of U as well as the probability with which each value occurs. For example,

$$P(U = 2) = P(X = 0, Y = 2) + P(X = 1, Y = 1) = \tfrac{1}{8} + \tfrac{1}{8} = \tfrac{1}{4}.$$

In this way, we obtain the entries in the following probability table for the random variable $U = X + Y$:

u	0	1	2	3	4
$P(U = u)$	1/8	1/4	1/4	1/4	1/8

From this table, we calculate the mean of U:

$$E(U) = E(X + Y) = 0(\tfrac{1}{8}) + 1(\tfrac{1}{4}) + 2(\tfrac{1}{4}) + 3(\tfrac{1}{4}) + 4(\tfrac{1}{8}) = 2.$$

From the marginal probability functions of X and Y, also given in Table 27, we find that

$$E(X) = 0(\tfrac{1}{2}) + 1(\tfrac{1}{2}) = \tfrac{1}{2}, \qquad E(Y) = 0(\tfrac{1}{8}) + 1(\tfrac{3}{8}) + 2(\tfrac{3}{8}) + 3(\tfrac{1}{8}) = \tfrac{3}{2}.$$

Observe that $E(X + Y) = E(X) + E(Y)$, a result that we will soon establish for *all* random variables X and Y.

If we define $z(x, y)$ as the product rather than the sum of x and y, then $V = z(X, Y) = XY$ is a random variable whose probability table is similarly found:

v	0	1	2	3
$P(V = v)$	1/2	1/8	1/4	1/8

Now we compute the mean of V,

$$E(V) = E(XY) = 0(\tfrac{1}{2}) + 1(\tfrac{1}{8}) + 2(\tfrac{1}{4}) + 3(\tfrac{1}{8}) = 1.$$

Observe that $E(XY) \neq E(X)E(Y)$.

The reader should note that what we do to determine the probability function of $z(X, Y)$ is collect all possible pairs of X and Y values that lead to the same value of $z(X, Y)$. But it is more convenient not to do this when we want to compute the mean of $z(X, Y)$. The following result tells us how to compute $E[z(X, Y)]$ directly from the joint probability table of X and Y *without* first determining the probability function of $z(X, Y)$. The proof is similar to that of Theorem 2.1, and we leave it for the problems.

Theorem 5.1. Let X and Y be random variables with joint probability function h. Then

(5.1) $$E[z(X, Y)] = \sum_{\text{all } j,k} z(x_j, y_k)h(x_j, y_k).$$

In words, we find $E[z(X, Y)]$ by moving from cell to cell in the joint probability table of X and Y, multiplying the value of $z(X, Y)$ corresponding to each cell by the probability appearing in that cell, and then adding these products for all cells.

Example 5.2. Refer to Example 5.1 and let us illustrate the use of Formula (5.1) by recalculating the means of $X + Y$ and XY. We find directly from Table 27, moving across the first row and then the second,

$$E(X + Y) = 0(\tfrac{1}{8}) + 1(\tfrac{1}{4}) + 2(\tfrac{1}{8}) + 3(0) + 1(0) + 2(\tfrac{1}{8})$$
$$+ 3(\tfrac{1}{4}) + 4(\tfrac{1}{8})$$
$$= 2, \text{ as before.}$$

There is of course no need to write down terms that have zero factors. Indeed, any cell in the joint probability table for which either $z(x_j, y_k) = 0$ or $h(x_j, y_k) = 0$ can be skipped in computing $E[z(X, Y)]$. For example, we note that XY has the value 0 for five of the eight cells in Table 27. Hence we skip these and find, as in Example 5.1,

$$E(XY) = 1(\tfrac{1}{8}) + 2(\tfrac{1}{4}) + 3(\tfrac{1}{8}) = 1.$$

Theorem 5.1 enables us to prove the following extremely important and often-used results.

Theorem 5.2. Let X and Y be any random variables defined on a sample space S. Then

(5.2) $$E(X + Y) = E(X) + E(Y).$$

In words, *the mean of the sum of two random variables is equal to the sum of their means.*

Proof. According to Formula (5.1) we have

$$E(X + Y) = \sum_{\text{all } j,k} (x_j + y_k)h(x_j, y_k)$$
$$= \sum_{\text{all } j,k} x_j h(x_j, y_k) + \sum_{\text{all } j,k} y_k h(x_j, y_k).$$

In the first term on the right, we sum over rows and then add the row-sums; in the second term, we sum over columns and then add the column-sums. We recall (4.6) and (4.7) and find

$$E(X + Y) = \sum_{j=1}^{M} x_j \sum_{k=1}^{N} h(x_j, y_k) + \sum_{k=1}^{N} y_k \sum_{j=1}^{M} h(x_j, y_k)$$
$$= \sum_{j=1}^{M} x_j f(x_j) + \sum_{k=1}^{N} y_k g(y_k)$$
$$= E(X) + E(Y).$$

Combining this result with Formula (2.7), we see that for any constants a and b,

$$(5.3) \qquad E(aX + bY) = aE(X) + bE(Y).$$

Still more general is the following theorem.

Theorem 5.3. Let n be any positive integer. If X_1, X_2, $\cdots$, X_n are any random variables defined on a sample space S, and if a_1, a_2, $\cdots$, a_n are any constants, then*

$$(5.4) \qquad E(a_1X_1 + a_2X_2 + \cdots + a_nX_n)$$
$$= a_1E(X_1) + a_2E(X_2) + \cdots + a_nE(X_n).$$

Proof. The result is true for $n = 1$ and $n = 2$ by Formula (5.3). The theorem is proved (by mathematical induction) as soon as we show that if the theorem is true for any positive integer, say $n = k$, then it is also true for the next integer, $n = k + 1$. Let us therefore assume that (5.4) is true for $n = k$. That is, letting $Y = a_1X_1 + \cdots + a_kX_k$, we are assuming

$$E(Y) = a_1E(X_1) + \cdots + a_kE(X_k).$$

The key idea of the proof is the observation that the sum of $k + 1$ random variables can be thought of as the sum of two random variables to which (5.3) can be applied. In particular,

$$E(a_1X_1 + \cdots + a_kX_k + a_{k+1}X_{k+1}) = E(Y + a_{k+1}X_{k+1})$$
$$= E(Y) + a_{k+1}E(X_{k+1})$$
$$= a_1E(X_1) + \cdots + a_kE(X_k)$$
$$+ a_{k+1}E(X_{k+1}).$$

But this last equality shows that (5.4) is true for $n = k + 1$, and so the proof is complete.

---

* Strictly speaking, the sum $a_1X_1 + a_2X_2 + \cdots + a_nX_n$ appearing in Formula (5.4) has been defined only if $n = 1$ or $n = 2$. We make the natural definition that the sum for any positive integer n is the random variable whose value at each $o_i \in S$ is the number $a_1X_1(o_i) + a_2X_2(o_i) + \cdots + a_nX_n(o_i)$.

Example 5.3. We apply (5.4) to derive a useful identity:

$$E[(X - \mu_X)(Y - \mu_Y)] = E(XY - \mu_X Y - \mu_Y X + \mu_X \mu_Y)$$
$$= E(XY) - \mu_X E(Y) - \mu_Y E(X) + \mu_X \mu_Y.$$

Except for sign the last three terms are equal. Hence

(5.5) $$E[(X - \mu_X)(Y - \mu_Y)] = E(XY) - \mu_X \mu_Y.$$

We turn now to some results leading to a formula for the variance of a sum of random variables.

Theorem 5.4. Let X and Y be *independent* random variables defined on a sample space S. Then

(5.6) $$E(XY) = E(X)E(Y).$$

In words, *the mean of the product of two independent random variables is equal to the product of their means.*

Proof. By Theorem 5.1 we write

$$E(XY) = \sum_{\text{all } j,k} x_j y_k h(x_j, y_k).$$

But the assumed independence of X and Y means, according to (4.11), that $h(x_j, y_k) = f(x_j)g(y_k)$ for all j and k. Hence

$$E(XY) = \sum_{\text{all } j,k} x_j y_k f(x_j)g(y_k)$$
$$= \sum_{j=1}^{M} x_j f(x_j) \sum_{k=1}^{N} y_k g(y_k)$$
$$= E(X)E(Y).$$

It is very important to note that *the converse of Theorem 5.4 is false.* As the following example shows, it is possible for (5.6) to be true for random variables X and Y that are dependent.

Example 5.4. Suppose X has probability table

x	-1	0	1
$P(X = x)$	1/4	1/2	1/4

Let $Y = X^2$. Then X and Y are surely dependent, since the value of X determines the value of Y. This dependence is obvious from the joint probabilities of X and Y, as given in Table 31.

TABLE 31

x \ y	0	1	$P(X = x)$
-1	0	1/4	1/4
0	1/2	0	1/2
1	0	1/4	1/4
$P(Y = y)$	1/2	1/2	1

Nevertheless, the reader can quickly check that $E(X) = 0$, $E(Y) = \frac{1}{2}$, and $E(XY) = E(X^3) = 0$, so that (5.6) holds. Recalling that (5.6) did not hold for the dependent random variables in Example 5.1, we conclude that (5.6) holds for *all* pairs of *independent* random variables and *some* but *not all* pairs of *dependent* random variables.

It is convenient to record here the following corollary:

(5.7) $E[(X - \mu_X)(Y - \mu_Y)] = 0$ if X, Y are independent.

This follows immediately from the identity in (5.5) if we apply Theorem 5.4.

We are now able to state a rule for finding the variance of the sum of two *independent* random variables.

Theorem 5.5. Let X and Y be *independent* random variables defined on a sample space S. Then

(5.8) $\mathrm{Var}(X + Y) = \mathrm{Var}(X) + \mathrm{Var}(Y)$.

In words, *the variance of the sum of two independent random variables is equal to the sum of their variances.*

Proof. By definition of variance, we have

$$\mathrm{Var}(X + Y) = E([(X + Y) - E(X + Y)]^2)$$
$$= E([(X - \mu_X) + (Y - \mu_Y)]^2),$$

where we have rearranged terms in the bracket after using (5.2). Now we perform the indicated squaring operation to obtain

$$\mathrm{Var}(X + Y) = E[(X - \mu_X)^2 + 2(X - \mu_X)(Y - \mu_Y) + (Y - \mu_Y)^2]$$
(5.9) $$= E[(X - \mu_X)^2] + 2E[(X - \mu_X)(Y - \mu_Y)]$$
$$+ E[(Y - \mu_Y)^2],$$

this last equality resulting from the use of (5.4). The middle term on the right vanishes according to (5.7). The other two terms on the right are, by Definition 3.1, precisely $\text{Var}(X)$ and $\text{Var}(Y)$. We have therefore completed the proof.

Now if X and Y are independent, then so are aX and bY for any constants a and b. (This obvious fact is technically a consequence of Theorem 4.2.) Hence we can apply Theorem 5.5 to aX and bY, and so find that

$$\text{Var}(aX + bY) = \text{Var}(aX) + \text{Var}(bY).$$

Now we use (3.11) to conclude that for any numbers a and b, if X and Y are *independent*, then

(5.10) $\text{Var}(aX + bY) = a^2\,\text{Var}(X) + b^2\,\text{Var}(Y).$

Still more general is the following result, whose proof we leave for the problems.

Theorem 5.6. Let n be any positive integer and suppose X_1, X_2, $\cdots$, X_n are *independent* random variables defined on a sample space S. Then for any constants $a_1, a_2, \cdots, a_n$ we have

(5.11) $\text{Var}(a_1X_1 + a_2X_2 + \cdots + a_nX_n) = \sum_{i=1}^{n} a_i^2\,\text{Var}(X_i).$

In particular (if $a_1 = a_2 = \cdots = a_n = 1$), the variance of the sum of any finite number of *independent* random variables is equal to the sum of their variances. Note that the corresponding result for the mean holds for any random variables, independent or dependent.

Example 5.5. A deck of cards numbered $1, 2, \cdots, n$ is shuffled and placed face down on the table. As each card is turned, a subject tries to guess what number it will be. Suppose the subject does not remember from one card to the next and calls his guesses independently and at random; i.e., he has the same probability $1/n$ for a correct guess at each trial (guess) of this n-trial experiment, and the trials are independent. (One way of doing this would be for the subject to have a duplicate deck and make each of his guesses by selecting one card at random from his deck. Note that our independence assumption means that the subject draws each card from the *full* duplicate deck; i.e., he is drawing a random sample of n cards *with replacement* from the deck of n cards. The corresponding problem in which his guesses are determined by drawing a random sample *without replacement* is discussed in Problem 6.8 of the next section.)

Let X denote the random variable whose value for any outcome of the n-trial experiment is the number of correct guesses made by the subject. We shall find the mean and variance of X by expressing X as a sum of n random variables, and then using Formulas (5.4) and (5.11).

For $k = 1, 2, \cdots, n$, let

$$(5.12) \qquad X_k = \begin{cases} 0 & \text{if the } k\text{th guess is wrong } \left(\text{probability } 1 - \dfrac{1}{n}\right) \\ 1 & \text{if the } k\text{th guess is correct } \left(\text{probability } \dfrac{1}{n}\right). \end{cases}$$

Then $X = X_1 + X_2 + \cdots + X_n$, since the value of the sum is equal to the number of 1's in the sum, and hence is precisely the number of correct guesses made by the subject, or the value of X. Now for $k = 1, 2, \cdots, n$, we find

$$(5.13) \qquad E(X_k) = 0\left(1 - \frac{1}{n}\right) + 1\left(\frac{1}{n}\right) = \frac{1}{n}.$$

Hence, by (5.4),

$$E(X) = E(X_1 + X_2 + \cdots + X_n) = E(X_1) + E(X_2) + \cdots + E(X_n)$$
$$= \frac{1}{n} + \frac{1}{n} + \cdots + \frac{1}{n}$$
$$= 1.$$

Thus we see that the mean number of correct guesses is 1, and therefore does not depend upon n, the number of cards in the deck.

To compute $\text{Var}(X)$, we note that X_k is determined by the kth trial of the experiment. Since the trials are independent, we conclude by Theorem 4.5 that $X_1, X_2, \cdots, X_n$ are independent random variables. Hence (5.11) is applicable. Since for $k = 1, 2, \cdots, n$,

$$\text{Var}(X_k) = E(X_k^2) - [E(X_k)]^2$$
$$(5.14) \qquad\qquad = \frac{1}{n} - \left(\frac{1}{n}\right)^2 = \frac{n-1}{n^2},$$

we obtain

$$\text{Var}(X) = \text{Var}(X_1 + X_2 + \cdots + X_n)$$
$$= \text{Var}(X_1) + \text{Var}(X_2) + \cdots + \text{Var}(X_n)$$
$$= n\,\frac{n-1}{n^2} = 1 - \frac{1}{n}.$$

In Chapter 5, we shall determine the probability function of X, but note that our method allows the calculation of the mean and variance of X without knowing this probability function. (Cf. Problem 3.11, Part 2, which is the special case of this problem when there are only three cards in the deck.)

We turn now to an application of our theorems to experiments made up of a number of independent repetitions of the same trial. Such experiments and random variables associated with them can be interpreted in a number of ways and, in particular, supply a mathematical model for repeated measurement in the sciences and for sampling with replacement in statistics.

Suppose a bowl contains N chips, each chip having a number on it. Some chips may have the same number on them, and we let $x_1, x_2, \cdots, x_M$ ($M \leq N$) be all the *different* numbers in the bowl. Suppose the number x_j occurs f_j times, so that the relative frequency with which this number appears in the bowl or the *proportion* of all chips having this number is f_j/N. It follows that

$$(5.15) \qquad \sum_{j=1}^{M} f_j = N \quad \text{or} \quad \sum_{j=1}^{M} \frac{f_j}{N} = 1.$$

If one chip is selected at random from the bowl and we denote by X the random variable whose value is the number on this chip, then the probability function of X is given by the following table:

x	x_1	x_2	$\cdots$	x_M
$P(X = x)$	f_1/N	f_2/N	$\cdots$	f_M/N

Thus we have a special probability function whose value for any x_j is the proportion of chips with x_j on them among all chips in the bowl:

$$(5.16) \qquad f(x_j) = P(X = x_j) = \frac{f_j}{N} \quad \text{for } j = 1, 2, \cdots, M.$$

The reader can verify that our definitions of mean and variance, when applied to this particular random variable X, yield the formulas

$$(5.17) \qquad \mu_X = E(X) = \frac{1}{N} \sum_{j=1}^{M} x_j f_j,$$

$$(5.18) \qquad \sigma_X^2 = \text{Var}(X) = \frac{1}{N} \sum_{j=1}^{M} (x_j - \mu_X)^2 f_j,$$

$$(5.19) \qquad \sigma_X^2 = \text{Var}(X) = \frac{1}{N} \sum_{j=1}^{M} x_j^2 f_j - \mu_X^2,$$

the last equality arising by use of (3.10).

It is customary in this context to say that we have a *population* of N chips and then to call μ_X and σ_X^2 the *population mean* and the *population variance* of X. As we know (Theorem 1.3), we can consider X defined on the sample space $S = \{x_1, x_2, \cdots, x_M\}$ whose simple events are assigned probabilities as given by the probability function of X, i.e., $P(\{x_j\}) = f_j/N$ for $j = 1, 2, \cdots, M$.

From the population of N chips, we now draw n chips, replacing each before the next draw. We consider this as an experiment made up of n independent trials, each trial being defined by the sample space S and the trials being independent by virtue of our assumption that we are sampling *with* replacement. Our mathematical counterpart for this n-trial sampling experiment is the sample space given by the Cartesian product set $S \times S \times \cdots \times S$ (n S's), together with an assignment of probabilities in accordance with the product rule discussed in Section II.9.

For $k = 1, 2, \cdots, n$, let X_k be the random variable whose value is the number on the kth chip drawn from the bowl. Thus we have n random variables $X_1, X_2, \cdots, X_n$ defined on the Cartesian product sample space. Since each trial is an exact duplicate of any other, these random variables all have the same probability function. Furthermore, since X_k is determined by the kth trial and the trials are independent, it follows that $X_1, X_2, \cdots, X_n$ are independent random variables. We summarize by saying that *the random variables $X_1, X_2, \cdots, X_n$ are independent and identically distributed, each with mean μ_X and variance σ_X^2.*

It is thus clear that our sampling experiment is completely determined as soon as we know the common probability function of the X_k's. For then we are given the possible sample values that can arise in each trial, together with their probabilities. In other words, it is meaningful to talk of a population specified by the probability function of a random variable X. Indeed, for this reason sampling with replacement is often referred to as sampling from a probability function.

The random variable $\overline{X}$ given by

$$(5.20) \qquad \overline{X} = \frac{X_1 + X_2 + \cdots + X_n}{n}$$

is called the *sample mean* of X. The value of $\overline{X}$ for any selection of n chips is just the arithmetic mean or average of the numbers on the chips. An experimenter who has incomplete knowledge of the composition of the bowl could nevertheless draw his random sample from the population and obtain a value of the sample mean $\overline{X}$. For example, if we think of the chips as corresponding to N people in a given population and the number on each chip as the income of the corresponding person, then the value of $\overline{X}$ is just the average of the n incomes selected in the random sample. Or, if the numbers on the chips are N possible measurements of some quantity, say the length of a bar to the nearest thousandth of a centimeter or the time to the nearest tenth of a second that it takes a rat to complete a maze, then the value of $\overline{X}$ is just the average of n such measurements. Before studying the random variable $\overline{X}$ in general, we pause to present a particular example to help fix these ideas.

Example 5.6. Suppose our population is specified by the probability table

x	-1	0	2
$P(X = x)$	.1	.5	.4

The reader can easily verify that the population mean and variance are given by

$$\mu_X = .7, \qquad \sigma_X^2 = 1.21.$$

TABLE 32

Sample	Probability of Drawing This Sample	Value of $\overline{X}$ For This Sample
$-1, -1$	.01	-1
$-1, 0$	.05	$-\frac{1}{2}$
$-1, 2$	.04	$\frac{1}{2}$
$0, -1$	.05	$-\frac{1}{2}$
$0, 0$	.25	0
$0, 2$	.20	1
$2, -1$	.04	$\frac{1}{2}$
$2, 0$	.20	1
$2, 2$	.16	2

In Table 32, we list all possible samples of size $n = 2$ taken with replacement from this population, the probability of obtaining each sample, and the corresponding value of the sample mean $\overline{X}$. We thus find the following probability table for $\overline{X}$, the sample mean:

x	-1	$-\frac{1}{2}$	0	$\frac{1}{2}$	1	2
$P(\overline{X} = x)$	.01	.10	.25	.08	.40	.16

And the reader can again verify that $\mu_{\overline{X}}$, the mean of $\overline{X}$, and $\sigma^2_{\overline{X}}$, the variance of $\overline{X}$, are given by

$$\mu_{\overline{X}} = .7, \qquad \sigma^2_{\overline{X}} = .605.$$

We observe that

$$\mu_{\overline{X}} = \mu_X \quad \text{and} \quad \sigma^2_{\overline{X}} = \frac{\sigma^2_X}{2};$$

i.e., the mean of the sample mean $\overline{X}$ is equal to the population mean of X, and the variance of the sample mean is equal to the population variance of X divided by the sample size. In other words, although the values of X and of $\overline{X}$ have the same average, the values of $\overline{X}$ spread less about this common average than do the values of X.

A similar procedure for samples of size $n = 3$ (there are now 27 possible samples) yields the following probability table for the sample mean $\overline{X}$. (Note that we do not complicate our notation by explicitly indicating the sample size when we write the symbol for the sample mean. It is therefore important to keep the sample size clearly in mind when writing $\overline{X}$.)

x	-1	$-\frac{2}{3}$	$-\frac{1}{3}$	0	$\frac{1}{3}$	$\frac{2}{3}$	1	$\frac{4}{3}$	2
$P(\overline{X} = x)$	.001	.015	.075	.137	.120	.300	.048	.240	.064

We compute the mean and variance of $\overline{X}$ and find

$$\mu_{\overline{X}} = .7, \qquad \sigma^2_{\overline{X}} = .403\cdots,$$

so that for samples of size 3,

$$\mu_{\overline{X}} = \mu_X \quad \text{and} \quad \sigma^2_{\overline{X}} = \frac{\sigma^2_X}{3}.$$

We see that X and $\overline{X}$ again have the same mean, but as compared with the values of X (which can be considered values of $\overline{X}$ for samples of size 1) or the values of $\overline{X}$ for samples of size 2, the values of $\overline{X}$ for samples of size 3 show less spread about the common mean μ_X. This fact corresponds to our intuitive feeling that we improve our estimate of the population mean as we take averages based on larger and larger samples from the population. We return to this point and make it precise in the theorems that follow.

With the results of this example before us, it should come as no surprise that the following general theorem holds.

Theorem 5.7. Let n be any positive integer and let $X_1, X_2, \cdots, X_n$ be n independent, identically distributed random variables, each with mean μ_X and variance σ_X^2. If

$$(5.21) \qquad \overline{X} = \frac{X_1 + X_2 + \cdots + X_n}{n},$$

then

$$(5.22) \qquad \mu_{\overline{X}} = \mu_X \quad \text{and} \quad \sigma_{\overline{X}}^2 = \frac{\sigma_X^2}{n}.$$

In words, for *sampling with replacement* from a population given by the probability function of a random variable X, *the mean of the sample mean $\overline{X}$ is equal to the population mean of X, and the variance of the sample mean is equal to the variance of X divided by the sample size.*

Proof. We first apply (5.4) to obtain

$$\mu_{\overline{X}} = E\left(\frac{X_1 + \cdots + X_n}{n}\right) = \frac{1}{n} E(X_1 + \cdots + X_n)$$

$$= \frac{1}{n}[E(X_1) + \cdots + E(X_n)] = \frac{1}{n}(n\mu_X) = \mu_X.$$

Similarly, applying (5.11) with $a_1 = \cdots = a_n = 1/n$, we find

$$\sigma_{\overline{X}}^2 = \mathrm{Var}\left(\frac{X_1 + \cdots + X_n}{n}\right)$$

$$= \frac{1}{n^2}[\mathrm{Var}(X_1) + \cdots + \mathrm{Var}(X_n)]$$

$$= \frac{1}{n^2}(n\sigma_X^2) = \frac{\sigma_X^2}{n}.$$

Note that the standard deviation of the sample mean is the standard deviation of X divided by the *square root* of the sample size:

$$(5.23) \qquad\qquad \sigma_{\bar{X}} = \frac{\sigma_X}{\sqrt{n}}.$$

Thus, as the sample size n increases, the values of the sample mean $\bar{X}$ tend to become more concentrated about the mean μ_X.

Observe that Theorem 5.7 was proved *without* finding the probability function of $\bar{X}$. For applications of this theorem to statistical problems, it becomes important to know more about $\bar{X}$ than its mean and standard deviation. Unfortunately, we must leave these interesting matters at this point, for they lead to probability problems that cannot be formulated using finite sample spaces.

We can however use Theorem 5.7 to prove the following result which is a special form of the so-called *law of large numbers*.

Theorem 5.8. Let a population be specified by a random variable X with mean μ_X and standard deviation σ_X. Let $\bar{X}$ be the mean of a random sample of size n drawn with replacement from this population. Let c be any positive number. Then as n increases without bound,

$$(5.24) \qquad\qquad P(\mu_X - c \leq \bar{X} \leq \mu_X + c)$$

approaches 1. In other words, by choosing the sample size n sufficiently large, the probability that the value of the sample mean differs from the population mean by at most c can be made as close to 1 as we like. Or, more colloquially, since c can be taken as small as we please: by choosing the sample size sufficiently large, we can be as sure as we like (short of certainty) that the value of the sample mean will be as near the population mean as we like.

Proof. To the random variable $\bar{X}$ we apply Chebyshev's inequality (3.15) and find that

$$P(|\bar{X} - \mu_{\bar{X}}| > c) < \frac{\sigma_{\bar{X}}^2}{c^2}.$$

We use (5.22) to write $\mu_{\bar{X}}$ and $\sigma_{\bar{X}}^2$ in terms of the population mean and variance. Then

$$P(|\bar{X} - \mu_X| > c) < \frac{\sigma_X^2}{nc^2}.$$

Hence

$$(5.25) \qquad\qquad P(|\bar{X} - \mu_X| \leq c) > 1 - \frac{\sigma_X^2}{nc^2}.$$

But as n increases, the quantity σ_X^2/nc^2 decreases and approaches zero. Hence $1 - \dfrac{\sigma_X^2}{nc^2}$ approaches 1 as n gets larger and larger, and so $P(|\overline{X} - \mu_X| \leq c)$, which is just the probability in (5.24), can be made as close to 1 as we like by choosing n sufficiently large. This completes the proof.

Example 5.7. Let the value of X corresponding to each person in a certain population be that person's annual income in thousands of dollars. Suppose $\mu_X = 6.5$ and $\sigma_X = 2.1$. A random sample of n persons is drawn with replacement from this population and a value of $\overline{X}$, the average income of these n persons, is obtained. We want the probability to be greater than .9 that this value differs from the population mean by at most .5. How large must the sample size n be?

We seek the smallest value of n for which

(5.26) $P(|\overline{X} - \mu_X| \leq .5) > .9.$

Putting $c = .5$ and $\sigma_X = 2.1$ in (5.25) we find

$$P(|\overline{X} - \mu_X| \leq .5) > 1 - \frac{17.64}{n}.$$

Hence (5.26) will be true if $17.64/n$ is less than .1 or if $n > 176.4$. The desired closeness of the sample mean income and the population mean income is therefore achieved with a sample of size $n = 177$. (This is a most conservative figure, since it applies no matter what probability function $\overline{X}$ has. In more advanced work, one derives the approximate form of the probability function of $\overline{X}$, and it is then possible to show that a sample of size $n = 50$ will suffice in this example.)

We conclude by showing how the law of large numbers can be used to supply a theoretical counterpart to our intuitive feeling that if an event A occurs f times in n identical trials and if n is large, then f/n, the proportion of times A occurs, should be near the probability $P(A)$ of the event A. We let the random variable X have the value 1 if the event A occurs, and have the value 0 otherwise. Thus X has the following probability table:

x	0	1
$P(X = x)$	$1 - P(A)$	$P(A)$

We note that $\mu_X = P(A)$. Also $\overline{X}$, the mean of a random sample of size n drawn with replacement from the population specified by the random variable X, is just the *proportion* of times the event A occurs. (For $X_1 + \cdots + X_n$ is the *number* of times A occurs and $\overline{X}$ equals this number divided by n.) According to Theorem 5.8, by taking n sufficiently large, the probability can be made arbitrarily close to 1 that the proportion of times A occurs will be as close as we like to the probability of A. (This fact, due to James Bernoulli, dates back to 1713.) It is in this form that we find support for the interpretation of probabilities as proportions in a large number of repeated independent trials.[*]

PROBLEMS

5.1. Suppose X and Y have the following joint probability table:

x \ y	1	2	3	$P(X = x)$
1	.1	.1	0	.2
2	.1	.2	.3	.6
3	.1	.1	0	.2
$P(Y = y)$	.3	.4	.3	1

(a) Determine the probability function of $X + Y$, and thus compute $E(X + Y)$. Check your answer by using (5.2).

(b) Determine the probability function of XY and thus compute $E(XY)$. Then check your answer by using (5.1) to find $E(XY)$ directly from the joint probability table.

(c) Show that (5.6) is true but that X and Y are dependent.

5.2. Let X and Y have the joint probability table given in the preceding problem. In each of the following parts, a function z is defined by giving its value $z(x, y)$ for any real numbers x and y. Determine the probability function of the random variable $z(X, Y)$ and calculate $E[z(X, Y)]$ from the probability function, and also by using (5.1).

---

[*] There are other interpretations of probability. See for example, the discussions in L. J. Savage, *The Foundations of Statistics*, John Wiley and Sons, Inc., 1954, especially pp. 3–4, 56–68, and in E. Nagel, *Principles of the Theory of Probability*, International Encyclopedia of Unified Science, Vol. 1, No. 6, University of Chicago Press. 1939.

(a) $z(x, y) = \min(x, y) = \begin{cases} x & \text{if } x \leq y \\ y & \text{if } x > y \end{cases}$

(b) $z(x, y) = \max(x, y) = \begin{cases} y & \text{if } x \leq y \\ x & \text{if } x > y \end{cases}$

(c) $z(x, y) = x/y$

(d) $z(x, y) = y/x$

(e) $z(x, y) = x^2 + y^2$

(f) $z(x, y) = \sqrt{x^2 + y^2}$

5.3. Which of the following are true for all random variables X and Y defined on a sample space? For those that are true for some but not all X and Y, find a pair X, Y for which the statement is true and another pair for which it is false. (Cf. the preceding problem.)

(a) $E[\min(X, Y)] = \min[E(X), E(Y)]$

(b) $E[\max(X, Y)] = \max[E(X), E(Y)]$

(c) $E(X/Y) = E(X)/E(Y)$

(d) $E(X/Y) = 1/E(Y/X)$

(e) $E(X^2 + Y^2) = E(X^2) + E(Y^2)$

(f) $[E(\sqrt{X^2 + Y^2})]^2 = E(X^2 + Y^2)$

5.4. In the text, corollary (5.7) is proved using the identity in (5.5). Prove the corollary without using this identity. (*Hint:* Use Theorem 4.2.)

5.5. X has mean 50 and standard deviation 12. Y has mean 30 and standard deviation 5. X and Y are independent. Find the mean and standard deviation of (a) $X + Y$, (b) $X - Y$, (c) $3X + 2Y$. (*Note:* $\sigma_{X+Y} \neq \sigma_X + \sigma_Y$.)

5.6. Start with the definition of $\text{Var}(X)$ given in (3.6) and use the theorems of the present section to prove that $\text{Var}(X) = E(X^2) - \mu_X^2$. (Cf. Theorem 3.2 and its proof.)

5.7. (a) Interpret the result $E(X + b) = E(X) + b$ established in Section 2 as a special case of Theorem 5.2. What then is the random variable Y?

(b) Interpret the result $E(aX) = aE(X)$ established in Section 2 as a special case of Theorem 5.4. What then is the random variable Y and why (as needed to apply Theorem 5.4) are X and Y independent?

5.8. Prove Theorem 5.1.

5.9. Generalize Theorem 5.4 by proving that if $X_1, X_2, \cdots, X_n$ are independent (n any positive integer), then

$$E(X_1 X_2 \cdots X_n) = E(X_1)E(X_2) \cdots E(X_n).$$

5.10. Let V_1, V_2, V_3 be independent random variables. Define $X = V_1 + V_2$ and $Y = V_1 + V_3$. Show that

$$E(XY) - E(X)E(Y) = \text{Var}(V_1).$$

5.11. (a) Let X_1, X_2, $\cdots$, X_n be independent random variables. Show that if k is any positive integer less than n and $Y_k = a_1X_1 + \cdots + a_kX_k$, then Y_k and X_{k+1} are independent.

(b) Prove Theorem 5.6 by mathematical induction.

5.12. A random sample of size n is drawn with replacement from a population and we find the sample mean $\bar{X}$ has mean $\mu_{\bar{X}}$ and standard deviation $\sigma_{\bar{X}}$. What happens to $\mu_{\bar{X}}$ and $\sigma_{\bar{X}}$ if the sample size is quadrupled?

5.13. Let a population be specified by the following probability table of the random variable X:

x	0	1	2
$P(X = x)$	1/4	1/2	1/4

(a) Find μ_X and σ_X, the population mean and standard deviation.

(b) List all possible samples of size 2 drawn with replacement from the population and determine the probability function of $\bar{X}$, the sample mean for samples of size 2. Compute $\mu_{\bar{X}}$ and $\sigma_{\bar{X}}^2$ from this probability function, and thus check (5.22).

(c) Repeat part (b) for samples of size 3.

5.14. Verify formulas (5.17)–(5.19).

5.15. The incomes of ten people are given in the following frequency table.

Income x_i	Number of People with This Income f_i
$3500	3
5000	4
7500	2
9000	1

(a) Use formulas (5.17)–(5.19) to compute μ_X and σ_X^2, the mean and variance of the incomes in this population. (*Note:* The computations are simplified if you "code" the incomes, for example, by letting $Y = (X - 5000)/500$. The three people with $3500 incomes ($X$-values) are thus given coded incomes (Y-values) of -3, the four people with \$5000 incomes are given coded incomes of 0, etc. Since the Y-values are small numbers, it is relatively easy to compute μ_Y and σ_Y^2 by using (5.17)–(5.19) with x_j replaced by y_j. By using the coding equation relating Y and X, you can easily find μ_X and σ_X^2 from μ_Y and σ_Y^2.)

(b) Determine the probability function of $\overline{X}$, the sample mean, based on samples of size 2 drawn with replacement from the population of ten people. Then compute $\mu_{\overline{X}}$ and $\sigma_{\overline{X}}^2$, and thus check (5.22).

5.16. Two samples of size n_1 and n_2 respectively are drawn with replacement from a population specified by the probability function of a random variable X with mean μ_X and standard deviation σ_X. Let $\overline{X}_1$ and $\overline{X}_2$ denote the respective sample means and suppose these means are independent random variables. Show that

$$E(\overline{X}_1 - \overline{X}_2) = 0,$$

$$\mathrm{Var}(\overline{X}_1 - \overline{X}_2) = \sigma_X^2 \left(\frac{1}{n_1} + \frac{1}{n_2} \right).$$

5.17. Let $X_1, X_2, \cdots, X_n$ be independent, identically distributed random variables, each with mean μ_X and standard deviation σ_X. (That is, a sample of size n is drawn with replacement from a population given by the probability function of X.) In the text we defined the random variable $\overline{X}$, the sample mean, and found its mean and variance. The *sample variance* can also be defined. It is a random variable, denoted by S^2, and given by

$$S^2 = \frac{1}{n-1} \sum_{k=1}^{n} (X_k - \overline{X})^2.$$

To find $E(S^2)$ proceed as follows:

(a) By writing $X_k - \overline{X} = (X_k - \mu_X) - (\overline{X} - \mu_X)$, show that

$$S^2 = \frac{1}{n-1} \sum_{k=1}^{n} (X_k - \mu_X)^2 - \frac{n}{n-1} (\overline{X} - \mu_X)^2.$$

(b) Show that

$$E[(X_k - \mu_X)^2] = \mathrm{Var}(X_k), \qquad E[(\overline{X} - \mu_X)^2] = \mathrm{Var}(\overline{X}).$$

(c) Conclude that

$$E(S^2) = \sigma_X^2,$$

that is, *the mean of the sample variance is equal to the population variance.*

6. Covariance and correlation; the sample mean (cont.)

Suppose we are given the joint probability table of two random variables X and Y defined on the same sample space S. Each of these variables has a mean and a variance, but the joint probability table is not needed to compute μ_X, σ_X^2, and μ_Y, σ_Y^2; these numbers are determined by the (marginal) probability functions of X and Y. In this section, we define some numbers that measure how the possible values of X are related to the possible values of Y; such numbers will depend upon the joint probability function of X and Y.

We are led to our first definition by reviewing the proof of Theorem 5.5. There we showed that

$$(6.1) \quad \text{Var}(X + Y) = \text{Var}(X) + \text{Var}(Y) + 2E[(X - \mu_X)(Y - \mu_Y)],$$

and since X and Y were assumed to be independent, we invoked (5.7) to conclude that the last term in (6.1) vanishes. But (6.1) is a result worth having, since it holds for dependent as well as independent random variables. We therefore want now to study the last term in (6.1), a term that we so hurriedly skipped over in the preceding section. As usual, a special symbol and name are introduced.

Definition 6.1. Let X and Y be random variables defined on a sample space S. The *covariance* of X and Y, denoted by $\text{Cov}(X, Y)$, is defined as the number given by

$$(6.2) \qquad \text{Cov}(X, Y) = E[(X - \mu_X)(Y - \mu_Y)]$$

or equivalently, because of the identity in (5.5),

$$(6.3) \qquad \text{Cov}(X, Y) = E(XY) - \mu_X\mu_Y.$$

Using this notation we rewrite (6.1) and obtain

$$(6.4) \qquad \text{Var}(X + Y) = \text{Var}(X) + \text{Var}(Y) + 2\,\text{Cov}(X, Y).$$

Let us also note here that Definition 6.1 treats X and Y symmetrically, i.e.,

$$(6.5) \qquad \text{Cov}(X, Y) = \text{Cov}(Y, X),$$

and since $X - \mu_X$ and $Y - \mu_Y$ each have mean zero,

$$(6.6) \qquad \text{Cov}(X - \mu_X, Y - \mu_Y) = \text{Cov}(X, Y).$$

Many problems require a generalization of (6.4) to more than two random variables. For example, by the definition of variance we find

$$\text{Var}(X_1 + X_2 + X_3) = E([X_1 + X_2 + X_3 - E(X_1 + X_2 + X_3)]^2)$$

If we use (5.4) and write μ_j for $E(X_j)$, then

(6.7) $\text{Var}(X_1 + X_2 + X_3)$
$$= E([(X_1 - \mu_1) + (X_2 - \mu_2) + (X_3 - \mu_3)]^2).$$

Let us now recall from algebra (or by using the multinomial theorem of Chapter 3) that

$$(a_1 + a_2 + a_3)^2 = a_1^2 + a_2^2 + a_3^2 + 2a_1a_2 + 2a_1a_3 + 2a_2a_3$$
$$= \sum_{j=1}^{3} a_j^2 + 2 \sum_{\substack{\text{all } j,k \\ j<k}} a_j a_k,$$

where the last sum is understood to include the $\binom{3}{2} = 3$ cross-products $a_j a_k$ with $1 \leq j < k \leq 3$. Applying this to (6.7) by putting $a_j = X_j - \mu_j$, we find

$$\text{Var}(X_1 + X_2 + X_3)$$
$$= E[\sum_{j=1}^{3} (X_j - \mu_j)^2 + 2 \sum_{\substack{\text{all } j,k \\ j<k}} (X_j - \mu_j)(X_k - \mu_k)].$$

Now using (5.4) again and the definition of variance and covariance, we obtain

(6.8) $$\text{Var}(X_1 + X_2 + X_3) = \sum_{j=1}^{3} \text{Var}(X_j) + 2 \sum_{\substack{\text{all } j,k \\ j<k}} \text{Cov}(X_j, X_k).$$

The derivation just completed, if applied to the sum of n random variables, leads to the following general result. We leave writing out the proof as an exercise for the reader.

Theorem 6.1. If $X_1, X_2, \cdots, X_n$ are any random variables ($n > 1$), then

(6.9) $$\text{Var}(X_1 + \cdots + X_n) = \sum_{j=1}^{n} \text{Var}(X_j) + 2 \sum_{\substack{\text{all } j,k \\ j<k}} \text{Cov}(X_j, X_k),$$

the last sum including the $\binom{n}{2}$ terms $\text{Cov}(X_j, X_k)$ with subscripts satisfying $1 \leq j < k \leq n$.

In the preceding section (pp. 221–226) we discussed some questions involving sampling *with replacement* from a population specified by the probability function of a random variable. In particular, we derived Formulas (5.22) for the mean and variance of the sample mean $\overline{X}$ based on random samples of size n drawn with replacement. We are now able to prove the corresponding results for the important case where random samples of size n are drawn *without replacement* from a finite population of N elements.

We suppose, as before, that we have a population of N chips and that each has a number on it. Let $x_1, x_2, \cdots, x_M$ be *all* the *different* numbers, and suppose x_j appears on f_j chips for $j = 1, 2, \cdots, M$. Then, precisely as in (5.16)–(5.19) of the preceding section, we define the random variable X and the population mean μ_X and population variance σ_X^2.

From this population of N chips we draw one chip at random, then another at random from the remaining $N - 1$ chips, and so on until a random sample of size n $(n \leq N)$ is drawn without replacement from the population. We again let X_k be the random variable whose value is the number on the kth chip drawn. Thus the sample mean $\overline{X}$ is given by (5.20), as before. Our task is to compute the mean and variance of $\overline{X}$.

The random variables $X_1, X_2, \cdots, X_n$ are independent when the sample is drawn with replacement. What makes our present analysis more complicated is the fact that, in sampling without replacement, these random variables are dependent. For example, we have

$$(6.10) \qquad P(X_1 = x_j) = \frac{f_j}{N} \qquad \text{for } j = 1, 2, \cdots, M,$$

but knowing the outcome of the first draw changes the probability of getting the number x_j on the second draw:

$$(6.11) \quad P(X_2 = x_j | X_1 \neq x_j) = \frac{f_j}{N - 1}, \quad P(X_2 = x_j | X_1 = x_j) = \frac{f_j - 1}{N - 1}.$$

But, in spite of being dependent, the random variables $X_1, X_2, \cdots, X_n$ are, as in the preceding section, all *identically distributed with probability function the same as that of X*; i.e., for $k = 1, 2, \cdots, n$, we have

$$(6.12) \qquad P(X_k = x_j) = \frac{f_j}{N} \qquad \text{for } j = 1, 2, \cdots, M.$$

This means that in the absence of information about the preceding draws, the probability that the kth draw results in a chip bearing the number x_j is just the proportion of chips bearing this number among all chips in the population. For the first draw this is clear and recorded in (6.10). We now prove it is true for the second draw. By Formula (II.6.1),

$$P(X_2 = x_j) = \sum_{k=1}^{M} P(X_2 = x_j|X_1 = x_k)P(X_1 = x_k).$$

Because of (6.11), we must isolate the term with $k = j$. Then

$$P(X_2 = x_j) = P(X_2 = x_j|X_1 = x_j)P(X_1 = x_j)$$
$$+ \sum_{k=1}^{M}\!\!{}^* P(X_2 = x_j|X_1 = x_k)P(X_1 = x_k),$$

where the asterisk means that the sum does *not* include the term with $k = j$. Continuing,

$$P(X_2 = x_j) = \frac{f_j - 1}{N - 1}\frac{f_j}{N} + \sum_{k=1}^{M}\!\!{}^* \frac{f_j}{N - 1}\frac{f_k}{N}$$

$$= \frac{f_j}{N(N - 1)}\left(f_j - 1 + \sum_{k=1}^{M}\!\!{}^* f_k\right)$$

$$= \frac{f_j}{N(N - 1)}\left(-1 + \sum_{k=1}^{M} f_k\right)$$

$$= \frac{f_j}{N(N - 1)}(N - 1) = \frac{f_j}{N}, \text{ as claimed.}$$

We leave the proof that (6.12) is also true for $k = 3, 4, \cdots, n$ for the problems.

From (6.12) it follows that $E(X_k) = \mu_X$ for $k = 1, 2, \cdots, n$, and so we apply (5.4) to obtain

$$\mu_{\bar{X}} = E\left(\frac{X_1 + \cdots + X_n}{n}\right) = \frac{1}{n}[E(X_1) + \cdots + E(X_n)] = \mu_X,$$

as in the case of sampling with replacement. The fact that $X_1, X_2, \cdots, X_n$ are no longer independent does not influence the calculation of $\mu_{\bar{X}}$, since (5.4) holds for dependent as well as independent random variables.

It is in the calculation of the variance of $\bar{X}$ that the dependence of the X_k's complicates matters. We must use (6.9) and so need to compute $\text{Cov}(X_j, X_k)$. (We know that $\text{Var}(X_j) = \sigma_X^2$, as given in (5.18), since X_j has the same probability function as X for $j = 1, 2, \cdots, n$.) A saving grace is the fact that

$$\text{Cov}(X_j, X_k) = \text{Cov}(X_1, X_2) \quad \text{for all } j \neq k.$$

This equality follows from the observation (see Problem 6.6) that each pair of random variables taken from $X_1, X_2, \cdots, X_n$ has the same joint probability function as any other pair. It therefore suffices to compute $\text{Cov}(X_1, X_2)$, and it is to this task that we now turn our attention.

By the definition of covariance, we have

$$\text{Cov}(X_1, X_2) = E[(X_1 - \mu_X)(X_2 - \mu_X)]$$
$$= \sum_{\text{all } j,k} (x_j - \mu_X)(x_k - \mu_X)P(X_1 = x_j, X_2 = x_k).$$

Again we isolate the terms with $j = k$ and indicate their absence by placing an asterisk on the summation symbol. Then

$$\text{Cov}(X_1, X_2) = \sum_{j=1}^{M} (x_j - \mu_X)^2 P(X_1 = x_j, X_2 = x_j)$$
$$+ \sum_{\text{all } j,k}^{*} (x_j - \mu_X)(x_k - \mu_X)P(X_1 = x_j, X_2 = x_k)$$

$$(6.13) \qquad = \sum_{j=1}^{M} (x_j - \mu_X)^2 \frac{f_j}{N} \frac{f_j - 1}{N - 1}$$
$$+ \sum_{\text{all } j,k}^{*} (x_j - \mu_X)(x_k - \mu_X) \frac{f_j}{N} \frac{f_k}{N - 1}.$$

To evaluate this last sum, we use the following device. Since $E(X_j) = \mu_X$, we know that $E(X_j - \mu_X) = 0$, i.e.,

$$\sum_{j=1}^{M} (x_j - \mu_X)f_j = 0.$$

Now square both sides of this equation to find

$$\sum_{j=1}^{M} (x_j - \mu_X)^2 f_j^2 + \sum_{\text{all } j,k}^{*} (x_j - \mu_X)(x_k - \mu_X)f_j f_k = 0.$$

The second sum is therefore the negative of the first. We use this in (6.13) and obtain

$$\text{Cov}(X_1, X_2)$$

$$= \frac{1}{N(N - 1)} \sum_{j=1}^{M} (x_j - \mu_X)^2 f_j(f_j - 1) - \frac{1}{N(N - 1)} \sum_{j=1}^{M} (x_j - \mu_X)^2 f_j^2$$

$$= -\frac{1}{N(N - 1)} \sum_{j=1}^{M} (x_j - \mu_X)^2 f_j$$

$$= -\frac{\sigma_X^2}{N - 1},$$

the final equality following from (5.18). Now at last we can apply Formula (6.9) to find $\sigma_{\bar{X}}^2$:

$$\sigma_{\bar{X}}^2 = \text{Var}\left(\frac{X_1 + \cdots + X_n}{n}\right) = \frac{1}{n^2}\text{Var}(X_1 + \cdots + X_n)$$

$$= \frac{1}{n^2}\left[\sum_{j=1}^{n}\sigma_X^2 + 2\sum_{\substack{\text{all } j,k \\ j<k}}\left(-\frac{\sigma_X^2}{N-1}\right)\right]$$

$$= \frac{1}{n^2}\left[n\sigma_X^2 + 2\binom{n}{2}\left(-\frac{\sigma_X^2}{N-1}\right)\right]$$

$$= \frac{\sigma_X^2}{n}\left(1 - \frac{n-1}{N-1}\right)$$

$$= \frac{\sigma_X^2}{n}\left(\frac{N-n}{N-1}\right).$$

With this lengthy calculation, we have completed the proof of the following important theorem.

Theorem 6.2. From a population of N elements with mean μ_X and variance σ_X^2, a random sample of size n is drawn without replacement. Let $\overline{X}$ be the sample mean. Then

(6.14) $\mu_{\bar{X}} = \mu_X \quad \text{and} \quad \sigma_{\bar{X}}^2 = \frac{\sigma_X^2}{n}\left(\frac{N-n}{N-1}\right).$

Before discussing these results we give a numerical example.

Example 6.1. Suppose we have a population of $N = 5$ people and know the IQ score of each. Table 33 summarizes the available information concerning the population. By using Formulas (5.17)–(5.19),

TABLE 33

IQ Score x_j	Number of People with This IQ Score f_j
80	1
100	2
130	2

the reader can check that for this population the mean IQ score and the variance of the IQ scores are

(6.15) $\mu_X = 108 \quad \text{and} \quad \sigma_X^2 = 376,$

so that the standard deviation of IQ scores in the population is $\sigma_X = 19.4$, approximately.

A random sample of $n = 2$ people is drawn without replacement from this population. In Table 34 we list all possible results of this sampling experiment, together with the information needed to de-

TABLE 34

IQ Score of First Person Selected	IQ Score of Second Person Selected	Number of Ways of Drawing These Scores in the Stated Order	Value of $\overline{X}$ for This Sample
80	100	2	90
80	130	2	105
100	80	2	90
100	100	2	100
100	130	4	115
130	80	2	105
130	100	4	115
130	130	2	130

termine the probability function of $\overline{X}$, the sample mean. Note that we are selecting two people from five people, not two IQ scores from the three different IQ scores. This means, for example, that we can get the scores 80 and 100 in that order in two ways, since two people have IQ scores of 100. There are altogether $5 \cdot 4 = 20$ ways of selecting first one person and then another from the population. Thus the numbers in the third column of Table 34 add to 20.

We see that $\overline{X}$ has the value 90 for four of the 20 samples, so that $P(\overline{X} = 90) = \frac{4}{20} = .2$. In this way, we obtain the following probability table for $\overline{X}$, the sample mean:

$\overline{x}$	90	100	105	115	130
$P(\overline{X} = \overline{x})$	.2	.1	.2	.4	.1

And the reader should now compute the mean and variance of $\overline{X}$ from the table. He will find that

(6.16) $$\mu_{\overline{X}} = 108, \qquad \sigma_{\overline{X}}^2 = 141.$$

Note that we have here a particular example to which Theorem 6.2 applies. The results in (6.16) can be checked against those predicted by the theorem. The values of μ_X and σ_X^2 are given in (6.15) and with $N = 5$, $n = 2$ we find from (6.14) that

$$\mu_{\overline{X}} = \mu_X = 108,$$

$$\sigma_{\overline{X}}^2 = \frac{\sigma_X^2}{n}\left(\frac{N - n}{N - 1}\right) = \frac{376}{2}\left(\frac{3}{4}\right) = 141,$$

as expected.

Comparing the results obtained in Theorem 5.7 (sampling with replacement) and Theorem 6.2 (sampling without replacement) we draw the following conclusions.

(1) In both sampling with and without replacement, the values of the sample mean $\overline{X}$ have the right "aim" in the sense that their average value $\mu_{\overline{X}}$ is equal to the population mean μ_X.

(2) In both sampling with and without replacement, if the sample size is greater than 1, then the values of $\overline{X}$ show less spread than the values of X about their common mean μ_X; i.e.,

$$\sigma_{\overline{X}}^2 < \sigma_X^2 \qquad \text{if } n > 1.$$

This follows by observing that $\sigma_{\overline{X}}^2$ is obtained by multiplying σ_X^2 by the factor $\frac{1}{n}$ in one case, by the factor $\frac{1}{n}\frac{N - n}{N - 1}$ in the other case, and both factors are less than 1 if $n > 1$.

(3) In both sampling with and without replacement, the sample variance $\sigma_{\overline{X}}^2$ decreases as the sample size n increases. For the factors mentioned in (2) are not only less than 1, but also decrease as n increases. Furthermore, in sampling without replacement, if the sample exhausts the population $(n = N)$, then $\sigma_{\overline{X}}^2 = 0$. For if we draw into the sample all the members of the population, then all samples differ only in the order in which members are drawn. Hence all samples have the same mean; i.e., there is only one possible value of $\overline{X}$. In this case we know that the variance of $\overline{X}$ is zero.

(4) When sampling from the same population and for samples of fixed size $n > 1$, $\sigma_{\overline{X}}^2$ is smaller when the sample is drawn without replacement than when it is drawn with replacement. For the variances differ by the factor $\frac{N - n}{N - 1}$, which is less than 1 when $n > 1$.

(5) Also, since

$$\frac{N - n}{N - 1} = \frac{1 - \dfrac{n}{N}}{1 - \dfrac{1}{N}},$$

this factor is close to 1 whenever N is very large compared to n. For then n/N and $1/N$ are close to zero. Thus, if samples of size n are drawn without replacement from a population of N elements, and if the population size is very large compared to the sample size n, then $\sigma_{\overline{X}}^2$ is approximately equal to σ_X^2/n. This accounts for the fact that the simpler formulas (5.22) are often used in statistics when N is very large compared to n, even though the sample is drawn without replacement.

We cannot here go into the question of how to use our results (in conjunction with other information about the sample mean $\overline{X}$, especially about its probability function) if just one sample is drawn from a population and we then want to use the sample values of $X_1, X_2, \cdots, X_n$ to make inferences about the population. This question is of great practical importance and is discussed in detail in statistics textbooks and courses.

From the definition of $\mathrm{Cov}(X, Y)$ in (6.2), we can see that the covariance is a measure of the extent to which the values of X and Y tend to increase or decrease together. If X has values greater than its mean μ_X whenever Y has values greater than its mean μ_Y and X has values less than μ_X whenever Y has values less than μ_Y, then $(X - \mu_X)(Y - \mu_Y)$ has positive values and $\mathrm{Cov}(X, Y) > 0$. On the other hand, if values of X are above μ_X whenever values of Y are below μ_Y and vice versa, then $\mathrm{Cov}(X, Y) < 0$. If X and Y are independent, then we know by Theorem 5.4 that $\mathrm{Cov}(X, Y) = 0$.

By a suitable choice of two random variables, we can make their covariance any number we like. For example, if a and b are constants, then

$$\begin{aligned}
\mathrm{Cov}(aX, bY) &= E(aXbY) - E(aX)E(bY) \\
&= abE(XY) - (a\mu_X)(b\mu_Y),
\end{aligned}$$

from which it follows that

(6.17) $\mathrm{Cov}(aX, bY) = ab\, \mathrm{Cov}(X, Y).$

It is now clear that if $\mathrm{Cov}(X, Y) \neq 0$, then by varying a and b we can make $\text{Cov}(aX, bY)$ positive or negative, as small or as large as we please.

It is more convenient to have a measure of relation that cannot vary so widely. We shall prove shortly that the covariance of X^* and Y^*, where X^* and Y^* are the standardized random variables corresponding to X and Y (as defined in Theorem 3.4), can vary only between -1 and $+1$.

Now by (6.17),

$$\text{Cov}(X^*, Y^*) = \text{Cov}\left(\frac{X - \mu_X}{\sigma_X}, \frac{Y - \mu_Y}{\sigma_Y}\right)$$

$$= \frac{1}{\sigma_X \sigma_Y} \text{Cov}(X - \mu_X, Y - \mu_Y)$$

$$= \frac{\text{Cov}(X, Y)}{\sigma_X \sigma_Y},$$

this last equality following from (6.6). We are thus led to the following definition.

Definition 6.2. Let X^* and Y^* be the standardized random variables corresponding to X and Y. The covariance of X^* and Y^* is called the *correlation coefficient* of X and Y and is denoted by $\rho(X, Y)$. In symbols,

$$(6.18) \qquad \rho(X, Y) = \text{Cov}(X^*, Y^*) = \frac{\text{Cov}(X, Y)}{\sigma_X \sigma_Y}.$$

If $\sigma_X = 0$ or if $\sigma_Y = 0$, and consequently (6.18) does not apply, we define $\rho(X, Y) = 0$. The random variables X and Y are said to be *uncorrelated* if and only if $\rho(X, Y) = 0$; otherwise they are said to be *correlated*.

If $\sigma_X > 0$ and $\sigma_Y > 0$, then the correlation coefficient $\rho(X, Y)$ is zero if and only if $\text{Cov}(X, Y) = 0$. In the exceptional case when one or both of the random variables have standard deviation zero, we know (see Problem 4.10) that X and Y are independent and hence $\text{Cov}(X, Y) = 0$. Thus $\rho(X, Y) = 0$ and $\text{Cov}(X, Y) = 0$ are equivalent conditions: *X and Y are uncorrelated if and only if their covariance is zero.*

Before commenting on this definition, let us see how to compute a correlation coefficient.

Example 6.2. Let X and Y be random variables with joint probabilities as given in Table 27 on p. 199. In Example 5.1, we found that

$\mu_X = \frac{1}{2}$, $\mu_Y = \frac{3}{2}$, and $E(XY) = 1$. Thus $\text{Cov}(X, Y) = \frac{1}{4}$ and we know that X and Y are correlated. We leave for the reader the verification that $E(X^2) = \frac{1}{2}$ and $E(Y^2) = 3$ so that $\sigma_X^2 = \frac{1}{4}$ and $\sigma_Y^2 = \frac{3}{4}$. We now apply (6.18) to find

$$\rho(X, Y) = \frac{\frac{1}{4}}{(\frac{1}{2})(\sqrt{3}/2)} = \frac{1}{\sqrt{3}} = .58, \text{ approximately.}$$

Example 6.3. In Example 5.4, we defined two random variables X and Y and found that they were functionally dependent ($Y = X^2$), but that (5.6) was true; i.e., $\text{Cov}(X, Y) = 0$. We therefore have an example of *random variables that are uncorrelated but not independent*. We conclude that one must exercise great care in interpreting the covariance or the correlation coefficient as a measure of relationship between values of X and Y. In particular, the fact that the correlation coefficient is zero does *not* mean that X and Y are unrelated, for we have just seen that $\rho(X, Y) = 0$ but X and Y are as strongly related as they can be: knowing the value of X we are certain of the value of Y, since $Y = X^2$.

Although it is merely a rephrasing of Theorem 5.4, we emphasize the point made in the last example by the following statement.

Theorem 6.3. If X and Y are independent random variables, then they are uncorrelated, but *not* conversely.

We turn now to some properties of the correlation coefficient.

Theorem 6.4. The correlation coefficient of X and Y is a number between -1 and $+1$ inclusive, i.e.,

(6.19) $-1 \le \rho(X, Y) \le 1$.

Proof. Consider the variance of $X^* + Y^*$, where X^* and Y^* are the standardized random variables corresponding to X and Y, respectively. By (6.4),

$$\text{Var}(X^* + Y^*) = \text{Var}(X^*) + \text{Var}(Y^*) + 2\,\text{Cov}(X^*, Y^*).$$

But $\text{Var}(X^*) = \text{Var}(Y^*) = 1$ and $\text{Cov}(X^*, Y^*) = \rho(X, Y)$ by definition. Hence

(6.20) $\text{Var}(X^* + Y^*) = 2[1 + \rho(X, Y)].$

Since $\text{Var}(X^* + Y^*) \ge 0$, it follows that $-1 \le \rho(X, Y)$. Similarly, the reader can show that

(6.21) $\qquad \text{Var}(X^* - Y^*) = 2[1 - \rho(X, Y)]$

from which we conclude, again since the variance is nonnegative, that $\rho(X, Y) \leq 1$. Thus the theorem is proved. (For another proof see Problem 6.16.)

It is important to understand the meaning of the extreme values $\rho(X, Y) = \pm 1$. Now the strongest relation exists between X and Y when the value of Y is uniquely determined as soon as the value of X is known. In such a case, Y is some function of X, say $Y = g(X)$. This situation exists whenever each row of the joint probability table of X and Y has all entries but one equal to zero. As we saw in Example 6.3, Y can be a function of X and yet $\rho(X, Y) = 0$. In that example, $Y = X^2$, a quadratic function of X. But if Y is a *linear* function of X, then we can prove that $\rho(X, Y)$ must have one of its extreme values. And we shall also be able to prove that conversely, if $\rho(X, Y) = \pm 1$, then X and Y are linearly related. Before stating and proving these results, let us look at an example.

Example 6.4. Suppose the random variables X and Y have the joint probabilities given in Table 35.

TABLE 35

x \ y	1	3	5	P(X = x)
1	0	0	.2	.2
2	0	.5	0	.5
3	.3	0	0	.3
P(Y = y)	.3	.5	.2	1

Since each row contains exactly one nonzero entry, we know that Y is some function of X. We also observe that all nonzero probabilities occur on the diagonal along which the values of Y increase as the values of X decrease. On the probability graph in Figure 26 we see that the points (x, y) at which positive probabilities are indicated all lie on the dotted straight line with negative slope. In fact, the joint probability table was constructed assuming Y is the linear function of X given by $Y = -2X + 7$. But let us calculate the correlation

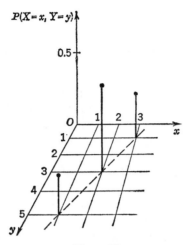

Figure 26

coefficient of X and Y without using this fact. It is easy to find directly from Table 35 that

$$\sigma_X = .7, \qquad \sigma_Y = 1.4, \qquad \text{Cov}(X, Y) = -.98.$$

Hence

$$\rho(X, Y) = \frac{-.98}{(.7)(1.4)} = -1.$$

Theorem 6.5. Let X be a random variable defined on a sample space S and suppose $\sigma_X > 0$. Let Y be a linear function of X; i.e., $Y = mX + b$, where m and b are numbers and $m \neq 0$. Then $\rho(X, Y) = +1$ if $m > 0$ and $\rho(X, Y) = -1$ if $m < 0$.

Proof. Since $Y = mX + b$, we have $\mu_Y = m\mu_X + b$. Hence $Y - \mu_Y = m(X - \mu_X)$, from which it follows that

$$\text{Cov}(X, Y) = E[m(X - \mu_X)^2] = m\sigma_X^2.$$

Also we know from (3.12) that $\sigma_Y = |m|\sigma_X$. Thus

$$\rho(X, Y) = \frac{\text{Cov}(X, Y)}{\sigma_X \sigma_Y} = \frac{m\sigma_X^2}{\sigma_X|m|\sigma_X} = \frac{m}{|m|}.$$

Since $m/|m|$ equals 1 if $m > 0$ and equals -1 if $m < 0$, the proof is complete.

Before we can prove the converse of Theorem 6.5, we must be careful to isolate a minor difficulty. We are going to want to prove that if $\rho(X, Y) = \pm 1$, then Y is a linear function of X; i.e., $Y = mX + b$ for some numbers m and b. What this means is that $Y(o_i) = mX(o_i) + b$ for each $o_i \in S$. But this is more than we can rightfully expect to prove. For suppose some simple event, say $\{o_1\}$, is assigned probability 0. Then we may as well forget about the element o_1, for $X(o_1)$ is not one of the possible values of X unless it is the value of X for some other element $o_i \in S$ for which $P(\{o_i\}) > 0$. In any case, the element o_1 could have been deleted from our sample space, since it plays no role in the construction of the joint proba-bility table of X and Y. Thus, changing the value $Y(o_1)$ cannot change the correlation coefficient of X and Y. This means that our best hope is to be able to prove that $Y(o_i) = mX(o_i) + b$ for all $o_i \in S$ except possibly for elements of S that together make up an event with probability 0. Let us introduce the following handy ter-minology for this state of affairs: We shall say that two random vari-ables are equal *with probability 1* whenever their values are equal for all elements of the sample space S except possibly for elements that together make up an event with probability 0.

Now we can state and prove the converse of Theorem 6.5.

Theorem 6.6. Let X and Y be random variables defined on a sample space S, and suppose $\rho(X, Y) = \pm 1$. Then Y is a linear function of X with probability 1. In fact, numbers $m > 0$, b, and c exist so that $Y = mX + b$ if $\rho = +1$ and $Y = -mX + c$ if $\rho = -1$, each with probability 1.

Proof. Suppose $\rho(X, Y) = +1$ and proceed from Equation (6.21). We see that $\operatorname{Var}(X^* - Y^*) = 0$ and it follows that $X^* - Y^*$ has one value that occurs with probability 1. This value must be the mean of $X^* - Y^*$ which is zero since the mean of each standardized random variable is zero. Thus $X^* = Y^*$ with probability 1 or

$$\frac{X - \mu_X}{\sigma_X} = \frac{Y - \mu_Y}{\sigma_Y}.$$

Simplification yields the desired result $Y = mX + b$ with

$$m = \frac{\sigma_Y}{\sigma_X} > 0 \quad \text{and} \quad b = \frac{\mu_Y \sigma_X - \mu_X \sigma_Y}{\sigma_X}.$$

The reader can complete the proof if $\rho(X, Y) = -1$ by starting with (6.20) and proceeding as above.

As these theorems hint, the correlation coefficient is a meaningful measure of relationship between values of X and Y only when this relationship is a linear one. For a fuller understanding of this fact, one must study the so-called *regression functions* of each random variable on the other. These functions are defined and some of their properties most often used in statistics are stated in Problem 6.20.

PROBLEMS

6.1. Prove Theorem 6.1.

6.2. From the population of $N = 10$ people whose incomes are given in the frequency table of Problem 5.15, a random sample of size $n = 2$ is drawn without replacement. If $\overline{X}$ is the sample mean income, then determine the probability function of $\overline{X}$, calculate $\mu_{\overline{X}}$ and $\sigma^2_{\overline{X}}$, and thus check Formulas (6.14). Compare the results with those of Problem 5.15, where the sample is drawn with replacement.

6.3. From the population of $N = 5$ people whose IQ scores are given in Table 33, a random sample of size n is drawn without replacement. Let $\overline{X}$ be the sample mean IQ score. Determine the probability function of $\overline{X}$, calculate $\mu_{\overline{X}}$ and $\sigma^2_{\overline{X}}$, and check Formulas (6.14) if the sample size is (a) $n = 3$, (b) $n = 4$, (c) $n = 5$. (The corresponding problem with $n = 2$ was solved in Example 6.1.)

6.4. Refer to Example 5.6 and suppose we have a population of $N = 10$ chips with relative frequencies as given in the probability table of X; i.e., one chip has -1 on it, five chips have 0's on them, and four chips have 2's on them. From this population, a random sample of size n is drawn without replacement. If $\overline{X}$ is the sample mean, then determine the probability function of $\overline{X}$, calculate $\mu_{\overline{X}}$ and $\sigma^2_{\overline{X}}$ if (a) $n = 1$, (b) $n = 2$, (c) $n = 3$, (d) $n = 9$, (e) $n = 10$. In each case check Formulas (6.14).

6.5. In a population of 10,000 families, annual income has mean $5000 and standard deviation $750. According to Chebyshev's inequality, within what interval will the sample mean $\overline{X}$ fall with probability at least $\frac{8}{9}$ if a random sample of size 100 is drawn without replacement from the population?

6.6. The following parts refer to the text discussion of sampling without replacement from a finite population.

(a) Prove (6.12) for $k = 3, 4, \cdots, n$ and thus complete the proof that $X_1, X_2, \cdots, X_n$ are identically distributed.

(b) Determine the joint probability function of X_1 and X_2, and also of X_1 and X_3. What is the joint probability function of X_j and X_k for any $j \neq k$?

(c) Show that $\rho(X_1, X_2) = -\dfrac{1}{N-1}$. Is this answer reasonable?

6.7. For sampling without replacement from a population of size N, you want to determine a sample size n_1 such that the standard deviation of the sample mean is half as big as it is for samples of size n. Show that

$$n_1 = \frac{4nN}{N + 3n},$$

provided the right hand side is an integer. (Note that n_1 and n are equal when $n = N$, and explain why this is reasonable.)

6.8. Refer to the card guessing experiment described in Example 5.5, but now suppose that the subject chooses at random a permutation of the numbers $1, 2, \cdots, n$ and then calls his guesses in the order specified by the selected permutation. As before, one way of doing this would be for the subject to have a duplicate deck. But now he makes his guesses by selecting a random sample of n cards, one by one *without replacement* from his deck.

Let X denote the random variable whose value is the number of correct guesses made by the subject, and thus write

$$X = X_1 + X_2 + \cdots + X_n,$$

where X_k has the value 0 or 1 according as the kth guess (trial) is wrong or right.

(a) Show that $X_1, X_2, \cdots, X_n$ are not independent.

(b) Prove that $X_1, X_2, \cdots, X_n$ are identically distributed, the probability of a correct guess at the kth trial being $1/n$ for $k = 1, 2, \cdots, n$.

(c) Show that $E(X) = 1$, as in Example 5.5.

(d) Prove that $\mathrm{Cov}(X_j, X_k) = \dfrac{1}{n^2(n-1)}$ for all $j \neq k$.

(e) Show that $\mathrm{Var}(X) = 1$, a somewhat higher variance than in Example 5.5.

6.9. Show that all other things being equal, the greater the correlation coefficient of two random variables, the greater the variance of their sum and the less the variance of their difference.

6.10. The average covariance of $X_1, X_2, \cdots, X_n$ is denoted by $\mathrm{Cov_{Av}}(X_1, \cdots, X_n)$ and defined as the sum of all $\mathrm{Cov}(X_j, X_k)$ with $1 \leq j < k \leq n$ divided by $\dbinom{n}{2}$, the number of such covariances.

(a) Show that

$$\text{Var}(\overline{X}) = \frac{1}{n^2} \sum_{j=1}^{n} \text{Var}(X_j) + \frac{n-1}{n} \text{Cov}_{\text{Av}}(X_1, \cdots, X_n).$$

(b) Suppose a number K (not depending on n) exists such that $\text{Var}(X_j) < K$ for all $j = 1, 2, \cdots, n$. Suppose further that as n increases without bound, $\text{Cov}_{\text{Av}}(X_1, \cdots, X_n)$ approaches some limiting value, say C. Show that then the variance of $\overline{X}$ also approaches C.

6.11. Let X and Y be the characteristic random variables (see the definition in Problem 1.14) of events A and B, respectively. Find $\rho(X, Y)$ and determine whether X and Y are independent if

(a) $P(A) = \frac{1}{4}, P(A|B) = \frac{1}{4}, P(B|A) = \frac{1}{2}$.
(b) $P(A) = \frac{1}{4}, P(A|B) = \frac{3}{4}, P(B|A) = \frac{1}{2}$.

6.12. Let X be the larger of the two numbers and Y be the sum of the numbers showing when two fair dice are rolled. Find $\rho(X, Y)$. (Cf. Problem 4.4.)

6.13. Let X be the number of empty cells and Y the number of objects in the first cell when three indistinguishable objects are randomly distributed into three numbered cells. Find $\rho(X, Y)$. (Cf. Problem 4.3.)

6.14. A fair die is rolled two independent times. Let X and Y denote the number of points showing on the first roll and the second roll, respectively. Define $U = X + Y$ and $V = X - Y$. Show that U and V are dependent random variables, but that $\rho(U, V) = 0$.

6.15. A fair coin is tossed four independent times. Let X be the number of heads obtained on the first two tosses and Y be the total number of heads. Find the joint probability table of X and Y and compute $\rho(X, Y)$.

6.16. (An alternate proof of Theorem 6.4.) Let $U = X - \mu_X$ and $V = Y - \mu_Y$.

(a) Note that $y = E([xU + V]^2) \geq 0$ for all real x.
(b) Expand and obtain

$$y = \sigma_X^2 x^2 + 2 \text{Cov}(X, Y)x + \sigma_Y^2 \geq 0.$$

(c) Interpret the inequality in (b) as showing that a certain parabola in the x-y plane lies entirely above the x-axis or has exactly one point of contact with the x-axis.
(d) Conclude that the quadratic equation $y = 0$ has either no real roots or two equal real roots. Hence the discriminant of the quadratic equation must be negative or zero. Thus find $-1 \leq \rho(X, Y) \leq 1$.

6.17. Suppose Y is a linear function of X. Let ρ_m be the value of $\rho(X, Y)$ when $Y = mX + b$. Draw a graph showing how ρ_m depends upon the slope m. (Plot m along the horizontal axis.)

6.18. Let X and Y be random variables and suppose a, b, c, d are any numbers provided only that $a \neq 0$, $c \neq 0$. Show that

$$\rho(aX + b, cY + d) = \frac{ac}{|ac|}\, \rho(X, Y).$$

[*Note:* This result shows that the absolute value of the correlation coefficient is not altered by a change in location of the origin or a change in scale on either x or y axis. This is a property expected of any reasonable measure of relationship. For example, the correlation between weight and height will have the same absolute value whether we measure height in inches, in feet, or in tenths of an inch above or below 68 inches. If a and c have opposite signs, then we see that $\rho(aX + b, cY + d) = -\rho(X, Y)$. Is this change of sign reasonable?]

6.19. Suppose X and Y each have only two possible values. Prove that, if X and Y are uncorrelated, then they are also independent.

6.20. The *conditional mean of Y given X* $= x_j$ is denoted by $E(Y \mid X = x_j)$ and defined for $j = 1, 2, \cdots, M$ by the equation

$$E(Y \mid X = x_j) = \sum_{k=1}^{N} y_k g(y_k \mid x_j).$$

(*Note:* Conditional probability functions are defined in Definition 4.3, p. 206.)

(a) Show that if X and Y are independent, then

$$E(Y \mid X = x_j) = E(Y) \qquad \text{for } j = 1, 2, \cdots, M.$$

(b) For any random variables X and Y, show that

$$E(Y) = \sum_{j=1}^{M} E(Y \mid X = x_j) f(x_j).$$

(c) The *conditional mean of X given Y* $= y_k$ is denoted by $E(X \mid Y = y_k)$ and defined for $k = 1, 2, \cdots, N$ by the equation

$$E(X \mid Y = y_k) = \sum_{j=1}^{M} x_j f(x_j \mid y_k).$$

State and prove the results analogous to those in parts (a) and (b), but now referring to the conditional mean of X given $Y = y_k$.

(d) The *regression function of Y on X* is defined as the function whose domain is the set of possible values of X and whose value at x_j is the conditional mean $E(Y \mid X = x_j)$ for $j = 1, 2, \cdots, M$. Similarly, the *regression function of X on Y* has the set of possible values of Y as domain, and its value at y_k is the conditional mean $E(X \mid Y = y_k)$ for $k = 1, 2, \cdots, N$. The regression graph of Y on X is a set of M points in the x-y plane, the point with x-coordinate x_j having y-coordinate $E(Y \mid X = x_j)$. Similarly, the regression graph of X on Y is the set of N points $(E(X \mid Y = y_k), y_k)$ for $k = 1, 2, \cdots, N$.

A regression function is said to be *linear* if all the points of the corresponding regression graph lie on a straight line. Otherwise, a regression function is said to be nonlinear.

   (i) For X and Y with joint probabilities given in Table 27 (p. 199), show that both regression functions are linear.

  (ii) For X and Y with joint probabilities given in Table 31 (p. 218), show that the regression function of Y on X is nonlinear, but the regression function of X on Y is linear.

 (iii) Construct a joint probability table so that both regression functions are nonlinear.

(e) Suppose the regression function of Y on X is linear; i.e., constants m and b exist such that for $j = 1, 2, \cdots, M$,

(*) $E(Y \mid X = x_j) = \sum_{k=1}^{N} y_k g(y_k \mid x_j) = mx_j + b.$

To evaluate m and b, proceed as follows. First multiply (*) by $f(x_j)$ and add the resulting equations for $j = 1, 2, \cdots, M$. Obtain $\mu_Y = m\mu_X + b$. Then multiply (*) by $x_j f(x_j)$ and add all M equations as before. Obtain $E(XY) = mE(X^2) + b\mu_X$. Solve these simultaneous linear equations and thus determine m and b. Finally, show that the linear regression function of Y and X can be written in the following form:

(**) $E(Y \mid X = x_j) = \mu_Y + \rho(X, Y) \dfrac{\sigma_Y}{\sigma_X} (x_j - \mu_X).$

Conclude that the points of the graph of the linear regression function lie on a straight line passing through the point (μ_X, μ_Y) and that this line is horizontal if and only if X and Y are uncorrelated.

(f) The experiment is performed and we are given the incomplete information that the value of X is x_j. We want to estimate the value of Y. Suppose we use a "least-squares" criterion; i.e., we seek an estimate, say c_j, such that the mean squared deviations of values of Y from the estimated value c_j will be as small as possible. In symbols, we seek the number c_j which minimizes

$$E[(Y - c_j)^2 \mid X = x_j] = \sum_{k=1}^{N} (y_k - c_j)^2 g(y_k \mid x_j).$$

Show that this least-squares estimate is the conditional mean of Y given $X = x_j$; i.e., show that $c_j = E(Y \mid X = x_j)$. [*Hint:* The proof follows immediately from the property of the mean stated in Problem 3.8.]

(g) For any pair of values (x_j, y_k), the error made by using the estimate $E(Y \mid X = x_j)$ in place of y_k is the difference $y_k - E(Y \mid X = x_j)$. The mean squared error, denoted by σ_e^2, is therefore the sum

$$\sigma_e^2 = \sum_{\text{all } j,k} [y_k - E(Y \mid X = x_j)]^2 h(x_j, y_k).$$

Show that if the estimate is given by the linear regression function (**), then

$$\sigma_e^2 = (1 - \rho^2)\sigma_Y^2,$$

and thus conclude that this mean squared error decreases and approaches zero as $\rho(X, Y)$ approaches either $+1$ or -1.

(h) State and prove the results analogous to those in parts (e)-(g), but now supposing the regression function of X on Y is linear and we are given that the value of Y is y_k.

SUPPLEMENTARY READING

In addition to the references listed at the end of Chapter 2, the following books are among the many sources of material on random variables and related topics. As with most of the previously mentioned references, only parts of these books can be read without some knowledge of the differential and integral calculus.

1. Adams, J. K., *Basic Statistical Concepts*, McGraw-Hill Book Company, Inc., 1955.

2. Brunk, H. D., *An Introduction to Mathematical Statistics*, Ginn and Company, 1960.

3. David, F. N., *Probability Theory for Statistical Methods*, Cambridge University Press, 1949.

4. Fraser, D. A. S., *Statistics: An Introduction*, John Wiley and Sons, Inc., 1958.

5. Lindgren, B. W. and G. W. McElrath, *Introduction to Probability and Statistics*, The Macmillan Company, 1959.

6. Mood, A. M., *Introduction to the Theory of Statistics*, McGraw-Hill Book Company, Inc., 1950.

7. Wilks, S. S., *Elementary Statistical Analysis*, Princeton University Press, 1948.

Chapter 5

BINOMIAL DISTRIBUTION
AND SOME APPLICATIONS

---

1. Bernoulli trials and the binomial distribution

Certain kinds of experiments and associated random variables occur time and again in the theory of probability and in its applications. They are therefore made the object of special study in which their properties are explored, values of frequently needed probabilities are tabulated, and so on. In this section, we describe a number of such experiments and random variables, paying special attention to the so-called binomial probability function. In the final sections of this chapter, we discuss two important problems of statistics in which this function plays a central role.

As we have seen in numerous examples throughout this book, many problems involve experiments made up of a number, say n, of individual trials. Each trial is itself really an arbitrary experiment, and is therefore defined in the mathematical theory by some sample space and assignment of probabilities to its simple events. The trials can be independent or dependent, and the simple events of the sample space for the n-trial experiment are assigned probabilities accordingly.

Although each trial may have many possible outcomes, we are often interested only in whether a certain result occurs or not. For example, a machine turns out parts which are classified defective or good; a card is selected from a standard deck and it is an ace or not an ace; two dice are rolled and the sum of the numbers showing is seven or is different from seven; a student selected from the senior class has a part-time job or has not; etc.

In order to have a convenient standard terminology for discussing all such trials, we shall call one of the two possible results of each trial a *success* and the other a *failure*. Which result is called a success is, of course, completely arbitrary—whether one calls a defective part a success or a failure, or a student with a part-time job a success or a failure, is a matter of taste as far as the theory goes. We must however make sure that we are consistent in our language in any one problem.

If when a trial is performed we are interested solely in whether a success or a failure results, then it is sensible to make the sample space defining the trial reflect this fact by containing just two elements, say S for success and F for failure. If the simple event $\{S\}$ is given probability p, then an acceptable assignment of probabilities is determined for every choice of the number p, provided only that $0 \leq p \leq 1$. Writing $q = 1 - p$ for convenience, we have

(1.1) $P(\{S\}) = p, \qquad P(\{F\}) = q, \qquad p + q = 1.$

As an example, consider drawing a card at random from a standard deck. Ordinarily we define as sample space a set containing 52 elements (one for each card) and assign probability $\frac{1}{52}$ to each simple event. But if we are interested only in whether or not the card is an ace, and we call drawing an ace a success and drawing any other face value a failure, then we prefer to use $\{S, F\}$ as sample space, with $p = \frac{1}{13}$ and $q = \frac{12}{13}$ as probability of success and failure, respectively.

Definition 1.1. Trials are called *Bernoulli trials* (after James Bernoulli, 1654–1705) if and only if they meet the following conditions:

(1) Each trial is defined by the sample space $\{S, F\}$; i.e., we consider that *each trial has only two outcomes:* either S (success) or F (failure).

(2) The same assignment of probabilities, as given in (1.1), is made to the simple events of each trial; i.e., *the probability of a success is the same on each trial* and is denoted by p.

(3) *The trials are independent.*

A sequence of any (not necessarily Bernoulli) trials can be thought of as a *process* in which outcomes of the individual trials are produced as the trials are performed. A process of this kind is called a *stochastic* (= probability) or *random* process, since the particular sequence of outcomes obtained depends upon chance. A random process made up of Bernoulli trials is called a *Bernoulli process*.

Example 1.1. Tossing a coin 100 independent times is interpreted to mean 100 Bernoulli trials in which each trial (toss of the coin) results in success (say, heads) or failure (tails), and the probability p of a head is the same for all 100 tosses. If the coin is fair, then $p = \frac{1}{2}$ and $q = \frac{1}{2}$; if the coin is biased, then $p \neq \frac{1}{2}$.

Example 1.2. Consider a manufacturing process in which a metal part is produced by an automatic machine. Suppose each part in a production run of 500 parts can be classified upon inspection as defective or good. We can think of the production of a part as a single trial which results in success (say, a defective part) or in failure (a good part). If we believe that the machine operation is just as likely to produce a defective on one trial as on any other, and if we also believe that the occurrence of a defective on any trial is made neither more nor less likely by the particular results obtained on the preceding trials, then it is reasonable to assume that the production run is a Bernoulli process with 500 trials. (The probability p of a defective on each trial is called the *average fraction defective* of the process.)

Of course, the Bernoulli process is a mathematical idealization of the actual production process. For example, if the machine setting wears down as the run proceeds, then the tendency of the machine to produce defectives will increase as time goes on and the probability p is therefore not the same for all 500 trials. It is clear that a real manufacturing process cannot be *exactly* represented by a Bernoulli process. Nevertheless, it is often closely approximated by such a process, and useful results are obtained by means of this idealization.

Example 1.3. The sample space for an experiment made up of three Bernoulli trials with probability p for success on each trial is the Cartesian product set $\{S, F\} \times \{S, F\} \times \{S, F\}$ containing $2^3 = 8$ three-tuples as elements. These three-tuples and the probabilities of the corresponding simple events, obtained by use of the product rule of Chapter 2, Section 9 (since the trials are independent), are listed in the first two columns of Table 36. The number of successes obtained in this experiment, denoted by S_3, is a random variable whose

TABLE 36

Outcome of Experiment	Probability of Corresponding Simple Event	Possible Value of S_3 k	$P(S_3 = k)$
FFF	$qqq = q^3$	0	q^3
FFS	$qqp = pq^2$		
FSF	$qpq = pq^2$	1	$3pq^2$
SFF	$pqq = pq^2$		
FSS	$qpp = p^2q$		
SFS	$pqp = p^2q$	2	$3p^2q$
SSF	$ppq = p^2q$		
SSS	$ppp = p^3$	3	p^3

possible values are 0, 1, 2, 3. The probability function of the random variable S_3 is determined in the last two columns of Table 36. Note that the probabilities in the last column are the terms in the binomial expansion of $(q + p)^3$. Since $p + q = 1$, it follows that the sum of these probabilities, as expected, is indeed 1.

The general argument about to be made is modeled on this last example. (The reader may find it helpful at this point to review Example 9.5 and Problems 9.1–9.7 in Chapter 2, where other special cases were presented.) The sample space for an experiment made up of n Bernoulli trials is the Cartesian product set

$$\{S, F\} \times \{S, F\} \times \cdots \times \{S, F\}$$

containing 2^n n-tuples as elements. Every n-tuple represents an outcome of the n-trial experiment and is made up of n symbols, each an S or an F. Since the trials are independent, the product rule of Chapter 2, Section 9 applies and, taking account of (1.1), we deduce that the probability of any simple event whose n-tuple contains k S's and $n - k$ F's (in any order) is $p^k q^{n-k}$ for $k = 0, 1, \cdots, n$. One such n-tuple is determined by selecting the k trials on which S's occur from among all n trials. Since this can be done in $\binom{n}{k}$ different ways, we conclude that there are $\binom{n}{k}$ n-tuples containing k S's and $n - k$ F's, and that the corresponding simple events all have the same probability, namely $p^k q^{n-k}$.

As in Example 1.3, we are interested in determining the probability function of the random variable whose value is the total number of successes obtained in the n-trial experiment. This random variable is denoted by S_n and clearly has possible values $0, 1, \cdots, n$. Now $S_n = k$, where k is any one of these possible values, is the event for which exactly k S's (and therefore $n - k$ F's) occur. This event is the union of the $\binom{n}{k}$ simple events determined by n-tuples with k S's and $n - k$ F's. As we observed, each such simple event has probability $p^k q^{n-k}$. Hence

$$(1.2) \qquad P(S_n = k) = \binom{n}{k} p^k q^{n-k} \qquad k = 0, 1, \cdots, n.$$

We have therefore proved the following result.

Theorem 1.1. Suppose an experiment consists of n Bernoulli trials with probability p for success on each trial. If S_n is the random variable whose value for any outcome of the experiment is the total number of successes obtained, then the probability function of S_n is given by (1.2).

For given values of n and p, the probability function defined by (1.2) is called the *binomial probability function* or the *binomial distribution* with parameters n and p*. Formula (1.2) thus defines not just one binomial distribution, but a whole family of binomial distributions, one for every possible pair of values for n and p. To show the dependence of the probabilities on the parameters, we shall write $b(k|n, p)$ for the probability in (1.2). Thus

$$(1.3) \qquad b(k|n, p) = P(S_n = k) = \binom{n}{k} p^k q^{n-k}$$

is the probability of exactly k successes, given the parameters n and p

---

* For the random variables considered in this volume the terms *probability distribution* and *probability function* are synonomous. We have avoided introducing the term probability distribution earlier due to possible confusion with the (cumulative) distribution function. But the reader should become familiar with the standard terminology, and we use it from now on. Note that one customarily shortens binomial probability distribution to binomial distribution.

of the binomial distribution; i.e., $b(k|n, p)$ is the probability of exactly k successes in n Bernoulli trials with probability p for success on each trial. The random variable S_n is said to be *binomially distributed* with parameters n and p when S_n has the probability distribution defined by (1.3).

The name *binomial distribution* arises from the fact that the probabilities $b(k|n, p)$ for $k = 0, 1, \cdots, n$ are the terms in the binomial expansion of $(q + p)^n$. (See Chapter 3, Section 2, for the binomial theorem and related identities involving binomial coefficients.) It follows since $p + q = 1$ that

$$(1.4) \qquad \sum_{k=0}^{n} b(k|n, p) = (q + p)^n = 1,$$

as required for a probability function.

Example 1.4. If a fair coin is tossed six times, the probability of getting *exactly* five heads is

$$b(5|6, \tfrac{1}{2}) = \binom{6}{5}\left(\frac{1}{2}\right)^5\left(\frac{1}{2}\right)^1 = \frac{6}{64} = .09375.$$

The probability of *at least* five heads is obtained by adding the probability of five heads and the probability of six heads. Since

$$b(6|6, \tfrac{1}{2}) = \binom{6}{6}\left(\frac{1}{2}\right)^6\left(\frac{1}{2}\right)^0 = \frac{1}{64} = .015625,$$

it follows that the probability of at least five heads is

$$P(S_6 \geq 5) = b(5|6, \tfrac{1}{2}) + b(6|6, \tfrac{1}{2}) = .109375,$$

where we write S_6 for the random variable denoting the total number of heads (successes) among the six tosses.

Example 1.5. Five percent of the metal parts produced by a machine are defective, the other 95 percent are good. How many parts must be produced in order for the probability of at least one defective to be $\tfrac{1}{2}$ or more? We assume that the production of parts is a Bernoulli process for which each trial (producing one part) results in a success (defective part) or failure (good part). The probability p for success on any trial is given as $p = .05$. We seek the smallest integer n such that $P(S_n \geq 1) \geq \tfrac{1}{2}$. Now

$$P(S_n \geq 1) = 1 - P(S_n = 0) = 1 - b(0|n, .05)$$
$$= 1 - \binom{n}{0}(.05)^0(.95)^n = 1 - (.95)^n,$$

so that we want the smallest integer n for which $1 - (.95)^n \geq \frac{1}{2}$ or $(.95)^n \leq \frac{1}{2}$. Using logarithms (Table 22, p. 141) we find $n \log (.95) \leq -\log 2$, from which $n \geq 13.5$ approximately. Hence $n = 14$ is the smallest lot size that can be used in order to have an even chance or better of finding at least one defective part in the lot.

For many applications, it is necessary to compute the probability not of exactly r successes, but of *at least* r or *at most* r successes. Since such cumulative probabilities are obtained by computing all the included individual probabilities and adding, this task soon becomes laborious. For example, to compute the probability of at least six successes in ten Bernoulli trials with $p = .3$, we must compute $b(k|10, .3)$ for $k = 6, 7, 8, 9, 10$ and then add these five probabilities. Fortunately, extensive tables are available to lighten the task of such computations.*

A small table of binomial probabilities is included here (Table 37) for our use. Wherever possible, examples and problems from now on will be formulated with numerical values for the parameters n and p that will allow the table to be used to find required probabilities. Note that we have tabulated $P(S_n \geq r)$, the probability of r *or more* (*at least* r) successes for $n = 1, 2, \cdots, 10, 20$ and for $p = .01, .05, .10, .20, .30, .40, .50$. For each pair of values for n and p and each possible value of r we read

$$P(S_n \geq r) = b(r|n, p) + b(r + 1|n, p) + \cdots + b(n|n, p)$$

directly from the table. (We do not include a row for $r = 0$, since $P(S_n \geq 0) = 1$ for all n and p.) We illustrate the use of this table in the following examples.

Example 1.6. In Example 1.4, we computed $P(S_6 \geq 5)$ for $p = \frac{1}{2}$ and found the answer .109375. In our table, for $n = 6, r = 5, p = .50$, we read .109, which agrees to three decimals with the exact answer. To find $P(S_6 = 5) = b(5|6, \frac{1}{2})$ we note that

$$P(S_6 = 5) = P(S_6 \geq 5) - P(S_6 \geq 6),$$

---

* See *Tables of the Cumulative Binomial Probability Distribution,* Annals of the Computation Laboratory of Harvard University, vol. XXXV, Harvard University Press, 1955; *Tables of the Binomial Probability Distribution,* National Bureau of Standards, Applied Mathematics Series, vol. 6, 1950; H. C. Romig, 50–100 *Binomial Tables,* John Wiley and Sons, Inc., 1953.

TABLE 37. Cumulative Binomial Probabilities

The entry is $P(S_n \geq r) = \sum_{k=r}^{n} b(k|n, p)$. Missing entries are less than .0005.

n	r	$p = .01$	$p = .05$	$p = .10$	$p = .20$	$p = .30$	$p = .40$	$p = .50$
1	1	.010	.050	.100	.200	.300	.400	.500
2	1	.020	.098	.190	.360	.510	.640	.750
	2		.002	.010	.040	.090	.160	.250
3	1	.030	.143	.271	.488	.657	.784	.875
	2		.007	.028	.104	.216	.352	.500
	3			.001	.008	.027	.064	.125
4	1	.039	.185	.344	.590	.760	.870	.938
	2	.001	.014	.052	.181	.348	.525	.688
	3			.004	.027	.084	.179	.312
	4				.002	.008	.026	.062
5	1	.049	.226	.410	.672	.832	.922	.969
	2	.001	.023	.081	.263	.472	.663	.812
	3		.001	.009	.058	.163	.317	.500
	4				.007	.031	.087	.188
	5					.002	.010	.031
6	1	.059	.265	.469	.738	.882	.953	.984
	2	.001	.033	.114	.345	.580	.767	.891
	3		.002	.016	.099	.256	.456	.656
	4			.001	.017	.070	.179	.344
	5				.002	.011	.041	.109
	6					.001	.004	.016
7	1	.068	.302	.522	.790	.918	.972	.992
	2	.002	.044	.150	.423	.671	.841	.938
	3		.004	.026	.148	.353	.580	.773
	4			.003	.033	.126	.290	.500
	5				.005	.029	.096	.227
	6					.004	.019	.062
	7						.002	.008
8	1	.077	.337	.570	.832	.942	.983	.996
	2	.003	.057	.187	.497	.745	.894	.965
	3		.006	.038	.203	.448	.685	.855
	4			.005	.056	.194	.406	.637
	5				.010	.058	.174	.363
	6				.001	.011	.050	.145
	7					.001	.009	.035
	8						.001	.004

TABLE 37. Cumulative Binomial Probabilities (cont.)

The entry is $P(S_n \geq r) = \sum_{k=r}^{n} b(k|n, p)$. Missing entries are less than .0005.

n	r	$p = .01$	$p = .05$	$p = .10$	$p = .20$	$p = .30$	$p = .40$	$p = .50$
9	1	.086	.370	.613	.866	.960	.990	.998
	2	.003	.071	.225	.564	.804	.929	.980
	3		.008	.053	.262	.537	.768	.910
	4		.001	.008	.086	.270	.517	.746
	5			.001	.020	.099	.267	.500
	6				.003	.025	.099	.254
	7					.004	.025	.090
	8						.004	.020
	9							.002
10	1	.096	.401	.651	.893	.972	.994	.999
	2	.004	.086	.264	.624	.851	.954	.989
	3		.012	.070	.322	.617	.833	.945
	4		.001	.013	.121	.350	.618	.828
	5			.002	.033	.150	.367	.623
	6				.006	.047	.166	.377
	7				.001	.011	.055	.172
	8					.002	.012	.055
	9						.002	.011
	10							.001
20	1	.182	.642	.878	.988	.999	1.000	1.000
	2	.017	.264	.608	.931	.992	.999	1.000
	3	.001	.075	.323	.794	.965	.996	1.000
	4		.016	.133	.589	.893	.984	.999
	5		.003	.043	.370	.762	.949	.994
	6			.011	.196	.584	.874	.979
	7			.002	.087	.392	.750	.942
	8				.032	.228	.584	.868
	9				.010	.113	.404	.748
	10				.003	.048	.245	.588
	11				.001	.017	.128	.412
	12					.005	.057	.252
	13					.001	.021	.132
	14						.006	.058
	15						.002	.021
	16							.006
	17							.001
	18							
	19							
	20							

since the event $(S_6 \geq 5)$ is the union of the mutually exclusive events $(S_6 = 5)$ and $(S_6 \geq 6)$. These cumulative probabilities are read directly from the table for $n = 6$, $r = 5$ and $n = 6$, $r = 6$, using the column headed $p = .50$. We find

$$P(S_6 = 5) = .109 - .016 = .093,$$

as compared to the exact answer .09375 computed in Example 1.4. The exact answer when rounded to three decimals is .094 rather than .093 as obtained from the table. Since each cumulative probability in the table is itself a rounded figure, such discrepancies are to be expected when subtracting tabular entries.

Example 1.7. To find $P(S_{10} \leq 3)$ when $p = .20$ we write

$$P(S_{10} \leq 3) = 1 - P(S_{10} \geq 4),$$

and read this cumulative probability under $p = .20$ and in the row labeled $n = 10$, $r = 4$. We get

$$P(S_{10} \leq 3) = 1 - .121 = .879.$$

Example 1.8. To use the table when $p > .50$, rephrase the problem in terms of $q = 1 - p$. For example, to find the probability of at least seven successes in ten Bernoulli trials with $p = .80$, we compute instead the equal probability of at most three failures in ten trials, but now entering the table with the probability appropriate to a failure, namely $p = .20$. This probability was computed in the preceding example.

Note that this method amounts to relabeling the two results of each trial so that S and F are interchanged. If the probability of a "success" is initially greater than .5, then after the relabeling it is less than .5 and the problem is reformulated in terms of this new language before using the table. (The formal identity used to justify this intuitively clear procedure is given in Problem 1.13.)

Example 1.9. Two teams, A and B, compete in a series of games. Each trial (play of one game in the series) can result in success (say, A wins) or failure (B wins). If we assume that the probability p that A wins is the *same* for all games in the series and that the games are *independent* trials, then a Bernoulli process serves as a mathematical model of the series competition.

The probability p is taken as a measure of the relative strength of the two teams. If $p > \frac{1}{2}$, team A is better than team B; if $p = \frac{1}{2}$, the teams are evenly matched; if $p < \frac{1}{2}$, team B is better than team A. How does the kind of series affect the probability of the better team to win the series? For example, a tie at the end of the regular baseball season between two National League teams is broken by a three-game series in which the team first to win two games is declared the pennant winner. The American League breaks ties by having the teams play a single game. World Series competition, however, is a seven-game series in which the team first to win four games is the winner. We feel intuitively that a superior team has a better chance of showing its superiority in a seven game series than in a three game series or in a single game against the same opponent.*

Although the World Series ends as soon as one team wins four games, we could imagine it continued to the full seven games. Winning the series is equivalent to winning *at least* four of the seven games, and we can therefore use our table of cumulative binomial probabilities to compute the probability of a team winning the series for various values of p. For example, if $p = .30$, then we enter the table for $n = 7$, $r = 4$ and read the probability .126 for team A, to win the series. If $p = .90$, then the probability that team A wins the series is not directly available from the table. Instead, we read the probability that team B wins (entering the table for $p = .10$ appropriate to the new meaning of a "success") and find .003. Hence, team A

TABLE 38

Probability of Team A Winning Single Game	Probability that Team A Wins an n-Game Series				
p	$n = 1$	$n = 3$	$n = 5$	$n = 7$	$n = 9$
0	0	0	0	0	0
.1	.100	.028	.009	.003	.001
.3	.300	.216	.163	.126	.099
.5	.500	.500	.500	.500	.500
.7	.700	.784	.837	.874	.901
.9	.900	.972	.991	.997	.999
1.0	1.000	1.000	1.000	1.000	1.000

* For a complete discussion of this and related points, see F. Mosteller, "The World Series Competition," *Journal of the American Statistical Association*, vol. 47 (1952), pp. 355–380.

wins the series with probability .997 if $p = .90$. In Table 38, we summarize these computations for various values of p and for series containing an odd number of games, the winner being required to win a majority of the games. These probabilities are graphed in Figure 27,

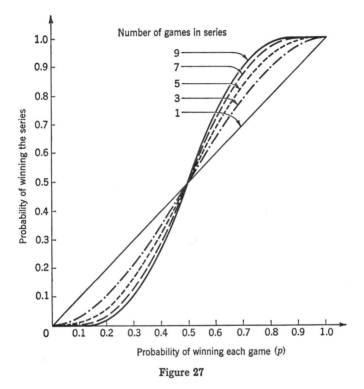

Figure 27

and we can see how increasing the number of games in the series decreases the probability of a poorer team winning (the graphs get lower if $p < .5$) and increases the probability of a better team winning (the graphs get higher if $p > .5$). The fact that all five graphs are very close together around $p = .5$ means that, if one team is only slightly better than the other (say $p = .51$), then a nine-game series is not very much more effective than a single game as a discriminator between the teams. In fact, with $p = .51$ it turns out that the better team wins a nine-game series with probability .525, which is only slightly higher than .51, the probability that it wins a single game. Put differently, this means that the poorer team will win the nine game series roughly 47.5% of the time in spite of the fact that it faces a superior opponent. Of course, one reduces the probability that the series will erroneously be won by the poorer team by increasing the number of games in the series. (Similar ideas appear in a variety of statistical problems, as we shall see in the next section.)

We turn now to a discussion of some properties of the random variable S_n. In particular, we want to determine the mean and variance of this binomially distributed random variable. Since the probability function of S_n is given in (1.2), we could use the definitions of mean and variance given in the preceding chapter to compute $E(S_n)$ and $\text{Var}(S_n)$. We would then have to evaluate the sums

$$(1.5) \qquad E(S_n) = \sum_{k=0}^{n} kb(k|n, p) = \sum_{k=0}^{n} k \binom{n}{k} p^k q^{n-k}$$

and

$$(1.6) \qquad E(S_n^2) = \sum_{k=0}^{n} k^2 b(k|n, p) = \sum_{k=0}^{n} k^2 \binom{n}{k} p^k q^{n-k},$$

from which we compute the variance of S_n by use of the formula

$$(1.7) \qquad\qquad \text{Var}(S_n) = E(S_n^2) - [E(S_n)]^2.$$

This way of calculating the mean and variance of S_n is direct and offers useful practice in the manipulation of binomial coefficients. But we choose to leave this for the problems and instead present an alternate derivation which gives added insight into the nature of the binomial distribution.

Suppose (Cf. Section 5 of the preceding chapter) we have a population specified by the random variable X whose probability table is as follows:

$$(1.8)$$

x	0	1
$P(X = x)$	q	p

X is here interpreted as the number of successes in a *single* trial, and we simulate a Bernoulli process by drawing random samples *with replacement* from this population, thinking of the occurrence of a success as corresponding to $x = 1$ and the occurrence of a failure as corresponding to $x = 0$. Indeed, if X_k is the kth sample value obtained, then the sum $X_1 + \cdots + X_n$ is precisely the number of ones in a random sample of size n or equivalently, the number of successes in n Bernoulli trials with probability p for success on each trial. Hence the sample mean $\overline{X}$ is related to S_n by the formula

$$(1.9) \qquad \overline{X} = \frac{S_n}{n} \quad \text{or} \quad S_n = n\overline{X},$$

and we can compute $E(S_n)$ and $\text{Var}(S_n)$ by using Theorem 5.7 of the preceding chapter, since from (1.9) we have

$$(1.10) \qquad E(S_n) = n\mu_{\overline{X}} \quad \text{and} \quad \text{Var}(S_n) = n^2\sigma^2_{\overline{X}}.$$

Now the population mean and variance are easily determined from (1.8) to be

$$(1.11) \qquad \mu_X = p, \quad \sigma^2_X = pq.$$

Hence

$$E(S_n) = n\mu_{\overline{X}} = n\mu_X = np$$

and

$$\text{Var}(S_n) = n^2\sigma^2_{\overline{X}} = n^2 \frac{\sigma^2_X}{n} = npq.$$

We have thus proved the following important result.

Theorem 1.2. A binomially distributed random variable with parameters n and p has mean np, variance npq, and standard deviation $\sqrt{npq}$.

Example 1.10. In 100 families containing four children, the number of families that had 0, 1, 2, 3, 4 girls were recorded as in the following frequency table:

Number of Girls in Family	0	1	2	3	4	Total
Number of Families	4	31	35	25	5	100

If the probability of giving birth to a girl is assumed constant, then how can we use these data to estimate this unknown probability? We think of the sexes of the children in each family as being determined by four Bernoulli trials with the probability p of success (female child) fixed but unknown. Using this theoretical binomial distribution, Theorem 1.2 tells us that the mean number of girls in a family of four children is $4p$. To estimate p we adopt the following procedure: *set the mean of the theoretical binomial probability distribution equal to the mean of the observed frequency distribution.*

The mean number of girls in the 100 families is

$$\frac{0(4) + 1(31) + 2(35) + 3(25) + 4(5)}{100} = 1.96.$$

Hence, according to the estimation procedure just stated, we equate $4p$ and 1.96 to obtain

$$\hat{p} = \frac{1.96}{4} = .49,$$

where we write $\hat{p}$ to denote an estimate of p based on the particular data given in this problem.

TABLE 39

Number of Girls in Family k	Observed Frequency	"Fitted" Binomial Probabilities $b(k\mid4, .49)$	Theoretically Expected Frequencies $100b(k\mid4, .49)$
0	4	.068	6.8
1	31	.260	26.0
2	35	.375	37.5
3	25	.240	24.0
4	5	.058	5.8

The binomial distribution with parameters $n = 4$ and $p = \hat{p} = .49$ is said to be "fitted" to the observed frequency distribution. From the fitted binomial distribution, we can compute the probabilities $b(k\mid4, \hat{p})$ and thus the theoretically expected frequencies $100b(k\mid4, \hat{p})$ for $k = 0, 1, 2, 3, 4$ and then can compare these with the actually observed frequencies to see how good a "fit" we have. The result is given in Table 39. How to test the "goodness of fit" between observed and theoretically expected frequencies as well as how to appraise the given estimation procedure as compared with other possible procedures are problems of great importance in statistics, but we cannot go into these matters here.

Using Theorem 1.2, the standardized random variable corresponding to S_n is seen to be

(1.12) $$S_n^* = \frac{S_n - np}{\sqrt{npq}}.$$

The event $-c \leq S_n^* \leq c$ is the same as

(1.13) $$np - c\sqrt{npq} \leq S_n \leq np + c\sqrt{npq}$$

and occurs when the number of successes in n Bernoulli trials differs from the mean np by no more than c standard deviations. According to Chebyshev's inequality, this probability is greater than $1 - (1/c^2)$, but we know that this estimate is not very helpful.

Much stronger results are available. Indeed, it can be shown by advanced methods that for large values of n, $P(-c \leq S_n^* \leq c)$ is closely approximated by the area under a certain bell-shaped curve known as the normal probability curve. For example, although Chebyshev's inequality tells us only that $P(-1 \leq S_n^* \leq 1) > 0$, the normal curve approximation tells us that this probability is about .68 if n is large. Similarly, we learn that $P(-2 \leq S_n^* \leq 2)$ is about .95 and $P(-3 \leq S_n^* \leq 3)$ is about .997, so that the number of successes in n Bernoulli trials is almost certain to be within three standard deviations of its mean if n is large. Unfortunately, we cannot do any more here than mention these results which are of such great practical and theoretical significance in probability. We do however give one illustrative example.

Example 1.11. A coin is tossed 400 times and falls heads 210 times. If the coin is fair, is it unlikely to get this many heads? The number of heads is binomially distributed with parameters $n = 400$ and, we assume, $p = \frac{1}{2}$. Hence the mean number of heads is 200 and the standard deviation is $\sqrt{400(\frac{1}{2})(\frac{1}{2})} = 10$. The probability that the number of heads is between 190 and 210 is approximately .68, since this is a one standard deviation interval on either side of the mean. To get as many as 210 heads is therefore not at all unlikely, on the assumption that the coin is fair. Similarly, we find that with a fair coin it is almost certain (probability .997) that the number of heads will fall within three standard deviations, or 30, on either side of the mean, i.e., in the range 170 to 230. Obtaining a number of heads less than 170 or more than 230 would therefore throw grave doubts on the hypothesis that the coin is fair.

PROBLEMS

1.1. If the production of parts by a machine is regarded as a Bernoulli process with process average defective equal to $p = .20$, is it more likely to have (a) no defectives among ten parts, or (b) at most one defective among 20 parts?

1.2. In a 20-question true-false examination, suppose a student tosses a fair coin to determine his answer to each question. If the coin falls heads, he answers "true"; if it falls tails, he answers "false." Find the probability that he answers at least 12 questions correctly and thus passes the exam.

1.3. The probability of having no ace in a bridge hand is approximately .30. What is the probability that a person who plays ten hands of bridge will never receive an ace?

1.4. From the cumulative probabilities given in Table 37, determine the probability function of a binomially distributed random variable with parameters (a) $n = 10$ and $p = .3$, (b) $n = 10$ and $p = .7$, (c) $n = 10$ and $p = .5$.

1.5. Let X be a binomially distributed random variable with mean 12 and variance 4.8. Find (a) $P(X > 5)$, (b) $P(5 < X < 10)$, (c) $P(X \leq 10)$.

1.6. A man is to throw a fair coin a certain number of independent times and is to receive a prize if he throws exactly five heads. At the outset, he is to choose the number of throws he will make. What number should he choose in order to maximize his chances of winning the prize? What then are the odds for his winning the prize?

1.7. For $n = 20$ Bernoulli trials, determine

(a) $P(S_{20} > 12)$ for $p = 0.7$.
(b) $P(10 \leq S_{20} \leq 14)$ for $p = 0.6$.
(c) The value of p for which $P(S_{20} \geq 8) = .50$. (*Hint:* Interpolate between two values found in the table.)

1.8. What is the probability of throwing exactly nine heads exactly twice in five throws of ten fair coins? (*Hint:* Use the binomial distribution twice.)

1.9. How many Bernoulli trials with probability .01 for success must be performed in order that the probability of at least one success be $\frac{1}{2}$ or more?

1.10. We are given the information that n Bernoulli trials resulted in exactly k successes. Show that the conditional probability of a success on any particular trial is k/n.

1.11. In order to decide whether to accept or reject a very large lot of items offered for sale, the buyer takes a sample of 20 items at random from the lot and tests them. If at most one defective is found, he accepts the entire lot; if more than one defective is discovered in the sample, he rejects the lot.

(a) Find the probability that the buyer accepts the lot if in fact it contains a proportion of defectives equal to p, where p assumes the values in Table 37.

(b) Graph the probability that the buyer accepts the lot against the proportion of defectives, showing the probability of acceptance on the vertical axis. (This is called an *operating characteristic curve* or OC curve for the single-sample decision rule adopted by the buyer.)

(c) Draw the operating characteristic curve for the following alternative single-sample decision rule: a sample of only ten items is drawn at random from the lot and tested. The lot is accepted if no defectives are found and rejected otherwise.

(d) Where in your analysis of this problem have you used the fact that the lot is very large?

1.12. (a) Prove the following *recursion formula* for binomial probabilities:
$$b(k + 1|n, p) = \frac{n - k}{k + 1} \frac{p}{q}\, b(k|n, p).$$

(b) Denote by m the unique integer for which
$$(n + 1)p - 1 < m \le (n + 1)p.$$
If $(n + 1)p$ is not an integer, show that as k goes from 0 to n, $b(k|n, p)$ increases up to a maximum value which occurs for $k = m$ and then decreases. But if $m = (n + 1)p$, then show that $b(k|n, p)$ increases up to $b(m - 1|n, p)$ which is equal to $b(m|n, p)$, and then decreases.

(c) Use Table 37 to compute the binomial probabilities for $n = 4$, $p = .4$ and $n = 5$, $p = .4$. For these special cases, check the assertions made in (b).

(d) The number m defined in (b) is called the *most probable number of successes* in n Bernoulli trials with probability of success equal to p. Determine $b(m|n, p)$ for $n = 20$, $p = .10$ and $n = 20$, $p = .50$. Does the most probable number of successes occur with high probability?

1.13. Show that
(a) $b(k|n, p) = b(n - k|n, 1 - p)$
(b) $\sum_{k=r}^{n} b(k|n, p) = 1 - \sum_{k=n-r+1}^{n} b(k|n, 1 - p).$

Interpret these formulas in words and show how they are used in relation to Table 37.

1.14. Show that

$$b(k|n + 1, p) = pb(k - 1|n, p) + qb(k|n, p)$$

and interpret this formula in words. Show how the formula can be used to extend Table 37 to $n = 11$.

1.15. (a) Compute $P(-c \leq S_n^* \leq c)$ for $c = 1, 2, 3$ if $n = 5$ and $p = .20$, and compare with the corresponding normal curve approximations.
(b) Repeat part (a), but with $n = 10$ and then $n = 20$.

1.16. Compute $E(S_n)$ and $\text{Var}(S_n)$ by evaluating the sums in (1.5) and (1.6).

1.17. The function G whose value for every real number t is given by

$$G(t) = \sum_{k=0}^{n} b(k|n, p)t^k$$

is called the *generating function* of the binomial distribution with parameters n and p (or of the random variable S_n). Show that $G(t) = (q + pt)^n$. [*Note:* Let those readers who know some differential calculus show from the definition of G that $G'(1) = E(S_n)$ and $G''(1) + G'(1) = E(S_n^2)$, where G' and G'' are the first and second derivatives of G, respectively. By computing these derivatives from the explicit expression for G obtained in this problem, derive formulas for the mean and variance of S_n.]

1.18. Consider a finite population of N objects, each of which is assigned the number 0 or 1, there being Nq 0's and Np 1's, where $p + q = 1$. For example, the objects can be manufactured parts that are good (0) or defective (1). If a random sample of size n is drawn with replacement, then we have seen that the number of 1's obtained is binomially distributed with parameters n and p. But now suppose that the sample of size n is drawn *without replacement*, and let Y_n be the random variable whose value is the number of 1's in the sample. The probability function of Y_n will depend on n, p, and N, and we indicate this dependence by writing

$$P(Y_n = k) = h(k|n, p, N).$$

Y_n is said to have a *hypergeometric* probability distribution with parameters n, p, and N.

(a) Show that

$$h(k|n, p, N) = \frac{\binom{Np}{k}\binom{Nq}{n-k}}{\binom{N}{n}}.$$

(*Note:* Recall the convention concerning binomial coefficients made in Formula (2.10) of Chapter 3.)

(b) Show that the sum of the probabilities $h(k|n, p, N)$ taken over all possible values of Y_n is equal to 1, as required of a probability function. (*Hint:* Use Formula (2.11) of Chapter 3.)

(c) Our notation has been chosen so that the population of N objects has the relative frequency table given in (1.8). If $\overline{X}$ is the sample mean obtained in selecting the random sample, show that $Y_n = n\overline{X}$. Now use Theorem 6.2 of the preceding chapter to conclude that the mean and variance of Y_n are given by

$$E(Y_n) = np$$
$$\text{Var}(Y_n) = npq \left(\frac{N - n}{N - 1} \right).$$

(d) Show that as the population size N increases without bound, the hypergeometric distribution with parameters n, p, and N approaches the binomial distribution with parameters n and p. In symbols.

$$h(k|n, p, N) \to b(k|n, p) \qquad \text{as } N \to \infty.$$

The importance of this limit theorem lies in the fact that when n/N is small enough, binomial probabilities can be used as approximations to hypergeometric probabilities. (*Hint:* Write out the binomial coefficients in (a) and thus show that $h(k|n, p, N)$ is equal to

$$\binom{n}{k} \frac{p \left(p - \dfrac{1}{N} \right) \cdots \left(p - \dfrac{k - 1}{N} \right) q \left(q - \dfrac{1}{N} \right) \cdots \left(q - \dfrac{n - k - 1}{N} \right)}{1 \left(1 - \dfrac{1}{N} \right) \cdots \left(1 - \dfrac{n - 1}{N} \right)}.$$

Now note what happens to each factor as $N \to \infty$.)

(e) Suppose a sample of size n is drawn without replacement from N objects of which Np are defective and Nq are good. In practice, one often knows N but the proportion defective p is unknown. If one obtains k defectives in the sample, then what is a reasonable estimate for this unknown proportion p? Since Np must be an integer, p is necessarily of the form j/N for some choice of j from among the integers $0, 1, \cdots, N$. Estimating p is therefore equivalent to finding an estimate for the integer j.

Now the probability of getting exactly k defectives depends only on j once k, n, and N are fixed. Let us write

$$h_j = h(k|n, \frac{j}{N}, N)$$

for this probability. The method known as *maximum-likelihood estimation* directs us to find that value of j, say $\hat{j}$, such that h_j is as large as possible. In other words, the probability of getting the experimental outcome actually obtained (i.e., exactly k defectives) is maximized if $j = \hat{j}$. The number $\hat{p} = \hat{j}/N$ is then called the maximum likelihood estimate of the unknown proportion defective p. To find $\hat{p}$, proceed as follows:

(i) Show that

$$\frac{h_j}{h_{j-1}} = \frac{j(N - j + 1 - n + k)}{(j - k)(N - j + 1)}.$$

(ii) Show that $\dfrac{h_j}{h_{j-1}}$ is greater than 1 if $j < \dfrac{k(N + 1)}{n}$ and is less than 1 if $j > \dfrac{k(N + 1)}{n}$.

(iii) Conclude that if $\hat{j}$ is the greatest integer less than or equal to $\dfrac{k(N + 1)}{n}$, then the maximum likelihood estimate of p is given by $\hat{p} = \hat{j}/N$.

(f) Repeat the preceding problem, but now assume the sample is drawn with replacement, so that the binomial distribution applies. Show that the maximum likelihood estimate of p is given by $\hat{p} = k/n$, the actual proportion defective found in the sample.

2. Testing a statistical hypothesis

In this section, we illustrate how the binomial distribution is used in a problem of statistical inference. We cannot here go into the general theory of hypothesis testing in statistics. Instead, we analyze one particular example in detail, in order to point out the highlights of the *method* of testing hypotheses.

The Committee for the Re-Election of Smith as Mayor is meeting well ahead of election day to discuss campaign strategy. Smith, as the incumbent, is felt to have the edge on his opponent, but the committee wants some more information about this advantage as a guide to deciding whether to plan a very vigorous and expensive campaign, or a less vigorous and less expensive campaign. Since Smith's opponent is going all out to win, and will undoubtedly reduce Smith's advantage during the campaign, the committee decides that they will raise funds for the more expensive campaign if 60% or less of the population is in favor of Smith, but that they will relax and wage the less expensive campaign if Smith has more than 60% of the voters on his side.

Let us denote by p the actual proportion of all voters in favor of Smith. If p were known, then the committee would have no problem. It would (according to its agreed-upon plan) decide on one course of action if $p \leq .60$ and on the alternative course of action if $p > .60$. *But p is unknown,* and some evidence will have to be obtained in order to choose among the two possible types of campaigns.

It is customary in statistics to say that there are two *hypotheses,* namely $p \leq .60$ and $p > .60$, and the procedure by which a choice is made between these hypotheses is called a *test* of one of the hypotheses *against* the other. The hypothesis that is tested is called the *null hypothesis;* the other is then called the *alternate hypothesis.* Although the committee will choose to accept one hypothesis or the other, it is customary to say instead that the committee's choice is between *acceptance* or *rejection* of the null hypothesis. We shall shortly make some comments about which hypothesis is to be taken as the null hypothesis, but for now let us make the following agreement:

Null Hypothesis: $p \leq .60$;
Alternate Hypothesis: $p > .60$.

The decision to accept or reject the null hypothesis will be based on the result of an experiment in which a certain number of people, say n, are randomly selected from the population of voters and then asked whether they are for or against candidate Smith. We cannot here discuss the very important practical problem of how to design a sample survey or opinion poll of this kind. But we shall assume that the selection of people is made in such a way that the process of sampling can reasonably be idealized as a Bernoulli process in which each trial (asking one of the selected people his voting intention) results in a success (will vote for Smith) or a failure (will not vote for Smith), the probability of a success on each trial being p, the proportion of people in the entire population who favor Smith. (Since the sample is ordinarily drawn without replacement, this theoretical model will be appropriate only if the sample size is very small compared to the number of voters in the entire population.)

Let us suppose that a sample of $n = 20$ people is drawn at random from the population and that each person is asked his voting intention. (We take so small a sample for illustrative purposes only; larger samples are discussed later.) For these $n = 20$ Bernoulli trials, let X denote the number of successes (people who say they favor Smith) obtained. Then X is binomially distributed with parameters $n = 20$ and p, but p is *unknown*. Our null and alternate hypotheses are thus statements about a parameter of a probability distribution. Such hypotheses are called *statistical hypotheses*.

The committee decision to accept or reject the null hypothesis will be based on the outcome of the poll of the 20 people in the sample, in particular, on the value of X obtained. Roughly speaking, the committee will act on the assumption that p is low (and therefore accept the null hypothesis) if the value of X is small, and it will act on the assumption that p is high (and therefore reject the null hypothesis) if the value of X is large. But this is quite vague, and it is clear that what we need is a rule that unequivocally prescribes the committee's decision for each possible outcome of the poll. Consider for the moment the following example of such a *decision rule:*

> Reject the null hypothesis if and only if
> at least 13 of the 20 people in the sample
> say they are in favor of candidate Smith.

Note that the decision rule is completely described by giving the values of X that result in rejection of the null hypothesis. These values are called the *critical* set of values of X for the given decision rule. If the observed outcome falls in the critical set, the null hypothesis is rejected; otherwise the null hypothesis is accepted.

Now the null hypothesis is, as a matter of fact, either true or false. And our decision rule leads either to acceptance or rejection of the null hypothesis. Hence the following possibilities can arise by use of this decision rule:

(1) The null hypothesis is actually true and the value of X does not fall in the critical set; i.e., the null hypothesis is accepted.

(2) The null hypothesis is actually true and the value of X falls in the critical set; i.e., the null hypothesis is rejected.

(3) The null hypothesis is actually false and the value of X does not fall in the critical set; i.e., the null hypothesis is accepted.

(4) The null hypothesis is actually false and the value of X falls in the critical set; i.e., the null hypothesis is rejected.

Now in (1) and (4) the action taken is the correct one, since the committee does indeed want to accept the null hypothesis when it is true and reject the null hypothesis when it is false. But in (2) and (3) the action taken is incorrect: case (2) is said to be an *error of the first kind* or a type I error; case (3) is said to be an *error of the second kind* or a type II error.

If the committee makes an error of the first kind, then it will with a false sense of confidence wage a mild and less expensive campaign, even though Smith has no more than 60% of the voters on his side. This error, although it saves money, may lead to the defeat of Smith at the polls. If the committee makes an error of the second kind, then it will with a false sense of urgency wage a very expensive campaign, even though Smith has the support of more than 60% of the voters. This error leads to spending money for an expensive campaign which the committee, *if* it knew that Smith has such support, would regard as unnecessary.

Since the committee is dedicated to Smith's re-election at all costs, the consequences of an error of the first kind are considered much more serious than the consequences of an error of the second kind. This fact accounts for our choice of $p \leq .60$ as the null hypothesis and $p > .60$ as the alternate hypothesis, rather than vice versa. For it is customary to formulate the null hypothesis so that rejecting it when it is true (error of first kind) is more serious than accepting it when it is false (error of second kind). For example, in testing a new drug there are the two hypotheses "drug is toxic" and "drug is not toxic." The former would be taken as the null hypothesis, since rejecting it when it is true will lead to deaths of patients, whereas accepting this hypothesis when it is false will have the less undesirable consequences of loss of money by the manufacturer and unnecessary waste of the drug. Of course, in cases where the two kinds of errors are of the same importance, it is immaterial which of the two hypotheses is called the null hypothesis.

To study the decision rule stated above (for which $n = 20$ and the null hypothesis is rejected if and only if the value of X is at least 13) it is convenient to define the function π whose value for each possible value of the parameter p is the probability of rejecting the null hypothesis. That is,

$$\pi(p) = P(X \geq 13) = \sum_{k=13}^{20} b(k|20, p).$$

The function π is called the *power function* of the given decision rule.

The reader should check the following values of this power function by referring to Table 37. (We write $0+$ for a positive number less than .0005, and $1-$ for a number greater than .9995 but less than 1.)

p	0	.10	.20	.30	.40	.50	.60	.70	.80	.90	1
$\pi(p)$	0	$0+$	$0+$	.001	.021	.132	.416	.772	.968	$1-$	1

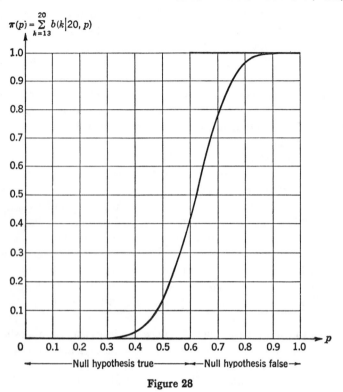

$$\pi(p) = \sum_{k=13}^{20} b(k|20, p)$$

Figure 28

The graph of this power function is shown in Figure 28 where we also indicate the graph (consisting of two horizontal line segments) of the power function for an *ideal* decision rule defined as a rule for which the probabilities of errors of both the first and second kind are zero. Since we are plotting the probability of rejecting the null hypothesis, we find in Figure 28 that this ideal power function has the value 0 when the null hypothesis is true and has the value 1 when the null hypothesis is false. For any value of p satisfying $p \le .60$, the difference in heights of the actual graph and the ideal graph is the probability of an error of the first kind; for any value of p satisfying $p > .60$, the difference in heights of the two graphs is the probability of an error of the second kind.

For the decision rule whose power function is graphed in Figure 28, we observe that as p increases from $p = 0$ to $p = .6$, the probability of an error of the first kind increases from 0 to .416. Similarly, as we move to the left from $p = 1$, the probability of an error of the second kind increases from 0 to .584 as we approach the borderline value $p = .6$.

The committee is not very happy with this decision rule, for it involves high error probabilities. For example, even if candidate Smith is favored by only 50% of the voting population, the probability is .132 that there will be at least 13 people in favor of Smith among the 20 people in the sample, thus leading the committee to plan a weak campaign when a strong one is clearly required. And if the percentage favoring candidate Smith is less than but near 60%, the committee is appalled to find that the decision rule will lead to a wrong decision roughly 40% of the time it is used. And even if candidate Smith is comfortably in the lead with, let us say, 70% of the voters on his side, the sample of 20 will contain *less* than 13 in favor of Smith with probability .228 and the decision rule will then lead the committee to erroneously wage a strong expensive campaign.

The committee therefore asks whether it is possible to formulate a decision rule for which errors of the first and second kind are both smaller than for the rule already mentioned. Let us see what happens if we keep the sample size fixed at $n = 20$. Then the only sensible rules that the committee will consider will be of the form:

> Reject the null hypothesis (that $p \leq .60$)
> if and only if X, the number in favor of
> Smith among the 20 people in the sample,
> is *at least* some specified number, say c.

Each choice of the number c determines one decision rule. We have already discussed the rule with $c = 13$. In order to compare the various possible rules, one determines the power function of each decision rule by using the definition of $\pi(p)$ as the probability of rejecting the null hypothesis when the parameter value is p. That is,

$$\pi(p) = P(X \geq c) = \sum_{k=c}^{20} b(k|20, p).$$

We have used our table of binomial probabilities to compute $\pi(p)$ for $c = 15, 16, 17$. These values are given in Table 40 and graphs of the corresponding power functions are drawn in Figure 29.

From either the table or the graphs we see that as c increases, the probability of an error of the first kind *decreases* for all p satisfying $p \leq .60$, i.e., the graphs move down *toward* the ideal graph for which

TABLE 40

p		0	.10	.20	.30	.40	.50	.60	.70	.80	.90	1
$\pi(p)$	$c = 15$	0	0+	0+	0+	.002	.021	.126	.416	.804	.989	1
	$c = 16$	0	0+	0+	0+	0+	.006	.051	.238	.630	.957	1
	$c = 17$	0	0+	0+	0+	0+	.001	.016	.107	.411	.867	1

the error probability is zero. But, at the same time the probability of an error of the second kind *increases;* i.e., the graphs move down *away from* the ideal graph for all p satisfying $p > .60$.

The committee thus learns that with sample size 20 it cannot simul-

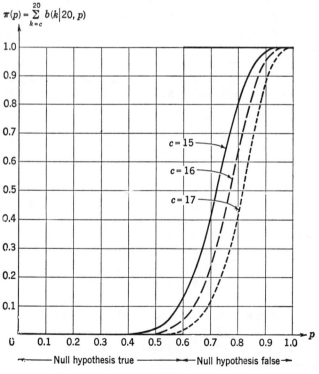

$$\pi(p) = \sum_{k=c}^{20} b(k \mid 20, p)$$

Figure 29

taneously lower *both* the probability of making an error of the first kind and the probability of making an error of the second kind. The customary statistical procedure in this circumstance is to concentrate on the errors of the first kind since, as we have noted earlier, they are presumably more serious than those of the second kind. The committee chooses a number, ordinarily denoted by α, which is the *maximum* probability of an error of the first kind that it will tolerate. In actual practice the number α is often chosen as one of the numbers .01, .05, or .10. Having picked α, the particular decision rule is adopted which not only meets the requirement that the maximum probability of an error of the first kind does not exceed α, but which in addition yields the lowest possible probabilities of errors of the second kind.

For example, suppose the committee chooses $\alpha = .02$. Then in Figure 29 we seek that value of c for which the height of the power curve for $p = .60$ (which gives the maximum probability of an error of the first kind) does not exceed $\alpha = .02$, but is as close to .02 as possible. We find from the figure or from the values in Table 40, that $c = 17$ has the required properties. Thus the committee's choice of $\alpha = .02$ dictates the use of the decision rule for which $c = 17$; i.e., the committee determines the value of X, the number of voters for candidate Smith in the sample of size 20, and rejects the null hypothesis if and only if $X \geq 17$. Although the committee now has a very high probability of making an error of the second kind (wasting money on an unnecessary strong campaign) it does have assurance that there is only at most a 2% chance for an error of the first kind (not waging a vigorous campaign when Smith needs it to win).

Now suppose that the committee is not willing to risk such large chances of an error of the second kind. From Table 40, for example, we find that $P(X \geq 17) = .107$ when $p = .70$. Thus there is almost a 90% chance of accepting the null hypothesis (and therefore wasting money on an unnecessary strong campaign), even when Smith has 70% of the voters on his side. What can be done to maintain the maximum risk level given by $\alpha = .02$ but at the same time to lower the risks of errors of the second kind?

With the sample size fixed at $n = 20$, there is nothing that can be done. But if larger samples are permitted, then risks of errors of both first and second kind can be controlled. We illustrate this point by considering samples of size $n = 50$, $n = 100$, and $n = 300$.

Our decision procedure is stated in general terms as follows:

A sample of n people is drawn from the population of all voters. Let X be the random variable whose value is the number (among the n selected people) who are in favor of Smith. Reject the null hypothesis (that $p \leq .60$) if and only if $X \geq c$, where c is determined so that the maximum probability of an error of the first kind does not exceed some prescribed value α (we suppose the committee chooses $\alpha = .02$) and so that probabilities of errors of the second kind are as small as possible.

From our previous discussion, the reader can see that to determine c we proceed as follows: First put $p = .60$, since it is for this value that the probability of an error of the first kind is largest. *Any* number c for which $P(X \geq c)$ does not exceed α, i.e., for which

$$(2.1) \qquad \sum_{k=c}^{n} b(k|n, .60) \leq \alpha,$$

will determine a decision rule whose maximum probability of an error of the first kind is at most α. To also minimize the probability of making an error of the second kind, we select the *smallest* value of c satisfying the inequality in (2.1). Put differently, we choose c as the smallest number in the set

$$(2.2) \qquad \{x \mid \sum_{k=x}^{n} b(k|n, .60) \leq \alpha\}.$$

(We are assuming that α is chosen so that $b(n|n, .60) \leq \alpha$. The set in (2.2) therefore contains the number n and so is not empty.) This set, containing the values of X for which the null hypothesis is rejected, is called the *critical set* of values of X for the given decision rule.

The reader should refer to Table 37 and check that with $\alpha = .02$ and $n = 20$, the value of c determined in this way is $c = 17$, as we have already seen in Table 40 and Figure 29. Using more extensive tables of cumulative binomial probabilities, we similarly find that with $\alpha = .02$, the smallest value of c satisfying (2.1) is $c = 38$ for $n = 50$, $c = 71$ for $n = 100$, and $c = 198$ for $n = 300$. We therefore have four decision rules, all determined by the committee's setting of $\alpha = .02$ as the maximum tolerable probability of an error of the first kind. Values of the power function of these four rules are given in Table 41. For comparison with Figure 29, we graph the three power functions for sample sizes $n = 50$, 100, and 300 in Figure 30.

TABLE 41

p		.50	.60	.70	.80	.90	1
$\pi(p)$	$n = 20, \quad c = 17$	.001	.016	.107	.411	.867	1
	$n = 50, \quad c = 38$	0+	.013	.223	.814	.999	1
	$n = 100, \quad c = 71$	0+	.015	.462	.989	1−	1
	$n = 300, \quad c = 198$	0+	.019	.941	1−	1−	1

As expected, the risk of making an error of the second kind goes down as the sample size increases; i.e., for each $p > .60$, as n increases the curves move up toward the ideal graph for which the probability of an error of the second kind is zero. From Table 41 we read that

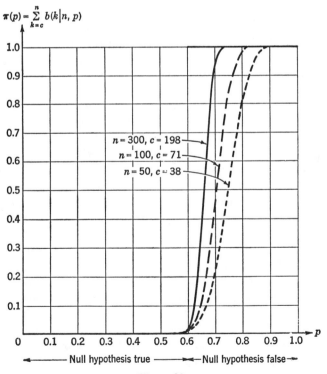

$$\pi(p) = \sum_{k=c}^{n} b(k|n, p)$$

$n = 300, c = 198$

$n = 100, c = 71$

$n = 50, c = 38$

← Null hypothesis true ——►◄—Null hypothesis false→

Figure 30

with $n = 300$, the probability of finding at least 198 persons in favor of Smith when $p = .70$ is .941, so that the probability of an error of the second kind is reduced to .059. Thus we have demonstrated how the committee can maintain its maximum tolerable probability of an error of the first kind at $\alpha = .02$ and can also control the risks of errors of the second kind by sampling a sufficiently large number of people from among the entire population of voters.

For the remainder of our discussion, in order that we may use Table 37, we return to the simple decision rule with $n = 20$, $c = 17$. Since c is chosen as the *smallest* number in the set defined in (2.2), we know that

$$(2.3) \quad P(X \geq x) = \sum_{k=x}^{20} b(k|20, .60) \leq .02 \qquad \text{for all } x \geq 17,$$

and

$$(2.4) \quad P(X \geq x) = \sum_{k=x}^{20} b(k|20, .60) > .02 \qquad \text{for all } x < 17.$$

We are now able to see that although it is helpful in understanding the method of testing hypotheses to determine decision rules and power functions, it is in practice unnecessary to do so if all one wants to do is decide whether the experimental evidence leads to acceptance or rejection of the null hypothesis.

The result of the poll of 20 voters is the occurrence of the event $X = x$, where x is an integer from 0 (none in favor of Smith) to 20 (all in favor of Smith). The larger the value of X, the more unfavorable is the result of the poll to the null hypothesis that $p \leq .60$. The number $P(X \geq x)$, calculated for the borderline value $p = .60$, is the probability of getting a value of X *at least as unfavorable to the null hypothesis as the one actually observed* and is called the *statistical significance* or the *descriptive level of significance* of the observed event $X = x$.

According to (2.3) and (2.4), if the descriptive level of significance of $X = x$ is less than or equal to $\alpha = .02$, then the null hypothesis is rejected, since x must then be greater than or equal to 17; if the descriptive level of significance of $X = x$ is greater than $\alpha = .02$, then the null hypothesis is accepted, since x is then less than 17. A value of X that leads to rejection of the null hypothesis is said to be *significant at the level* $\alpha = .02$ (or at the 2% level of significance); a value of X that leads to acceptance of the null hypothesis is not significant at the level $\alpha = .02$. *Testing the significance of the observed value of X at the level* α (i.e., computing $P(X \geq x)$ for $p = .60$ and comparing it with α) is therefore a way of determining the action to be taken *without* first finding the decision rule and its power function. For this reason, tests of statistical hypotheses are often called *tests of significance.*

To illustrate these ideas, suppose the committee has decided to sample $n = 20$ people and has set $\alpha = .02$ as the maximum tolerable probability of an error of the first kind. The 20 people are polled and the event $X = 16$ occurs; i.e., 16 people are in favor of Smith. With $p = .60$, we find from Table 37 that

$$P(X \geq 16) = .051.$$

Since $.051 > .02$, the observed event $X = 16$ is not significant at the 2% level of significance and the committee therefore accepts the null hypothesis. Note that if the committee had set $\alpha = .06$, say, then this same value of X would be significant at the 6% level of significance (since $.051 < .06$) and would therefore lead to rejection of the null hypothesis. By increasing α, the committee increases its chances of rejecting the null hypothesis. It also, of course, increases the chances of making an error of the first kind.*

We conclude by reminding the reader that we have discussed only one particular problem and that null hypotheses, decision rules, and tests of significance will generally assume different forms in different problems. Nevertheless, an *understanding* of this section should enable the reader to solve a variety of problems of the sort treated here where the binomial distribution applies. *This* hypothesis can be tested by trying the problems that follow.

PROBLEMS

2.1. Consider the decision problem discussed in the text, and suppose the committee chooses to base its decision on a sample of $n = 20$ people. As in the text, let α denote the maximum tolerable probability of an error of the first kind.

---

* The obvious fact that errors *can* be made when using tests of significance means that these tests are fraught with danger and must be interpreted with great caution. For a particularly impressive discussion of this point, with special reference to the field of psychology, see T. D. Sterling, "Publication Decisions and Their Possible Effects on Inferences Drawn From Tests of Significance—or Vice Versa," *Journal of the American Statistical Association*, vol. 54 (1959), pp. 30–34.

(a) What decision rule is determined if $\alpha = 0$? What then is the probability of an error of the second kind for all $p > .60$? Draw the graph of the power function for this rule.

(b) What decision rule is determined if the committee insists that the probability of an error of the second kind must be zero for all $p > .60$? What then is the value of α? Draw the graph of the power function of this decision rule.

(c) What decision rule is determined if $\alpha = .10$? For this decision rule, what is the probability of an error of the second kind if $p = .70$? if $p = .80$? What is the probability of an error of the first kind if $p = .50$?

(d) The committee has decided to use $\alpha = .10$ as in part (c) and finds that 75% of the people in the sample are in favor of Smith. What is the descriptive level of significance of this observed event? Does the committee wage a very expensive or a less expensive campaign?

(e) The committee decides to use $\alpha = .01$. How many people in the sample of 20 must be in favor of Smith before the observed event is significant at the level .01?

2.2. Suppose the committee decides on a sample of $n = 10$ people and sets $\alpha = .05$. Determine the decision rule that should be used and draw the graph of its power function.

2.3. In a study of the effects of stress,* 20 college students were taught to tie a bowline knot by two different methods. Half the subjects learned method A first and the other half learned method B first. Later—after an active day and an evening final examination—each subject was asked to tie the knot. The prediction was that stress would induce regression, i.e., the subjects would tend to revert to the first-learned method of tying the knot. Each subject was classified as a success (used knot-tying method he learned first) or a failure (used method he learned last). Assume that the experiment can be idealized as a set of 20 Bernoulli trials with (unknown) probability p for success on each trial.

Suppose the null hypothesis expresses the fact that there is *no* regression and that under stress it is equally likely to use either of the two methods of tying the knot. (It is cases of this kind that explain the use of the word "null": the null hypothesis asserts that stress has *no* effect.) The alternate hypothesis will then state that regression *does* occur; i.e., it is more probable that under stress the first-learned method is used than the second-learned method.

(a) Formulate these hypotheses in terms of p and determine the decision

---

* Barthol, R. P. and N. D. Ku, "Regression Under Stress to First Learned Behavior," *Journal of Abnormal and Social Psychology*, vol. 59 (1959), pp. 134–36.

rule if $\alpha = .05$; i.e., if an error of the first kind has probability at most .05.

(b) Suppose 15 of the 20 subjects use the first-learned method of tying the knot. What is the descriptive level of significance of this observed outcome? Is it significant at the 5% level? Is it significant at the 1% level?

2.4. Determine a decision rule to test the null hypothesis $p = .20$ against the alternate hypothesis $p = .60$, assuming a sample of size $n = 10$ and a maximum tolerable probability for an error of the first kind equal to .05. What is the actual probability of an error of the first kind for your test? What is the probability that you will incorrectly accept the null hypothesis when $p = .60$?

2.5. The production manager of a company submits a report recommending hiring of additional repairmen. His conclusions are based on the assumption that, on the average, 20% of the machines in the shop will require maintenance on any given day; i.e., the probability is .20 that a machine observed for a period of a day (a machine-day) will need the services of a repairman. The president of the company is interested in testing this assumption, since the conclusions of the report will be different if the assumed 20% is either too high or too low. Suppose (unrealistically, but in order to be able to use Table 37) that only 20 machine-days are observed and the president is willing to take at most a 10% risk of rejecting the assumption if it is true ($\alpha = .10$).

(a) Formulate a null and alternate hypothesis, explaining how the binomial distribution applies (i.e., define trial, success, failure, etc.)

(b) Determine a reasonable decision rule for testing the null hypothesis.

(c) Of the 20 machine-days observed, seven required services of a repairman. What is the descriptive level of significance of this event? Is it significant at the .10 level?

(d) Draw the graph of the power function of the decision rule in (b) and on the same figure also draw the corresponding graph for an *ideal* decision rule for which the probabilities of both kinds of errors are zero.

2.6. A hair tonic manufacturer claims that his product will cure baldness at least 70% of the time it is used according to instructions. Formulate null and alternate hypotheses to test this claim. Determine a decision rule assuming the maximum tolerable probability of an error of the first kind is $\alpha = .05$. Use a sample of size $n = 20$.

3. An example of decision-making under uncertainty

Analyses of the type discussed in the preceding section can be carried still further and made more realistic if we assign relative values to the losses that will arise when various kinds of errors are made. We should also take note of the fact that sampling involves certain expenses and that in some practical situations larger samples may cost more to obtain than the consequent reduction in probabilities of errors is worth. In short, statistical investigations are undertaken as a basis for action, and decisions should therefore be made in the light of all their relevant consequences.

We shall illustrate this approach by discussing a particularly simple problem that can be solved with the mathematical skills we have now accumulated.*

Before each production run, a machine used to produce a certain part must be adjusted by an operator. Five hundred parts are produced in each such run, and each part is classified as either good or defective. On the basis of his experience with the machine, the manufacturer is willing to assume that the production of the 500 parts can be thought of as a Bernoulli process in which the probability of a defective, denoted by p and called the average fraction defective, is the same on each of the 500 trials.

Two delicate adjustments must both be made perfectly by the operator before each run in order to have $p = .01$, which is the very best that the machine can do, because of mechanical limitations. But if only one of these adjustments is properly made, then the average fraction defective becomes $p = .10$, and if the operator happens to make neither adjustment properly then $p = .20$. We therefore have three possible "states" for the machine. On the basis of records of past production runs made by the operator, the manufacturer estimates that the operator will have both adjustments right 80% of the time, miss exactly one adjustment 15% of the time, and miss

---

* This problem, with changes in numerical values, is one treated in great detail by somewhat different methods in R. Schlaifer, *Probability and Statistics for Business Decisions*, McGraw-Hill Book Company, Inc., 1959, especially Chapters 22 and 33. I here express my appreciation to Professor Schlaifer for permission to use this material. I am also indebted to Professor Howard Raiffa for introducing me to this kind of decision problem and for the particular method of solution used in the text.

both adjustments 5% of the time. These data are summarized in Table 42.

TABLE 42

State of Machine	Probability of a Defective Part Given This State	Probability of This State
I (Both adjustments right)	.01	.80
II (Only one adjustment right)	.10	.15
III (Neither adjustment right)	.20	.05

Each of the 500 parts produced by the machine, whether good or defective, is used in assembling the final product, but *a defective part requires special hand fitting which costs $5.00 per part*. This means that a faulty machine setup (i.e., machine in states II or III) can lead to a fairly high cost of using defective parts.

However, the manufacturer can reduce these costs by calling in a master mechanic before the production run. If this is done, the machine is certain to be properly adjusted and therefore will be in state I, where the average fraction defective is equal to its minimum value $p = .01$. *This special use of the master mechanic, however, costs $50.* Thus, if the regular operator has put the machine in state I (as he does most often), then this $50 would be a total loss. On the other hand, if the operator has missed one or both of the adjustments, then the saving in cost of using defective parts more than offsets the extra cost of hiring the master mechanic.

Finally, the manufacturer considers that he may be able to reduce his average costs by inspecting a sample of the product after the operator prepares the machine, but before beginning the actual production run. He might, for example, make a sample run of ten parts and then inspect each part. When the number of defectives among the ten parts in the sample is "high," he would call the master mechanic; when it is "low," he would order the regular run of 500 parts to be made without readjustment. *But sample production and inspection cost $2.00 per part.*

There are two possible decisions that the manufacturer can make:

1. Order the production run to proceed after the machine is prepared by the regular operator without calling the master mechanic; we shall call this a decision to *proceed*.

2. Call the master mechanic and have him readjust the machine so that its average fraction defective is certain to be $p = .01$; we shall call this a decision to *readjust*.

The manufacturer, whose aim is to make the average cost of the entire production run as low as possible, asks the following questions:

1. Should the decision to proceed or readjust be made without a sample run or on the basis of evidence accumulated (at a price due to inspection costs) in a sample run?

2. If a sample run is *not* indicated, then which of the two decisions should be made?

3. If a sample run *is* indicated, then how large a sample should be taken; i.e., how many parts should be made by the machine? (We assume that all of these parts will be inspected.) And what *decision rule* should then be adopted; i.e., for each possible outcome (as measured by the number of defective parts discovered in the sample), which of the two decisions should be made?

By answering these questions, we give the manufacturer a rule for action in the face of uncertainty (since the actual state of the machine is unknown). Moreover, this rule will be optimum in that it minimizes the average cost of the entire production run.

We first investigate the costs involved in each decision, assuming no sample run is made. If the decision is to proceed and we suppose for the moment that the state of the machine is given (i.e., p is known), then the mean number of defectives produced in the run is $500p$, and so the mean cost of defectives is $500p \times \$5.00 = 2500p$ dollars. We compute this mean cost for each possible state of the machine in the third column of Table 43. Similarly, if the decision is to readjust, then the master mechanic adjusts the machine (makes $p = .01$) and the mean number of defectives produced in the run is

TABLE 43

State of Machine	Probability of a Defective Part Given This State p	Mean Cost of Defectives Given This State When Decision Is to		Loss Due to Proceeding Rather Than Taking Better Decision if State Were Known	Loss Due to Readjusting Rather Than Taking Better Decision if State Were Known	Probability of This State
		Proceed	Readjust			
I	.01	$ 25*	$75	$ 0	$50	.80
II	.10	250	75*	175	0	.15
III	.20	500	75*	425	0	.05

$500p = 5$ parts. Hence, no matter what state the machine was put in by the regular operator, the mean cost of defectives when the decision is to readjust is $5 \times \$5.00 = \$25.00$, plus the $50.00 cost of the master mechanic, or a total of $75.00. This cost is listed in the fourth column of Table 43.

Note that if the machine is in state I, then the better decision (i.e., the one with the lower mean cost) is the decision to proceed. But if the machine is in state II or III, then the better decision is to readjust. The asterisks in Table 43 indicate the mean costs of the better decision if the state of the machine were known; i.e., if *perfect* information concerning the quality of the adjustments made by the regular operator were available to the manufacturer. But such perfect information is *not* available. Thus, if the machine is known to be in state I, then the better decision is to proceed and this action involves a mean cost of $25. Since this is the better decision for state I, the loss due to proceeding in the absence of perfect information happens to be zero; but the loss due to readjusting rather than taking the better decision is the mean cost of readjusting ($75) minus the mean cost of the better decision ($25), and hence is $50. Similarly, if the machine is in state II, then the better decision is to readjust. Hence, the loss due to proceeding rather than taking this better decision is the mean cost of proceeding ($250) minus the mean cost of the better decision ($75) or $175. In this way, we compute the losses given in the fifth and sixth columns of Table 43.

The loss due to proceeding is therefore $0, $175, or $425 with probability .80, .15, and .05 respectively, these being the given probabilities of the three states of the machine. Hence we find:

$$\text{Mean loss due to proceeding} = 0(.80) + 175(.15) + 425(.05)$$
$$= \$47.50.$$

Similarly, we find:

$$\text{Mean loss due to readjusting} = 50(.80) + 0(.15) + 0(.05)$$
$$= \$40.00.$$

We conclude that if no sample run is ordered, then the mean loss due to the decision to proceed is $7.50 more than the mean loss due to the decision to readjust. The manufacturer should therefore decide to readjust; i.e., the master mechanic should be called in before the production run, thus making average costs $7.50 less per run than with the alternate decision to proceed.

Although the decision to readjust is the better decision, we have computed that the mean cost due to readjusting is $40.00. Hence $40.00 is the *mean cost of uncertainty* in this problem, and the manufacturer could therefore afford to pay any price up to $40.00 for the certain knowledge (never available in practice) of the value of p. In other words, his costs would average $40.00 less per run if he had information that allowed him to make the better decision for each state, i.e., if he knew the true value of p and took the decision to readjust *only* when $p = .10$ or $p = .20$ (when it paid to readjust).

The state of uncertainty is somewhat reduced by evidence accumulated in a sample run. But such evidence costs $2.00 for each part produced by the machine and inspected. We turn now to the problem of determining whether the mean loss of $40.00 just computed can be lowered by ordering a sample run before making a decision to proceed or readjust.

The decision rules we allow are of the following form:

> Order a sample run of n parts. Let X be the random variable whose value is the number of defectives among the n parts in the sample. Make the decision to readjust if $X \geq c$, where c is some specified number. Make the decision to proceed if $X < c$.

Each choice of n and c determines one such rule, which we call the (n, c) decision rule. For example, the $(5, 1)$ decision rule requires that a sample of $n = 5$ parts be produced; the decision to readjust is then made if and only if the number of defectives turns out to be at least 1.

We now demonstrate how to compute the mean loss from the use of such a decision procedure. We shall for the moment concentrate on explaining the construction of Table 44 which concerns the $(5, 1)$ decision rule. *A similar analysis applies to any (n, c) rule, no matter what the values of n and c.*

TABLE 44

(1)	(2)	(3)	(4)	(5)	(6)	(7)	(8)
State of Machine	Probability of a Defective Part Given This State	Probability of Decision to Readjust Using (5, 1) Rule	Probability of Decision to Proceed Using (5, 1) Rule	Probability of Wrong Decision	Loss Due to Wrong Decision	Mean Loss Due to Wrong Decision	Mean Total Loss Due to Wrong Decision
I	.01	.049	.951	.049	$ 50	$ 2.45	$ 12.45
II	.10	.410	.590	.590	175	103.25	113.25
III	.20	.672	.328	.328	425	139.40	149.40

Mean loss due to use of (5, 1) decision rule = 12.45(.80) + 113.25(.15) + 149.40(.05)
= $34.42.

Columns (1) and (2) of Table 44 are clear. Column (3) is obtained directly from the cumulative binomial probabilities in Table 37. Under the (5, 1) rule, the decision to readjust is made when X, the number of successes (defectives) in the $n = 5$ Bernoulli trials making up the sample run, is at least 1. If $p = .01$, we find $P(X \geq 1) = .049$. Similarly, we read directly from the binomial tables that $P(X \geq 1)$ equals .410 if $p = .10$ and equals .672 if $p = .20$. Thus column (3) is completed. The probability that the (5, 1) rule will lead to a decision to proceed is 1 minus the probability that it will lead to a decision to readjust. Hence, the entries in column (4) in Table 44 are obtained directly from those in column (3).

The probability of a wrong decision, entered in column (5), is merely the probability that the decision rule leads to readjustment if $p = .01$ (when the better decision is known to be to proceed) and to proceeding if $p = .10$ or $p = .20$ (when the better decision is known to be to readjust).

The loss due to a wrong decision has been computed in Table 43, and so column (6) of Table 44 is easily completed.

The entries in column (5) are multiplied by the entries in column (6) to give the mean loss due to a wrong decision; i.e., this mean loss (for given p) is the product of the loss and the probability with which it is sustained.

Finally, to the mean loss entered in column (7) we add the cost of the sample, which is \$2 for each of the five parts sampled. The entries in column (8) are therefore merely \$10 more than those in column (7).

Since we are given (in Table 42) the probabilities of the three possible states, we can compute the overall mean loss due to the use of the (5, 1) decision rule. This we do in the lower part of Table 44. Since this mean loss is \$34.42 and is therefore *less* than the mean loss of \$40.00 due to a decision to readjust without a sample run, we see immediately that *it does pay to order a sample run*. The only remaining question is therefore the choice of the best possible decision rule.

To find the best decision rule, we must find the best pair of values of n, the sample size, and c, the smallest number of defectives leading to a decision to readjust. (Of course, "best" is interpreted as lowest mean loss for the production run of 500 parts.) We proceed by first finding the best value of c, given the sample size n; we then compare different values of n when each is used with the value of c that is best for it. From this point on, we only state results. The reader can verify each of our statements by carrying out computations similar to those used in constructing Table 44. (We omit problems at the end of this section, since there is ample opportunity to test one's understanding by checking our results.)

By keeping the sample size fixed at $n = 5$ and varying c, we find the following mean losses due to use of $(5, c)$ decision rules:

Decision Rule	Mean Loss
(5, 1)	$34.42
(5, 2)	49.82
(5, 3)	56.03

It is clear that for samples of size $n = 5$, the best value of c is $c = 1$. In fact, similar computations show that for each of the sample sizes $n = 4, 5, 6, 7, 8, 9$ the best value of c is $c = 1$. (This is a peculiarity of our particular problem and is not generally true.) We thus obtain the following mean losses for decision rules with various sample sizes, each computed with the value $c = 1$ that is best for it.

Decision Rule	Mean Loss
(4, 1)	$35.49
(5, 1)	34.42
(6, 1)	33.87
(7, 1)	33.73
(8, 1)	33.94
(9, 1)	34.45

We note that rule $(7, 1)$ has the lowest mean loss. It is therefore the decision rule preferred by the manufacturer. He orders a run of $n = 7$ sample parts. If at least one of these seven parts is defective, he spends the $50 required to have the master mechanic readjust the machine. If no defectives are found among the seven parts, then he orders the run to proceed without readjustment. His mean cost of uncertainty is thereby reduced to $33.73 from the $40.00 cost resulting from his best decision (namely, to always call the master mechanic) in the absence of a sample run.

SUPPLEMENTARY READING

The binomial distribution is discussed to a greater or lesser extent in the probability and statistics books included in reading lists at the end of preceding chapters. Some idea of statistical applications will be obtained by also consulting the following books, in addition to the references mentioned in footnotes.

1. Ackoff, R. L., *The Design of Social Research*, University of Chicago Press, 1953.

2. Bross, I. D. J., *Design for Decision*, The Macmillan Company, 1953.

3. Cowden, D. J., *Statistical Methods in Quality Control*, Prentice-Hall, Inc., 1957.

4. Dodge, H. F. and H. G. Romig, *Sampling Inspection Tables*, 2nd edition, John Wiley and Sons, Inc., 1959.

5. Mosteller, F., "Applications," pp. xxxiv–lxi, in *Tables of the Cumulative Binomial Probability Distribution*, Annals of the Computation Laboratory of Harvard University, vol. XXXV, Harvard University Press, 1955.

6. Sprowls, R. C., *Elementary Statistics for Students of Social Science and Business*, McGraw-Hill Book Company, Inc., 1955.

7. Wallis, W. A., and H. V. Roberts, *Statistics, A New Approach*, The Free Press, 1956.

Note. Now that you are at the end of this book, you can review some of the things you have learned and prepare the way for continued study of probability by reading the following articles.

Curtiss, J. H., "Elements of a Mathematical Theory of Probability," *Mathematics Magazine*, vol. 26 (1953), 233–254.

Halmos, P. R., "The Foundations of Probability," *American Mathematical Monthly*, vol. 51 (1944), 493–510.

Robbins, H., "The Theory of Probability," Chap. XI in *Insights Into Modern Mathematics*, Twenty-third Yearbook, National Council of Teachers of Mathematics, Inc., 1957.

ANSWERS TO ODD-NUMBERED PROBLEMS

Chapter 1

1.1. (a) Finite, one element; (c) Infinite; (e) Finite, four elements, $\{1 \to 2 \to 4 \to 3 \to 2, \ 1 \to 2 \to 3 \to 4 \to 2, \ 2 \to 4 \to 3 \to 2 \to 1, \ 2 \to 3 \to 4 \to 2 \to 1\}$; (g) Finite, two elements, $\{2, 1\}$.

1.3. Consider numbers of the form $n^2 + x(n - 1)(n - 2)(n - 3)$, and find x such that this number is 94 when $n = 4$. $x = 13$.

1.5. (a) $\{(2, 3)\}$, the point of intersection of the two lines;
(b) $\emptyset$, for the two lines, being parallel, have no points in common;
(c) $\{(x, y) | x + y = 5\}$, the set of all points on the graph of the equation $x + y = 5$, for the two equations define the same line.

1.7. (a) $A = B$; (c) $A = B$; (e) $A = \{\tfrac{1}{2}, 2\} \neq B = \{0, \tfrac{1}{2}, 2\}$.

2.1. (a) The same number appears on each die; (c) The sum of the numbers is 4.

2.3. $B = \emptyset$.

2.5. (a) Correct; (b) Incorrect, for the only subsets of $\{\{1\}\}$ are $\emptyset$ and $\{\{1\}\}$; (c) Correct, for the elements of $\{1, \{1\}\}$ are 1 and $\{1\}$; (d) Correct.

2.7. (a) $\{(0, 2), (0, -2)\}$; (c) $\emptyset$; (e) Upper semicircle, including the points $(-2, 0)$ and $(2, 0)$.

2.11. $5 \cdot 4 = 20$.

2.13. $8 \cdot 8 \cdot 9 \cdot 10^4 = 5{,}760{,}000$.

2.15. (a) 6; (b) 9; (c) 3; (d) 6.

2.17. Assuming the coins distinguishable: 2^3, 2^4, 2^n ways.

2.19. (a) 169; (b) 338; (c) 169.

2.21. (a) 2^3, 3^3, n^3; (b) 2^r, 3^r, n^r.

3.1. $A' = \{b, c\}$, $B' = \{a, c\}$, $A \cup B = \{a, b\}$, $A \cap B = \emptyset$, $A' \cap B' = \{c\}$, $A' \cap (A \cup B) = \{b\}$.

3.3. (a) $\mathcal{U} = \{\text{HHH, HHT, HTH, THH, HTT, THT, TTH, TTT}\}$, where an element represents the outcome for the penny first, the nickel, and then the dime;
(b) $A' = \{\text{THH, THT, TTH, TTT}\}$,
$A \cup B = \{\text{HHH, HHT, HTT, HTH, TTT}\}$,
$A \cap C = \{\text{HHH, HHT, HTH}\}$,
$A' \cap C = \{\text{THH}\}$, $(A \cap B) \cap C = \{\text{HHH}\}$.

3.5. (a) (i) 54, (iii) 3;   (b) (1) $Y \cap C$, (3) $(Y \cup N)'$ or $Y' \cap N'$.

3.7. (a) $n(\mathcal{U}) = n(A) + n(A')$; (b) Let $B = A'$ and note that then $n(A \cap B)$ becomes $n(\emptyset) = 0$.

3.9. 4.

3.11. $\emptyset \subseteq (A \cap B) \cap C \subseteq B \cap A = A \cap B \subseteq B \subseteq A \cup B$
$\subseteq (A \cup B) \cup C = A \cup (B \cup C) \subseteq \mathcal{U} = \emptyset'$.

3.13. (a) $P \cup B = B$ or $P \cap B = P$ or $P \cap B' = \emptyset$;
(c) $(M \cap C) \cap W = \emptyset$;
(e) $P \cap (B \cap M) \neq \emptyset$;   (g) $(P \cap B) \cap (I' \cap W') \neq \emptyset$;
(i) $B \cup I = I$ or $B \cap I = B$ or $B \cap I' = \emptyset$;   (k) $B = I$.

4.1. (1a)

A	$\emptyset$	$A \cup \emptyset$
$\in$	$\notin$	$\in$
$\notin$	$\notin$	$\notin$

(4a)

A	A'	$\mathcal{U}$	$A \cup A'$
$\in$	$\notin$	$\in$	$\in$
$\notin$	$\in$	$\in$	$\in$

(9a)

A	B	C	$B \cap C$	$A \cup (B \cap C)$	$A \cup B$	$A \cup C$	$(A \cup B) \cap (A \cup C)$
$\in$	$\in$	$\in$	$\in$	$\in$	$\in$	$\in$	$\in$
$\in$	$\in$	$\notin$	$\notin$	$\in$	$\in$	$\in$	$\in$
$\in$	$\notin$	$\in$	$\notin$	$\in$	$\in$	$\in$	$\in$
$\in$	$\notin$	$\notin$	$\notin$	$\in$	$\in$	$\in$	$\in$
$\notin$	$\in$	$\in$	$\in$	$\in$	$\in$	$\in$	$\in$
$\notin$	$\in$	$\notin$	$\notin$	$\notin$	$\in$	$\notin$	$\notin$
$\notin$	$\notin$	$\in$	$\notin$	$\notin$	$\notin$	$\in$	$\notin$
$\notin$	$\notin$	$\notin$	$\notin$	$\notin$	$\notin$	$\notin$	$\notin$

Referring to Figure 10 of the text, we see that both sides are represented by R_1 & R_2 & R_3 & R_4 & R_5.

4.3. (a)

A	B	A'	B'	$A' \cap B'$	$(A' \cap B')'$	$A \cup B$
∈	∈	∉	∉	∉	∈	∈
∈	∉	∉	∈	∉	∈	∈
∉	∈	∈	∉	∉	∈	∈
∉	∉	∈	∈	∈	∉	∉

(e)

A	B	C	A'	B'	C'	$B \cap C$	$(A' \cap (B \cap C))$	$(A' \cap (B \cap C))'$	$A \cup B'$	$A \cup B' \cup C'$
∈	∈	∈	∉	∉	∉	∈	∉	∈	∈	∈
∈	∈	∉	∉	∉	∈	∉	∉	∈	∈	∈
∈	∉	∈	∉	∈	∉	∉	∉	∈	∈	∈
∈	∉	∉	∉	∈	∈	∉	∉	∈	∈	∈
∉	∈	∈	∈	∉	∉	∈	∈	∉	∉	∉
∉	∈	∉	∈	∉	∈	∉	∉	∈	∉	∈
∉	∉	∈	∈	∈	∉	∉	∉	∈	∈	∈
∉	∉	∉	∈	∈	∈	∉	∉	∈	∈	∈

4.5. (a) In Figure 9 of the text, both sides are represented by R_1 & R_2 & R_3;
(b) In Figure 9 of the text, both sides are represented by R_1 & R_2 & R_4;
(e) In Figure 10 of the text, both sides are represented by R_1 & R_2 & R_3 & R_4 & R_6 & R_7 & R_8.

4.7. (a) If (1) $C \cup B = B$ and (2) $B \cup W = W$, then $C \cup W = W$.
Proof: $C \cup W = C \cup (B \cup W) = (C \cup B) \cup W = B \cup W = W$,
by using (2), law 8a, (1) and (2), respectively.

4.9. $(A \cap B) \cap (C \cap D) = [(A \cap B) \cap C] \cap D$, by law 8b,
$$= [A \cap (B \cap C)] \cap D, \text{ by law 8b.}$$

4.11. (a) $\mathfrak{U}, \emptyset$;   (b) $\emptyset, \mathfrak{U}$;   (c) $A, A', \mathfrak{U}, \emptyset$;   (d) same as (c).

5.1. (a) $(1, 1)$, $(1, 2)$, $(2, 1)$, $(2, 2)$;   (c) $(1, 2)$, $(1, 3)$, $(2, 2)$, $(2, 3)$;
(e) $(2, 3)$;   (g) $(1, 2)$, $(1, 3)$, $(2, 2)$, $(2, 3)$.

5.3. (a) If $A = B$, then $A \times B = A \times A = B \times A$. The converse is false,
but if $A \times B \neq \emptyset$, i.e., neither A nor B is the null set, then the converse is true;
(c) $A \times B \subseteq C \times D$ if $A \subseteq C$ and $B \subseteq D$. *Proof:* We consider two cases: (1) If $A \times B = \emptyset$, then clearly $A \times B = \emptyset \subseteq C \times D$. (2) If $A \times B \neq \emptyset$, then let (a, b) be any element of $A \times B$. Since $A \subseteq C$ and $B \subseteq D$, we have $a \in C$ and $b \in D$. Thus $(a. b) \in C \times D$. We con-

clude that $A \times B \subseteq C \times D$. The converse is false, but if either $A = \emptyset$ and $B = \emptyset$, or $A \neq \emptyset$ and $B \neq \emptyset$, then the converse is true.

5.5. $(A \times \mathfrak{U}) \cap (\mathfrak{U} \times B) = (A \times \mathfrak{U}) \cap ((A \cup A') \times B)$

$\qquad = (A \times \mathfrak{U}) \cap ((A \times B) \cup (A' \times B))$, by Problem 5.4,

$\qquad = ((A \times \mathfrak{U}) \cap (A \times B)) \cup ((A \times \mathfrak{U}) \cap (A' \times B))$,

$\qquad\qquad\qquad\qquad\qquad\qquad$ by 9b of Theorem 4.1,

$\qquad = (A \times B) \cup \emptyset = A \times B$,

since $(A \times \mathfrak{U}) \cap (A \times B) = (A \times B)$ and $(A \times \mathfrak{U}) \cap (A' \times B) = \emptyset$.

5.7. (a) $a = d$, $b = e$, and $c = f$ implies that $(a, b, c) = (d, e, f)$.

Conversely, $(a, b, c) = (d, e, f)$ implies $((a, b), c) = ((d, e), f)$, which implies $(a, b) = (d, e)$ and $c = f$, which in turn implies that $a = d$ and $b = e$, proving the assertion.

(b) If the corresponding objects of the r-tuples are equal, the equality follows immediately. We now show that, conversely,

$(a_1, a_2, \cdots, a_r) = (b_1, b_2, \cdots, b_r)$ implies $a_1 = b_1$, $a_2 = b_2$, $\cdots$, $a_r = b_r$, for any integer $r > 1$. We know the result is true for $r = 2$ and $r = 3$ by part (a). Assume the result is true for the integer k, where $k > 1$. Using the definition of an ordered $(k + 1)$-tuple,

$$(a_1, a_2, \cdots, a_k, a_{k+1}) = (b_1, b_2, \cdots, b_k, b_{k+1}) \text{ means}$$
$$((a_1, a_2, \cdots, a_k), a_{k+1}) = ((b_1, b_2, \cdots, b_k), b_{k+1}).$$

This equality of ordered pairs implies that $a_{k+1} = b_{k+1}$ and $(a_1, a_2, \cdots, a_k) = (b_1, b_2, \cdots, b_k)$. But by the induction hypothesis, it follows that $a_1 = b_1$, $\cdots$, $a_k = b_k$, which establishes the result for $r = k + 1$. Hence we have shown that if the statement is true for $r = k$, where k is any integer greater than 1, then it is also true for $r = k + 1$. This, together with the result that it is true for $r = 2$, completes the proof for all integers $r > 1$.

Chapter 2

1.1. (a) The set D in (1.3);

(c) $S = \{PN, PD, PQ, NP, ND, NQ, DP, DN, DQ, QP, QN, QD\}$;

(e) $S = \{(0, 2), (1, 1), (2, 0)\}$, where $(0, 2)$, for example, represents the outcome of zero objects in cell 1 and two objects in cell 2;

(g) $S = \{FFF, FFM, FMF, FMM, MFF, MFM, MMF, MMM\}$;

(i) $S = \{0, 1, 2, \cdots, r\}$, the set of possible numbers of heads, or with more detail,

$$S = \{(x_1, \cdots, x_r) \mid x_i \in A, \ i = 1, 2, \cdots, r\},$$

where $A = \{H, T\}$.

1.3. 4.

1.5. All are suitable except (b) and (e).

1.7. $S = \{1b, 1w, 2b, 2w\}$ or $S = \{1b_1, 1w_1, 1w_2, 2b_2, 2b_3, 2w_3\}$.

2.1. (a) $E = \{A_s, K_s, \cdots, 2_s\}$;   (c) $E = \{A_s\}$.

2.3. (a) Let $A = \{1, 2, \cdots, 365\}$. Then

$E = \{3\} \times A \times A \times \cdots \times A \ (r - 1 \ A\text{'s})$,
$F = A \times \{28\} \times A \times \cdots \times A \ (r - 1 \ A\text{'s in all})$,
$E \cap F = \{3\} \times \{28\} \times A \times \cdots \times A \ (r - 2 \ A\text{'s})$;

(b) $n(E) = n(F) = 365^{r-1}$, $n(E \cap F) = 365^{r-2}$,   and $n(E \cup F) = 729(365)^{r-2}$ (cf. Example I.3.4).

2.5. The relations are readily seen from the following:
$S = \{(0, 2), (1, 1), (2, 0)\}$, $E = \{(0, 2)\}$, $F = \{(2, 0)\}$, $G = \{(0, 2)\}$.

3.1. (a) $\frac{1}{12}$;   (c) $\frac{1}{18}$.

3.3. (a) $S = \{$ABC, ABD, ABE, ABF, ACD, ACE, ACF, ADE, ADF, AEF, BCD, BCE, BCF, BDE, BDF, BEF, CDE, CDF, CEF, DEF$\}$, assign probability $\frac{1}{20}$ to each simple event;   (c) $\frac{1}{5}$;   (e) $\frac{1}{2}$.

3.5. (a) If $S = \{$HHH, HHT, HTH, THH, HTT, THT, TTH, TTT$\}$, and $\frac{1}{8}$ is assigned as the probability of each simple event, then $P(\text{exactly two tails}) = \frac{3}{8}$;

(c) If $S = \{(0, 2), (1, 1), (2, 0)\}$ and we assign $\frac{1}{3}$ as the probability of each of the three simple events, then $P(\text{one cell empty}) = \frac{2}{3}$. If we assign probabilities of $\frac{1}{4}$, $\frac{1}{2}$, and $\frac{1}{4}$ to the simple events $\{(0, 2)\}$, $\{(1, 1)\}$, and $\{(2, 0)\}$ respectively, then $P(\text{one cell empty}) = \frac{1}{2}$. The latter assignment is preferred;

(e) If $S = \{(x_1, x_2, \cdots, x_r) \mid x_i \ \epsilon \ A, i = 1, 2, \cdots, r\}$, where $A = \{$H, T$\}$, and we assign the same probability to each simple event of S, then $P(\text{all coins fall heads}) = (\frac{1}{2})^r$;

(g) If $S = \{$Sun., Mon., Tues., Wed., Thurs., Fri., Sat.$\}$ and we assign to each simple event the probability $\frac{1}{7}$, then $P(\text{13th day falls on Sunday}) = \frac{1}{7}$. But see *American Mathematical Monthly*, vol. 40 (1933), p. 607, for a demonstration that the 13th day is more likely to be Friday than any other day of the week.

3.7. $P(E \cap F) = (\frac{1}{365})^2$, $P(E \cup F) = 729/(365)^2$.

3.9. (a) 123, 132, 213, 231, 312, 321;

(c) $P(E_1) = P(E_2) = P(E_3) = \frac{1}{3}$, $P(E_1 \cup E_2) = \frac{1}{2}$, $P(E_1 \cap E_2) = \frac{1}{6}$, $P(E_1 \cap E_2 \cap E_3) = \frac{1}{6}$, $P(E_1 \cup E_2 \cup E_3) = \frac{2}{3}$.

4.1. $P(E) = \frac{11}{12}$, 11 to 1.

4.3. 5 to 4.

4.5. $\frac{1}{12} \leq P(F) \leq \frac{3}{4}$, the extreme values occurring when $E \cap F = \emptyset$ and $E \cap F = E$, respectively.

4.7. (a) $S = \{(x, y) \mid x \in D, y \in D, x \neq y\}$ where D is defined in Example 1.3., and we assign probability 1/2652 to each simple event of S; (b) $\frac{1}{26}$; (c) 25 to 1.

4.9. 0.8.

4.11. (a) $\frac{1}{4}$; (b) $\frac{1}{4}$; (c) If, for any $k = 1, 2, 3, \cdots$, an integer p is selected at random from among the first $2(10)^k$ positive integers, then the probability that p is divisible by either 6 or 8 is $\frac{1}{4}$.

4.13. (a) $P(E' \cup F') = 1 - P(E \cap F)$, the probability of not both E and F; (c) $P(E' \cup F) = 1 - P(E) + P(E \cap F)$, the probability of F or not E; (e) $P(E \cap F') = P(E) - P(E \cap F)$, the probability of E but not F.

4.15. If E_1 represents selecting a spade, E_2 an honor card, and E_3 a deuce, then
$P(E_1 \cup E_2 \cup E_3) = \frac{13}{52} + \frac{20}{52} + \frac{4}{52} - \frac{5}{52} - \frac{1}{52} - \frac{0}{52} + \frac{0}{52} = \frac{31}{52}$.

4.17. The theorem follows immediately from Formula (4.6), Definitions 4.2 and 3.3, and by noting that if E_1, E_2, and E_3 are mutually exclusive in pairs, then

$$E_1 \cap E_2 \cap E_3 = E_1 \cap (E_2 \cap E_3) = E_1 \cap \emptyset = \emptyset.$$

4.19. The theorem is true for $k = 2$ and $k = 3$. (Cf. Theorem 4.5 and Problem 4.17.) Assuming the theorem to be true for any k events (the induction hypothesis), we must show that it is true for $k + 1$ events. This, plus the fact that the theorem is true for $k = 2$, will complete the proof for all integers $k > 1$. Now

$$P(E_1 \cup E_2 \cup \cdots \cup E_k \cup E_{k+1}) = P((E_1 \cup E_2 \cup \cdots \cup E_k) \cup E_{k+1}).$$

But by Theorem I.4.2. and since $E_1, E_2, \cdots, E_{k+1}$ are mutually exclusive in pairs, it follows that

$$(E_1 \cup E_2 \cup \cdots \cup E_k) \cap E_{k+1} = \emptyset.$$

Hence, by Theorem 4.5 and the induction hypothesis,

$$P(E_1 \cup E_2 \cup \cdots \cup E_k \cup E_{k+1})$$
$$= P(E_1) + P(E_2) + \cdots + P(E_k) + P(E_{k+1}),$$

establishing the theorem for $k + 1$ events and thus completing the proof.

4.21. (a) $\dfrac{1}{1} - \dfrac{1}{2 \cdot 1} + \dfrac{1}{3 \cdot 2 \cdot 1} = \dfrac{2}{3}$;

(b) $\dfrac{1}{1} - \dfrac{1}{2 \cdot 1} + \dfrac{1}{3 \cdot 2 \cdot 1} - \dfrac{1}{4 \cdot 3 \cdot 2 \cdot 1} = \dfrac{5}{8}$;

(c) In general, the probability of at least one match is

$$\frac{1}{1!} - \frac{1}{2!} + \frac{1}{3!} - \cdots \pm \frac{1}{N!},$$

where $N!$ denotes the product of the first N positive integers.

5.1. (a) $\dfrac{\frac{1}{36}}{\frac{6}{36}} = \dfrac{1}{6}$;   (c) $\dfrac{\frac{3}{36}}{\frac{18}{36}} = \dfrac{1}{6}$.

5.3. $\dfrac{\frac{3}{27}}{\frac{21}{27}} = \dfrac{1}{7}$. (Our sample space contains ten elements, but the simple

events are *not* assigned equal probabilities.)

5.5. $\frac{246}{321} \cdot \frac{245}{320} = .59$, approximately.

5.7. (a) (i) $\frac{1}{2}$.   (ii) $\frac{1}{2}$;

(b) (i) $S = \{(x_1, \cdots, x_N) \mid x_i \in \{H, T\}, i = 1, 2, \cdots, N\}$; assign probability $(\frac{1}{2})^N$ to each simple event of S.

(ii) $2^{N-1}/2^N = \frac{1}{2}$.   (iii) $(\frac{1}{2})^N/(\frac{1}{2})^{N-1} = \frac{1}{2}$.

5.9. $\frac{1}{6}, \frac{2}{6}, \frac{3}{6}$.

5.11. $\frac{1}{3}$.

5.13. (a) .00359;   (b) $(1 - .00359)(.00380) = .00379$;

(c) $(1 - .00359)(1 - .00380)(.00396) = .00393$.

5.15. First derive the identity in Problem 5.16(f) and then use $P(E \cap F) > P(E)P(F)$, which follows from the given inequality.

5.17. $a, b, c, d,$ and e must satisfy the following equations:

$$a + b + c + d + e = 1, \quad a + b = \frac{1}{2}, \quad \frac{a}{a + b} = \frac{3}{8}, \quad \frac{c}{c + d + e} = \frac{2}{5},$$

$\dfrac{d}{c + d + e} = \dfrac{1}{5}.$ Solving these equations, we find the unique solution

$$a = \tfrac{3}{16}, \quad b = \tfrac{5}{16}, \quad c = \tfrac{1}{5}, \quad d = \tfrac{1}{10}, \quad \text{and} \quad e = \tfrac{1}{5}.$$

5.19. (a) Plan 1: $P(E) = \left(\dfrac{100 - x}{100}\right)\left(\dfrac{99 - x}{99}\right).$

Plan 2: $P(E) = 1 - \left(\dfrac{x}{100}\right)\left(\dfrac{x - 1}{99}\right).$

Plan 3: $P(E) = \left(\dfrac{100 - x}{100}\right)\left(\dfrac{99 - x}{99}\right) + 2\left(\dfrac{100 - x}{100}\right)\left(\dfrac{x}{99}\right)\left(\dfrac{99 - x}{98}\right).$

6.1. (a) Define $S = \{(x, y) \mid x \in C, y \in C, \text{ and } x \neq y\}$, where
$C = \{B_1, B_2, B_3, B_4, G_1, G_2\}$, the set of six children. Assign probability
$\frac{1}{30}$ to each simple event and note that there are ten elements in the
subset E for which the second child is a girl. Thus $P(E) = \frac{1}{3}$;
(b) $P(E) = (\frac{2}{3})(\frac{4}{5}) + (\frac{1}{3})(\frac{2}{5}) = \frac{1}{3}$.

6.3. The probabilities needed are: $P(E) = 0.254$, $P(E') = 0.746$,
$P(E_1|E) = \frac{150}{254}$, $P(E_2|E) = \frac{75}{254}$, $P(E_3|E) = \frac{25}{254}$, $P(E_4|E) = \frac{4}{254}$,
$P(E_1|E') = \frac{150}{746}$, $P(E_2|E') = \frac{175}{746}$, $P(E_3|E') = \frac{225}{746}$, $P(E_4|E') = \frac{196}{746}$.

6.5. $\frac{2}{3}$.

6.7. $\frac{28}{31} = .90$, approximately.

6.9. $\frac{3}{11}$.

6.11. $\frac{1}{2}$.

6.13. $E \cap E_i \subseteq E$, $i = 1, 2, \cdots, n$, by Definition I.3.1., demonstrating condition (i) of Definition 6.1. Also, use Theorem I.4.1. to show that for
$i \neq j$, $(E \cap E_i) \cap (E \cap E_j) = \emptyset$ follows from the hypothesis that
$E_i \cap E_j = \emptyset$. Thus condition (ii) holds. Finally, since $\{E_1, \cdots, E_n\}$ is
a partition of S, if x is any element of $E \subseteq S$, then there exists some
E_i such that $x \in E_i$. Then $x \in (E \cap E_i)$, which demonstrates condition (iii).

7.1. Independent events in (a) and (b), dependent in (c).

7.3. (a) $P(E)P(F) = (\frac{5}{16})(\frac{14}{16}) \neq P(E \cap F) = \frac{1}{4}$; (b) If $n = 3$, the events
are independent. (Cf. Example 7.3.) To prove the "only if" part, we
note that if $S = \{(x_1, \cdots, x_n) \mid x_i \in \{H, T\}, i = 1, 2, \cdots, n\}$ and we
assign probability $(\frac{1}{2})^n$ to each simple event of S, then

$$P(E) = \frac{n+1}{2^n}, \quad P(F) = \frac{2^n - 2}{2^n}, \quad \text{and} \quad P(E \cap F) = \frac{n}{2^n}.$$

If E and F are independent, then

$$\left(\frac{n+1}{2^n}\right)\left(\frac{2^n - 2}{2^n}\right) = \frac{n}{2^n},$$

which implies that $n + 1 = 2^{n-1}$ or $n = 3$.

7.5. (a) Let S be the set of 7460 females in the sample. $1/7460$;
(c) 0.143; (e) 0.014, approximately.

7.7. All independent.

7.9. Since $P(F) = 1$ and, by Theorem 4.2, $P(E \cup F) \geq P(F)$, it follows
that $P(E \cup F) = 1$. Now use Theorem 4.4 to show that

$$P(E)P(F) = P(E \cap F).$$

7.11. No. For counterexample, choose any event F with $P(F) = 1$ and let E and G be any dependent events. (Cf. Problem 7.9.)

7.13. (a) $P(S)P(A) = (\frac{1}{2})(\frac{1}{2}) \neq P(S \cap A) = \frac{7}{16}$; (b) 9.

8.1. $P(E_1)P(E_2) = (\frac{1}{6})(\frac{1}{6}) = P(E_1 \cap E_2),$
$P(E_1)P(E_3) = (\frac{1}{6})(\frac{1}{2}) = P(E_1 \cap E_3),$
$P(E_2)P(E_3) = (\frac{1}{6})(\frac{1}{2}) = P(E_2 \cap E_3),$
but $P(E_1)P(E_2)P(E_3) = \frac{1}{72} \neq P(E_1 \cap E_2 \cap E_3) = 0.$

8.3. $P((E_1 \cap E_2) \cap E_3) = P(E_1 \cap E_2 \cap E_3) = P(E_1)P(E_2)P(E_3)$, since we know Equation (8.3) holds. But then $P(E_1)P(E_2) = P(E_1 \cap E_2)$, and thus $P((E_1 \cap E_2) \cap E_3) = P(E_1 \cap E_2)P(E_3)$. That E_1 and E_3 are not necessarily independent may be seen by letting $E_2 = \emptyset$, and E_1 and E_3 be any dependent events.

8.5. Twice, with probability .46.

8.7. $P(E_1 \cap E_2) = P(E_1)P(E_2)$ by hypothesis. Consider the case $P(E_3) = 0$. Then since $0 \leq P(E_1 \cap E_3) \leq P(E_3) = 0$, it follows that
$$0 = P(E_1 \cap E_3) = P(E_1)P(E_3).$$
By an identical argument, $P(E_2 \cap E_3) = P(E_2)P(E_3)$. Similarly,
$$P(E_1 \cap E_2 \cap E_3) = P(E_1)P(E_2)P(E_3) = 0.$$
In the case $P(E_3) = 1$, since $P(E_3) \leq P(E_1 \cup E_3)$, it follows that $P(E_1 \cup E_3) = 1$. Then, by Theorem 4.4, $P(E_1 \cap E_3) = P(E_1)P(E_3)$. By the same argument $P(E_2 \cap E_3) = P(E_2)P(E_3)$. Also since $P(E_3) = 1$, it follows that $P(E_1 \cup E_2 \cup E_3) = 1$, and then, by using the result of Problem 4.14,
$$P(E_1 \cap E_2 \cap E_3) = P(E_1)P(E_2)P(E_3).$$

8.9. $\dfrac{1}{n+1}$. One needs to prove that if $E_1, E_2, \cdots, E_n$ are independent events, then $E_1', E_2', \cdots, E_n'$ are also independent.

8.11. .012.

9.1. (a) Sample space is $S \times S \times S$, where $S = \{\text{H, T}\}$, and $\frac{1}{8}$ is the probability assigned to each simple event of $S \times S \times S$;
(b) Using the same sample space as in (a), we assign the probability $p^k q^{3-k}$ to each simple event whose 3-tuple contains k H's and therefore $3 - k$ T's.

9.3. The sample space is $S \times S \times \cdots \times S$ (ten S's) where $S = \{\text{correct, incorrect}\}$. $(\frac{1}{6})^k(\frac{5}{6})^{10-k}$ is the probability assigned to each simple event whose 10-tuple contains exactly k "corrects."

$P(9 \text{ or } 10 \text{ correct}) = 51(\frac{1}{6})^{10} = .00000084$, approximately.

9.5. .784.

9.7. 7.

9.9. (a) $S_n \times S_{n-1}, \dfrac{1}{n(n-1)}$;   (b) B and D;   (d) $B, D, A,$ and Y.

10.1. (a) $u_1 = u_2 = \frac{1}{16}, 2v_1 = 2v_2 = \frac{6}{16}, w_1 = w_2 = \frac{9}{16}$;
   (c) $u_1 = u_2 = 1, 2v_1 = 2v_2 = w_1 = w_2 = 0$.

10.3. (a) $u_1 = \frac{1}{16}, 2v_1 = \frac{6}{16}, w_1 = \frac{9}{16}, u_2 = \frac{324}{5041}, 2v_2 = \frac{1908}{5041}, w_2 = \frac{2809}{5041},$
   $f_0 = \frac{3}{4}, f_1 = \frac{53}{71}, f_2 = 16{,}645/22{,}396$.

10.5. Substituting $f_n = 1/g_n$ in (10.12), we have

$$\frac{1}{g_{n+1}} = 1 - \frac{1}{1 + 1/g_n} = \frac{1}{g_n + 1}$$

and taking reciprocals we have (10.14).

Chapter 3

1.1. (a) 56;   (c) 126;   (e) 1,260.

1.3. (a) .13;   (c) .16.

1.5. (a) $\dbinom{n}{2}$;   (b) $\dbinom{n}{2} - n$.

1.7. 432,516.

1.9. (a) $\dbinom{n}{4}$;   (b) $3\dbinom{n}{3}$;   (c) $\dbinom{n}{2}$;

   (d) by solving $6\dbinom{n}{4} = \dbinom{n}{4} + 3\dbinom{n}{3} + \dbinom{n}{2}$, find $n = 6$.

1.11. (a) $\dfrac{(2)5!5!}{10!} = \dfrac{1}{126}$;   (b) Same as (a).

1.13. (a) $\frac{3}{19}$;   (b) $\frac{11}{19}$.

1.15. (a) .251;   (b) .215;   (c) .633.

1.17. $\dbinom{8}{2,\,3,\,3}\left(\dfrac{1}{6}\right)^2\left(\dfrac{1}{6}\right)^3\left(\dfrac{1}{6}\right)^3 = .00033$, approximately.

1.19. .19, approximately.

1.21. $p_0 = .16, p_1 = .31, p_2 = .29, p_3 = .16, p_4 = .06, p_5 = .02$, where p_n is the probability to two decimal places that sample contains exactly n defectives.

1.23. (a) $\dfrac{8\binom{7}{1,\,4,\,2}}{\binom{11}{1,\,4,\,4,\,2}} = \dfrac{4}{165}$;   (b) $\dfrac{6\binom{5}{2}}{\binom{9}{4,\,3,\,2}} = \dfrac{1}{21}$;   (c) $\dfrac{4\binom{3}{1}}{\binom{7}{4,\,2,\,1}} = \dfrac{4}{35}$.

1.25. We give the *number* of different poker hands of each kind. The required probability is this number divided by 2,598,960, the total number of poker hands. (a) 1,302,540;   (b) 123,552;   (c) 54,912;   (d) 10,200; (e) 5108;   (f) 3744;   (g) 624;   (h) 40.

1.27. (a) $S = S_1 \times S_1 \times S_1 \times S_1$, but the probabilities of simple events of S are *not* assigned according to the product rule. In general, knowledge of any hand changes the probability of any of the other hands having a certain makeup;

(b) $\dfrac{\binom{4}{1}\binom{48}{12}\binom{39}{13}\binom{26}{13}\binom{13}{13}}{\binom{52}{13}\binom{39}{13}\binom{26}{13}\binom{13}{13}} = \dfrac{\binom{4}{1}\binom{48}{12}}{\binom{52}{13}}$, this latter ratio being the

answer to (c);   (d) Refer to Formula II.9.8, using $P(E_1) = P(C_1)$.

1.29. (a) $\dfrac{4!\binom{13}{5}\binom{13}{4}\binom{13}{3}\binom{13}{1}}{\binom{52}{13}} = .13$, approximately;

(b) $\dfrac{\dfrac{4!}{2!}\binom{13}{4}\binom{13}{4}\binom{13}{3}\binom{13}{2}}{\binom{52}{13}} = .21$, approximately;

(c) $\dfrac{\dfrac{4!}{3!}\binom{13}{4}\binom{13}{3}^3}{\binom{52}{13}} = .11$, approximately.

1.31. From the preceding problem, the probability that the queen falls is
$$.407 + (\tfrac{1}{4})(.497) = .531.$$

Hence the odds are approximately 53 to 47, or 1.13 to 1.

2.1. (a) $p^5 + 5p^4q + 10p^3q^2 + 10p^2q^3 + 5pq^4 + q^5$;
(c) $a^4 - 12a^3b + 54a^2b^2 - 108ab^3 + 81b^4$.

2.3. (a) 1.072;   (b) 1.219;   (c) .922.

2.5. (a) $r\binom{n}{r} = \dfrac{rn!}{r!(n-r)!} = \dfrac{rn(n-1)!}{r(r-1)!(n-r)!} = n\binom{n-1}{r-1}$;

(c) $\binom{n-1}{r-1} + \binom{n-1}{r} = \dfrac{(n-1)!}{(r-1)!(n-r)!} + \dfrac{(n-1)!}{r!(n-r-1)!}$

$$= \dfrac{r(n-1)! + (n-r)(n-1)!}{r!(n-r)!} = \binom{n}{r}.$$

2.7. $\binom{n}{r+1} = \dfrac{n!}{(r+1)!(n-r-1)!} = \dfrac{n!(n-r)}{(r+1)r!(n-r)(n-r-1)!}$

$$= \dfrac{n-r}{r+1}\binom{n}{r}.$$

2.9. (a) $\binom{x}{r} = \dfrac{(x)_r}{r!} = \dfrac{x(x-1)(x-2)\cdots(x-r+1)}{r!}$. If x is an integer such that $0 \le x < r$, then a term of the numerator above is zero. Hence $\binom{x}{r} = 0$, as defined in Equation (2.10). If $x \ge r$ and an integer, then by multiplying the above expression by $\dfrac{(x-r)!}{(x-r)!}$, we have $\binom{x}{r} = \dfrac{x!}{r!(x-r)!}$, as previously defined.

2.11. (a) 1; (b) 252; (c) 12,600.

Chapter 4

1.1. (a) $f(x) = \frac{1}{8}, \frac{3}{8}, \frac{3}{8}, \frac{1}{8}$ for $x = 0, 1, 2, 3$, respectively, $f(x) = 0$ otherwise;
(b) $F(x) = 0$ if $x < 0$, $F(x) = \frac{1}{8}$ if $0 \le x < 1$, $F(x) = \frac{4}{8}$ if $1 \le x < 2$, $F(x) = \frac{7}{8}$ if $2 \le x < 3$, $F(x) = 1$ if $x \ge 3$.

1.3. (a) $f(x) = \frac{15}{70}, \frac{40}{70}, \frac{15}{70}$ for $x = 0, 1, 2$, respectively, $f(x) = 0$ otherwise;
(b) $F(x) = 0$ if $x < 0$, $F(x) = \frac{15}{70}$ if $0 \le x < 1$, $F(x) = \frac{55}{70}$ if $1 \le x < 2$, $F(x) = 1$ if $x \ge 2$.

1.5. (a) $k = \frac{1}{6}$; (b) $\frac{1}{2}, 1, \frac{1}{3}$; (c) 2; (d) $F(x) = 0$ if $x < 0$, $F(x) = \frac{1}{6}$ if $0 \le x < 1$, $F(x) = \frac{1}{2}$ if $1 \le x < 2$, $F(x) = 1$ if $x \ge 2$.

1.7. (b) $\frac{1}{2}, \frac{1}{4}, \frac{5}{12}, \frac{1}{2}, \frac{2}{3}, \frac{2}{3}, 1, \frac{1}{6}$;
(c) $f(x) = \frac{1}{4}, \frac{1}{4}, \frac{1}{6}, \frac{1}{3}$ for $x = -1, 1, 2, 3$, respectively, $f(x) = 0$ otherwise.

1.9. $f(x) = \dfrac{\binom{13}{x}\binom{39}{13-x}}{\binom{52}{13}}$ for $x = 0, 1, \cdots, 13$; one finds (with three decimal place accuracy)

$f(x) = .013, .080, .206, .286, .239, .125, .042, .009, .001$

for $x = 0, 1, 2, 3, 4, 5, 6, 7, 8$ respectively, and $f(x) = 0$ for all other x.

1.11. (a) The event $(X \leq b)$ is the union of the two mutually exclusive events $(X \leq a)$ and $(a < X \leq b)$. Hence $F(b) = F(a) + P(a < X \leq b)$ from which the result follows;

(c) The event $(a \leq X \leq b)$ is the union of the mutually exclusive events $(a \leq X < b)$ and $(X = b)$. Hence, using result in (b),

$$F(b) - F(a) + f(a) = P(a \leq X < b) + f(b).$$

1.13. (a) X_1 and X_2 are not equal, for they have different domains; but their probability functions are both given by f where $f(1) = f(2) = \frac{1}{2}$, $f(x) = 0$ if $x \neq 1$ or 2;

(b) Let $S = \{o_1, o_2, \cdots, o_n\}$ be *any* set with at least two elements. Make an acceptable assignment of probabilities to the simple events of S so that some one simple event, say $\{o_1\}$, has probability $\frac{1}{2}$. Define X by $X(o_1) = 1$, $X(o_j) = 2$ if $j \neq 1$. Then X has the probability function f defined in (a). We get a different random variable X with each choice of S and there are infinitely many sets from which to choose S.

2.1. (a) $\frac{3}{2}$;

(b) $E(Y) = (-1.5)(\frac{1}{8}) + (-.5)(\frac{3}{8}) + (.5)(\frac{3}{8}) + (1.5)(\frac{1}{8}) = 0$;

(c) $E(Z) = \frac{3}{4}$.

2.3. $E(X) = -.05$ of a dollar.

2.5. 1.

2.7. (a) Mean net profit (in cents) is $-80, 50, 100, 90$, when stock is 0, 1, 2, 3 flowers, respectively; (b) At least $1.00.

2.9. $b = 1.04$, $E(Y) = 0$.

2.11. $20.

2.13. $E(X^2) = \frac{329}{6}$, $[E(X)]^2 = 49$.

2.15. (a) $f(-e) = \frac{125}{216}$, $f(e) = \frac{75}{216}$, $f(2e) = \frac{15}{216}$, $f(3e) = \frac{1}{216}$;

(b) $E(X) = -17e/216$ or $-.08$, approximately, when $e = 1$.

2.17. $E(X) = 3$.

2.19. (a) $P(X_1 = k) = \frac{1}{10}$ for $k = 0, 1, \cdots, 9$;

$P(X_2 = k) = \frac{1}{20}$ for $k = -3, -2, \cdots, 3, 4$,

$P(X_2 = k) = \frac{3}{20}$ for $k = 5, 6$, $\quad P(X_2 = k) = \frac{1}{10}$ for $k = 7, 8, 9$;

(b) $E(X_1) = \$4.50$, $E(X_2) = \$4.25$, accept option 1.

2.21. (a) Unique if $p \neq F(x_k)$ for $k = 1, 2, \cdots, N$. If there is a possible value x_k of X for which $F(x_k) = p$, then there are infinitely many medians.

2.23. From the hypothesis it follows that if $a + d_j$ is a possible value of X for any number d_j, then $a - d_j$ is also a possible value and

$$f(a + d_j) = f(a - d_j).$$

Suppose there are p such pairs. Then

$$E(X) = af(a) + \sum_{j=1}^{p} (a + d_j)f(a + d_j) + \sum_{j=1}^{p} (a - d_j)f(a - d_j)$$

$$= a \left[f(a) + \sum_{j=1}^{p} f(a + d_j) + \sum_{j=1}^{p} f(a - d_j) \right] = a,$$

since the sum in brackets is the sum of $f(x_k)$ over all possible values x_k of X, and hence equals 1.

3.1. $\mu_X = \frac{3}{4}$, $\sigma_X^2 = \frac{3}{16}$, $\sigma_X = 0.43$, approximately;
$\mu_Y = 7000$, $\sigma_Y^2 = 4{,}500{,}000$, $\sigma_Y = 2121$, approximately;
$\mu_Z = \frac{7}{4}$, $\sigma_Z^2 = \frac{19}{16}$, $\sigma_Z = 1.09$, approximately.

3.3. $\sigma_X^2 = \frac{35}{6}$, $\sigma_X = 2.41$, approximately.

3.5. (a) $\text{Var}(X_k) = k/4$;   (b) $\text{Var}(X_k) = kp(1 - p)$.

3.7. (a) 2504;   (b) 16;   (c) 4;   (d) 4;   (e) 2.

3.9. Use Theorem 2.1 to find

$$E(Y) = \sum_{k=1}^{N} (a + bx_k + cx_k^2)f(x_k) = a + bE(X) + cE(X^2).$$

Now use (3.10).

3.11. (a) If f, g, h are probability functions of X for methods (1), (2), and (3), respectively, then $f(1) = 1$; $g(0) = \frac{8}{27}$, $g(1) = \frac{12}{27}$, $g(2) = \frac{6}{27}$, $g(3) = \frac{1}{27}$; $h(0) = \frac{2}{6}$, $h(1) = \frac{3}{6}$, $h(3) = \frac{1}{6}$;
(b) $E(X) = 1$ for each method;
(c) Standard deviations of X for the methods are 0, $\sqrt{6}/3$, and 1 respectively.

3.13. Mean absolute deviations are .8125, 2.3125, .8125, and 1.625 for X_1, X_2, X_3, and X_4, respectively.

3.15. (a) -1.5, 0, .3;   (b) 80, 90, 96, 113.

3.17. In each case, Chebyshev's inequality says probability is greater than $\frac{5}{9}$ for $z = 1.5$ and greater than $\frac{3}{4}$ for $z = 2$. The actual probabilities are, for $z = 1.5$ and $z = 2$ respectively: (a) 1, 1;   (b) $\frac{5}{8}$, $\frac{17}{18}$;   (c) $\frac{7}{8}$, 1.

3.19. $z = \sqrt{2}$, $\sqrt{10}$, $\sqrt{20}$, 10.

4.1.

z \ y	0	1	2	3	$P(Z = z)$
1	0	3/8	3/8	0	3/4
3	1/8	0	0	1/8	1/4
$P(Y = y)$	1/8	3/8	3/8	1/8	1

Y and Z are dependent.

4.3.

x \ y	0	1	2	3	$P(X = x)$
0	0	6/27	0	0	2/9
1	6/27	6/27	6/27	0	6/9
2	2/27	0	0	1/27	1/9
$P(Y = y)$	8/27	12/27	6/27	1/27	1

X and Y are dependent.

4.5. $h(x, y) = \dfrac{\binom{13}{x}\binom{13}{y}\binom{26}{13 - x - y}}{\binom{52}{13}}$ if x and y are any nonnegative integers for which $x + y \leq 13$, $h(x, y) = 0$ otherwise. X and Y are dependent since $h(13, 13) = 0$, but

$$P(X = 13) = P(Y = 13) = 1 \Big/ \binom{52}{13}$$

so that $h(13, 13) \neq f(13)g(13)$.

4.7. Independence follows from Theorem 4.2 by considering the four tosses as two independent trials of two tosses each.

4.9. (a)

x \ y	1	2	3	4	5	$P(X = x)$
1	1/25	1/25	1/25	1/25	1/25	1/5
2	0	1/20	1/20	1/20	1/20	1/5
3	0	0	1/15	1/15	1/15	1/5
4	0	0	0	1/10	1/10	1/5
5	0	0	0	0	1/5	1/5
$P(Y = y)$	12/300	27/300	47/300	77/300	137/300	1

(c) $P(Y = y | X = 3) = \frac{1}{3}$ for $y = 3, 4, 5$;

(d) $P(X = x | Y = 3) = \frac{12}{47}, \frac{15}{47}, \frac{20}{47}$ for $x = 1, 2, 3$, respectively;

(e) $\frac{7}{15}, \frac{103}{300}$.

4.11.
$$P(X \leq x, Z \leq z) = \begin{cases} 0 & \text{if } x < 0 \text{ or if } z < 1 \\ \frac{3}{8} & \text{if } 0 \leq x < 1 \text{ and } 1 \leq z < 3 \\ \frac{1}{2} & \text{if } 0 \leq x < 1 \text{ and } 3 \leq z \\ \frac{3}{4} & \text{if } 1 \leq x \text{ and } 1 \leq z < 3 \\ 1 & \text{if } 1 \leq x \text{ and } 3 \leq z \end{cases}$$

4.13. Using notation in Table 29, let $c_{ij} = f(x_j)/f(x_i)$ and show that if X and Y are independent, then for $k = 1, 2, \cdots, N$ we have

$$h(x_j, y_k) = c_{ij} h(x_i, y_k).$$

4.15. (c) Let X have exactly two possible values differing only in sign, say $+1$ and -1. Let Y be any random variable such that X and Y are dependent. Since X^2 has only one possible value, X^2 and Y^2 are independent.

5.1. (a) Let $f(x) = P(X + Y = x)$. Then $f(x) = .1, .2, .3, .4$ for $x = 2, 3, 4, 5$, respectively, and $f(x) = 0$ otherwise; $E(X + Y) = 4$;

(b) Let $g(x) = P(XY = x)$. Then $g(x) = .1, .2, .1, .2, .4$ for $x = 1, 2, 3, 4, 6$, respectively, and $g(x) = 0$ otherwise; $E(XY) = 4$.

5.3. (a) Not true for all X, Y. False for random variables in Problem 5.2(a). True if $Y = X$, for example;

(c) False for random variables in Problem 5.2(c). True if $Y = X$;

(e) True for all X, Y.

5.5. (a) 80 and 13;   (b) 20 and 13;   (c) 210 and $\sqrt{1396}$.

5.7. (a) $Y(o_k) = b$ for all $o_k \epsilon S$; i.e., Y is a constant function. Note that our notation does not distinguish a constant function from the number that is its constant value;

(b) Y is the function equal to a for all $o_k \epsilon S$. X and Y are independent by the result in Problem 4.10.

5.9. First generalize Theorem 5.1 to functions of n random variables, and thus show that

$$E(X_1 X_2 \cdots X_n) = \sum v_1 v_2 \cdots v_n h(v_1, v_2, \cdots, v_n)$$

where $h(v_1, v_2, \cdots, v_n) = P(X_1 = v_1, X_2 = v_2, \cdots, X_n = v_n)$ and the sum extends over all possible values v_1 of X_1, v_2 of X_2, $\cdots$, v_n of X_n. But by Definition 4.4,

$$h(v_1, v_2, \cdots, v_n) = f_1(v_1) f_2(v_2) \cdots f_n(v_n)$$

where f_k is the probability function of X_k. Hence (as in Theorem 5.4 where $n = 2$), the sum can be written as a product of the n sums $\sum v_k f_k(v_k)$ for $k = 1, 2, \cdots, n$. Since the kth sum extends over all possible values v_k of X_k, it is equal to $E(X_k)$ and the result follows. (*Note:* A proof by mathematical induction is also possible. In such a proof one needs to use the following fact: If X_1, X_2, $\cdots$, X_n are independent and if $Y = X_1 X_2 \cdots X_{n-1}$, then Y and X_n are independent. This result can be proved by a method similar to that used below in the solution to part (a) of Problem 5.11.)

5.11. (a) Let x and y be any numbers. Then

$$P(Y_k = y, X_{k+1} = x) = \sum P(X_1 = v_1, \cdots, X_k = v_k, X_{k+1} = x),$$

the summation extending over all values v_1 of X_1, $\cdots$, v_k of X_k such that

$$a_1 v_1 + \cdots + a_k v_k = y.$$

Now we invoke Definition 4.4 to obtain

$$P(Y_k = y, X_{k+1} = x) = \sum P(X_1 = v_1) \cdots P(X_k = v_k) P(X_{k+1} = x).$$

Since x is a constant with respect to this summation, the term $P(X_{k+1} = x)$ can be placed before the summation sign. The remaining sum is just $P(Y_k = y)$. Hence,

$$P(Y_k = y, X_{k+1} = x) = P(Y_k = y) P(X_{k+1} = x),$$

which proves the independence of Y_k and X_{k+1}.

(b) Let I be the set of positive integers for which the theorem is true. By Theorem 3.3 and (5.10), we know that $1 \epsilon I$ and $2 \epsilon I$. Now let us assume that $k \epsilon I$ and show that then $(k + 1) \epsilon I$ for any integer k.

When there are $k + 1$ independent random variables $X_1, \cdots, X_{k+1}$, then by part (a), Y_k and X_{k+1} are independent. Hence, by (5.10),

$$\mathrm{Var}(Y_k + a_{k+1}X_{k+1}) = \mathrm{Var}(Y_k) + a_{k+1}^2\,\mathrm{Var}(X_{k+1}).$$

Now use the induction hypothesis (that $k \in I$) to expand $\mathrm{Var}(Y_k)$ and thus show that $(k + 1) \in I$. This completes the proof.

5.13. (a) $\mu_X = 1$, $\sigma_X = 1/\sqrt{2}$; (b) The probability function of $\overline{X}$ for samples of size 2 is given by:

$\bar{x}$	0	1/2	1	3/2	2
$P(\overline{X} = \bar{x})$	1/16	4/16	6/16	4/16	1/16

$\mu_{\bar{X}} = 1$ and $\sigma_{\bar{X}}^2 = \frac{1}{4}$; (c) The probability function of $\overline{X}$ for samples of size 3 is given by:

$\bar{x}$	0	1/3	2/3	1	4/3	5/3	2
$P(\overline{X} = \bar{x})$	1/64	6/64	15/64	20/64	15/64	6/64	1/64

$\mu_{\bar{X}} = 1$, $\sigma_{\bar{X}}^2 = \frac{1}{6}$.

5.15. (a) $\mu_X = \$5450$, $\sigma_X^2 = 3{,}322{,}500$, $\sigma_X = \$1823$, approximately;

(b) The probability function of $\overline{X}$ is given by:

$\bar{x}$	3500	4250	5000	5500	6250	7000	7500	8250	9000
$P(\overline{X} = \bar{x})$	.09	.24	.16	.12	.22	.08	.04	.04	.01

$\mu_{\bar{X}} = \$5450$, $\sigma_{\bar{X}}^2 = 1{,}661{,}250$, $\sigma_{\bar{X}} = \$1289$, approximately.

6.1. Write μ_j for $E(X_j)$ and use the definition of variance together with (5.4) to obtain

$$\mathrm{Var}(X_1 + \cdots + X_n) = E([(X_1 - \mu_1) + \cdots + (X_n - \mu_n)]^2).$$

Now perform the indicated squaring of the sum in brackets and use Definitions 3.1 and 6.1 to complete the proof.

6.3. (a) Letting $f(\bar{x}) = P(\bar{X} = \bar{x})$ we have

$$f(\bar{x}) = .1, .4, .2, .1, .2$$

for $\bar{x} = \frac{280}{3}, \frac{310}{3}, \frac{330}{3}, \frac{340}{3}, \frac{360}{3}$, and $f(\bar{x}) = 0$ otherwise;
$E(\bar{X}) = 108$, $\mathrm{Var}(\bar{X}) = \frac{188}{3}$;
(c) $f(\bar{x}) = 1$ for $\bar{x} = 108$, $f(\bar{x}) = 0$ otherwise;
$E(\bar{X}) = 108$, $\mathrm{Var}(\bar{X}) = 0$.

6.5. $\sigma_{\bar{x}} = \$74.63$, approximately, and the required interval, extending three standard deviations on either side of the mean, is \$4776 to \$5224.

6.7. Set $\dfrac{\sigma_X^2}{n_1}\left(\dfrac{N - n_1}{N - 1}\right) = \dfrac{1}{4}\dfrac{\sigma_X^2}{n}\left(\dfrac{N - n}{N - 1}\right)$ and solve for n_1.

6.9. Show that $\mathrm{Var}(X \pm Y) = \sigma_X^2 + \sigma_Y^2 \pm 2\rho(X, Y)\sigma_X\sigma_Y$, from which the results are immediate.

6.11. (a)

x \ y	0	1	$P(X = x)$
0	3/8	3/8	3/4
1	1/8	1/8	1/4
$P(Y = y)$	1/2	1/2	1

$\rho(X, Y) = 0$,
X and Y independent.

(b)

x \ y	0	1	$P(X = x)$
0	17/24	1/24	3/4
1	1/8	1/8	1/4
$P(Y = y)$	5/6	1/6	1

$\rho(X, Y) = 2/\sqrt{15} = .52$
X and Y dependent.

6.13. $\rho(X, Y) = 0$. (Note that X and Y are dependent but uncorrelated.)

6.15.

x \ y	0	1	2	3	4	$P(X = x)$
0	1/16	2/16	1/16	0	0	1/4
1	0	2/16	4/16	2/16	0	2/4
2	0	0	1/16	2/16	1/16	1/4
$P(Y = y)$	1/16	4/16	6/16	4/16	1/16	1

$\rho(X, Y) = \sqrt{2}/2 = .71$, approximately.

6.17. $\rho_m = \begin{cases} -1 & \text{if } m < 0 \\ 0 & \text{if } m = 0 \\ 1 & \text{if } m > 0. \end{cases}$

6.19. Without loss of generality (see Problem 6.18), we can assume that X and Y each have possible values 0 and 1. Then

$$E(X) = P(X = 1), \quad E(Y) = P(Y = 1), \quad E(XY) = P(X = 1, Y = 1)$$

so that $\text{Cov}(X, Y) = 0$ implies

$$P(X = 1, Y = 1) = P(X = 1)P(Y = 1).$$

Show then that the other three joint probabilities must also be products of the corresponding marginal probabilities.

Chapter 5

1.1. Probabilities are .107 for (a) and .069 for (b).

1.3. .000006.

1.5. Using Theorem 1.2, find $p = .60$ and $n = 20$. Required probabilities are (a) .998 (b) .126 (c) .245.

1.7. (a) .772; (b) .746; (c) $p = .376$.

1.9. $n \geq 69$.

1.11. Corresponding to the values of p given in Table 37, the probabilities of accepting the lot are .983, .736, .392, .069, .008, .001, and .000 in part (a), and .904, .599, .349, .107, .028, .006, and .001 in part (c).

1.13. (a) By (1.3), $b(n - k|n, 1 - p) = \binom{n}{n-k}(1 - p)^{n-k}p^{n-(n-k)}$. Now recall that

$$\binom{n}{n-k} = \binom{n}{k};$$

(b) By (a), $\displaystyle\sum_{k=r}^{n} b(k|n, p) = \sum_{k=r}^{n} b(n - k|n, 1 - p)$

$$= \sum_{k=0}^{n-r} b(k|n, 1 - p).$$

1.15.

	$n = 5$	$n = 10$	$n = 20$	Normal Approx.
$P(-1 \leq S_n^* \leq 1)$	.409	.772	.598	.68
$P(-2 \leq S_n^* \leq 2)$	.942	.967	.956	.95
$P(-3 \leq S_n^* \leq 3)$	.993	.994	.997	.997

1.17. $G(t) = \sum_{k=0}^{n} \binom{n}{k}(pt)^k (q)^{n-k}$, which equals $(q + pt)^n$, by the binomial theorem.

2.1. (a) Accept null hypothesis if $X \geq 0$; i.e., accept no matter what result is obtained from the sample of 20. The probability of an error of the second kind is then 1.

(b) Reject the null hypothesis no matter what value of X occurs. Then $\alpha = 1$.

(c) Reject null hypothesis if and only if $X \geq 16$. From Table 40 with $c = 16$, find $1 - \pi(.70) = .762$, $1 - \pi(.80) = .370$, $\pi(.50) = .006$.

(d) .126, accept null hypothesis and wage very expensive campaign.

(e) $P(X \geq 17) = .016$ and $P(X \geq 18) = .004$; therefore at least 18 must favor Smith.

2.3. (a) Null hypothesis: $p = \frac{1}{2}$; Alternate hypothesis: $p > \frac{1}{2}$. From Table 37, find
$$P(X \geq 14) = .058, \quad P(X \geq 15) = .021.$$
Hence reject null hypothesis if and only if X, the number who revert to first-learned method, is at least 15.

(b) Significant at the 5% level, since $P(X \geq 15) = .021 < .05$. Not significant at the 1% level.

2.5. (a) Let a trial (observing a machine for a day) result in success (machine needs repair) or failure (machine does not need repair). Let p = probability of a success. Null hypothesis: $p = .20$; Alternate hypothesis: $p \neq .20$.

(b) Since mean number of successes is $np = 4$ if the null hypothesis is true, we reject null hypothesis if X, number of successes observed, is either too much larger or too much smaller than four; i.e., we reject null hypothesis if and only if either $X \leq 4 - d$ or $X \geq 4 + d$, where d denotes the smallest deviation from the mean that makes X "too much" larger or "too much" smaller than the mean. The number d is determined by requiring the probability of an error of the first kind to be no larger than .10 but as close to .10 as possible. This error probability is $P(X \leq 4 - d) + P(X \geq 4 + d)$, calculated for $p = .20$. If $d = 3$, this probability is greater than .10; if $d = 4$, it is less than .10 (from Table 37). Hence, reject null hypothesis if and only if $X = 0$ or $X \geq 8$. (This is called a two-tailed test.)

(c) Probability that X deviates from its mean in either direction by at least as much as the observed value does is $P(X \leq 1) + P(X \geq 7)$ or $.069 + .087 = .156$. Not significant at .10 level.

(d) Ideal decision rule has $\pi(p) = 0$ if $p = .20$, $\pi(p) = 1$ if $p \neq .20$.

INDEX

A CATALOG OF SELECTED

DOVER BOOKS
IN SCIENCE AND MATHEMATICS

A CATALOG OF SELECTED

DOVER BOOKS

IN SCIENCE AND MATHEMATICS

QUALITATIVE THEORY OF DIFFERENTIAL EQUATIONS, V.V. Nemytskii and V.V. Stepanov. Classic graduate-level text by two prominent Soviet mathematicians covers classical differential equations as well as topological dynamics and ergodic theory. Bibliographies. 523pp. 5⅜ × 8½. 65954-2 Pa. $10.95

MATRICES AND LINEAR ALGEBRA, Hans Schneider and George Phillip Barker. Basic textbook covers theory of matrices and its applications to systems of linear equations and related topics such as determinants, eigenvalues and differential equations. Numerous exercises. 432pp. 5⅜ × 8½. 66014-1 Pa. $9.95

QUANTUM THEORY, David Bohm. This advanced undergraduate-level text presents the quantum theory in terms of qualitative and imaginative concepts, followed by specific applications worked out in mathematical detail. Preface. Index. 655pp. 5⅜ × 8½. 65969-0 Pa. $13.95

ATOMIC PHYSICS (8th edition), Max Born. Nobel laureate's lucid treatment of kinetic theory of gases, elementary particles, nuclear atom, wave-corpuscles, atomic structure and spectral lines, much more. Over 40 appendices, bibliography. 495pp. 5⅜ × 8½. 65984-4 Pa. $11.95

ELECTRONIC STRUCTURE AND THE PROPERTIES OF SOLIDS: The Physics of the Chemical Bond, Walter A. Harrison. Innovative text offers basic understanding of the electronic structure of covalent and ionic solids, simple metals, transition metals and their compounds. Problems. 1980 edition. 582pp. 6⅛ × 9¼. 66021-4 Pa. $14.95

BOUNDARY VALUE PROBLEMS OF HEAT CONDUCTION, M. Necati Özisik. Systematic, comprehensive treatment of modern mathematical methods of solving problems in heat conduction and diffusion. Numerous examples and problems. Selected references. Appendices. 505pp. 5⅜ × 8½. 65990-9 Pa. $11.95

A SHORT HISTORY OF CHEMISTRY (3rd edition), J.R. Partington. Classic exposition explores origins of chemistry, alchemy, early medical chemistry, nature of atmosphere, theory of valency, laws and structure of atomic theory, much more. 428pp. 5⅜ × 8½. (Available in U.S. only) 65977-1 Pa. $10.95

A HISTORY OF ASTRONOMY, A. Pannekoek. Well-balanced, carefully reasoned study covers such topics as Ptolemaic theory, work of Copernicus, Kepler, Newton, Eddington's work on stars, much more. Illustrated. References. 521pp. 5⅜ × 8½. 65994-1 Pa. $11.95

PRINCIPLES OF METEOROLOGICAL ANALYSIS, Walter J. Saucier. Highly respected, abundantly illustrated classic reviews atmospheric variables, hydrostatics, static stability, various analyses (scalar, cross-section, isobaric, isentropic, more). For intermediate meteorology students. 454pp. 6⅛ × 9¼. 65979-8 Pa. $12.95

ASYMPTOTIC METHODS IN ANALYSIS, N.G. de Bruijn. An inexpensive, comprehensive guide to asymptotic methods—the pioneering work that teaches by explaining worked examples in detail. Index. 224pp. 5⅜ × 8½. 64221-6 Pa. $6.95

OPTICAL RESONANCE AND TWO-LEVEL ATOMS, L. Allen and J.H. Eberly. Clear, comprehensive introduction to basic principles behind all quantum optical resonance phenomena. 53 illustrations. Preface. Index. 256pp. 5⅜ × 8½. 65533-4 Pa. $7.95

COMPLEX VARIABLES, Francis J. Flanigan. Unusual approach, delaying complex algebra till harmonic functions have been analyzed from real variable viewpoint. Includes problems with answers. 364pp. 5⅜ × 8½. 61388-7 Pa. $7.95

ATOMIC SPECTRA AND ATOMIC STRUCTURE, Gerhard Herzberg. One of best introductions; especially for specialist in other fields. Treatment is physical rather than mathematical. 80 illustrations. 257pp. 5⅜ × 8½. 60115-3 Pa. $5.95

APPLIED COMPLEX VARIABLES, John W. Dettman. Step-by-step coverage of fundamentals of analytic function theory—plus lucid exposition of five important applications: Potential Theory; Ordinary Differential Equations; Fourier Transforms; Laplace Transforms; Asymptotic Expansions. 66 figures. Exercises at chapter ends. 512pp. 5⅜ × 8½. 64670-X Pa. $10.95

ULTRASONIC ABSORPTION: An Introduction to the Theory of Sound Absorption and Dispersion in Gases, Liquids and Solids, A.B. Bhatia. Standard reference in the field provides a clear, systematically organized introductory review of fundamental concepts for advanced graduate students, research workers. Numerous diagrams. Bibliography. 440pp. 5⅜ × 8½. 64917-2 Pa. $11.95

UNBOUNDED LINEAR OPERATORS: Theory and Applications, Seymour Goldberg. Classic presents systematic treatment of the theory of unbounded linear operators in normed linear spaces with applications to differential equations. Bibliography. 199pp. 5⅜ × 8½. 64830-3 Pa. $7.95

LIGHT SCATTERING BY SMALL PARTICLES, H.C. van de Hulst. Comprehensive treatment including full range of useful approximation methods for researchers in chemistry, meteorology and astronomy. 44 illustrations. 470pp. 5⅜ × 8½. 64228-3 Pa. $10.95

CONFORMAL MAPPING ON RIEMANN SURFACES, Harvey Cohn. Lucid, insightful book presents ideal coverage of subject. 334 exercises make book perfect for self-study. 55 figures. 352pp. 5⅜ × 8¼. 64025-6 Pa. $8.95

OPTICKS, Sir Isaac Newton. Newton's own experiments with spectroscopy, colors, lenses, reflection, refraction, etc., in language the layman can follow. Foreword by Albert Einstein. 532pp. 5⅜ × 8½. 60205-2 Pa. $9.95

GENERALIZED INTEGRAL TRANSFORMATIONS, A.H. Zemanian. Graduate-level study of recent generalizations of the Laplace, Mellin, Hankel, K. Weierstrass, convolution and other simple transformations. Bibliography. 320pp. 5⅜ × 8½. 65375-7 Pa. $7.95

NUMERICAL METHODS FOR SCIENTISTS AND ENGINEERS, Richard Hamming. Classic text stresses frequency approach in coverage of algorithms, polynomial approximation, Fourier approximation, exponential approximation, other topics. Revised and enlarged 2nd edition. 721pp. 5⅜ × 8½.
65241-6 Pa. $14.95

THEORETICAL SOLID STATE PHYSICS, Vol. I: Perfect Lattices in Equilibrium; Vol. II: Non-Equilibrium and Disorder, William Jones and Norman H. March. Monumental reference work covers fundamental theory of equilibrium properties of perfect crystalline solids, non-equilibrium properties, defects and disordered systems. Appendices. Problems. Preface. Diagrams. Index. Bibliography. Total of 1,301pp. 5⅜ × 8½. Two volumes. Vol. I 65015-4 Pa. $12.95
Vol. II 65016-2 Pa. $12.95

OPTIMIZATION THEORY WITH APPLICATIONS, Donald A. Pierre. Broad-spectrum approach to important topic. Classical theory of minima and maxima, calculus of variations, simplex technique and linear programming, more. Many problems, examples. 640pp. 5⅜ × 8½. 65205-X Pa. $13.95

THE MODERN THEORY OF SOLIDS, Frederick Seitz. First inexpensive edition of classic work on theory of ionic crystals, free-electron theory of metals and semiconductors, molecular binding, much more. 736pp. 5⅜ × 8½.
65482-6 Pa. $15.95

ESSAYS ON THE THEORY OF NUMBERS, Richard Dedekind. Two classic essays by great German mathematician: on the theory of irrational numbers; and on transfinite numbers and properties of natural numbers. 115pp. 5⅜ × 8½.
21010-3 Pa. $4.95

THE FUNCTIONS OF MATHEMATICAL PHYSICS, Harry Hochstadt. Comprehensive treatment of orthogonal polynomials, hypergeometric functions, Hill's equation, much more. Bibliography. Index. 322pp. 5⅜ × 8½. 65214-9 Pa. $9.95

NUMBER THEORY AND ITS HISTORY, Oystein Ore. Unusually clear, accessible introduction covers counting, properties of numbers, prime numbers, much more. Bibliography. 380pp. 5⅜ × 8½. 65620-9 Pa. $8.95

THE VARIATIONAL PRINCIPLES OF MECHANICS, Cornelius Lanczos. Graduate level coverage of calculus of variations, equations of motion, relativistic mechanics, more. First inexpensive paperbound edition of classic treatise. Index. Bibliography. 418pp. 5⅜ × 8½. 65067-7 Pa. $10.95

MATHEMATICAL TABLES AND FORMULAS, Robert D. Carmichael and Edwin R. Smith. Logarithms, sines, tangents, trig functions, powers, roots, reciprocals, exponential and hyperbolic functions, formulas and theorems. 269pp. 5⅜ × 8½. 60111-0 Pa. $5.95

THEORETICAL PHYSICS, Georg Joos, with Ira M. Freeman. Classic overview covers essential math, mechanics, electromagnetic theory, thermodynamics, quantum mechanics, nuclear physics, other topics. First paperback edition. xxiii + 885pp. 5⅜ × 8½. 65227-0 Pa. $18.95

CATALOG OF DOVER BOOKS

HANDBOOK OF MATHEMATICAL FUNCTIONS WITH FORMULAS, GRAPHS, AND MATHEMATICAL TABLES, edited by Milton Abramowitz and Irene A. Stegun. Vast compendium: 29 sets of tables, some to as high as 20 places. 1,046pp. 8 × 10½. 61272-4 Pa. $22.95

MATHEMATICAL METHODS IN PHYSICS AND ENGINEERING, John W. Dettman. Algebraically based approach to vectors, mapping, diffraction, other topics in applied math. Also generalized functions, analytic function theory, more. Exercises. 448pp. 5⅜ × 8¼. 65649-7 Pa. $8.95

A SURVEY OF NUMERICAL MATHEMATICS, David M. Young and Robert Todd Gregory. Broad self-contained coverage of computer-oriented numerical algorithms for solving various types of mathematical problems in linear algebra, ordinary and partial, differential equations, much more. Exercises. Total of 1,248pp. 5⅜ × 8½. Two volumes. Vol. I 65691-8 Pa. $14.95
Vol. II 65692-6 Pa. $14.95

TENSOR ANALYSIS FOR PHYSICISTS, J.A. Schouten. Concise exposition of the mathematical basis of tensor analysis, integrated with well-chosen physical examples of the theory. Exercises. Index. Bibliography. 289pp. 5⅜ × 8½.
65582-2 Pa. $7.95

INTRODUCTION TO NUMERICAL ANALYSIS (2nd Edition), F.B. Hildebrand. Classic, fundamental treatment covers computation, approximation, interpolation, numerical differentiation and integration, other topics. 150 new problems. 669pp. 5⅜ × 8½. 65363-3 Pa. $14.95

INVESTIGATIONS ON THE THEORY OF THE BROWNIAN MOVEMENT, Albert Einstein. Five papers (1905-8) investigating dynamics of Brownian motion and evolving elementary theory. Notes by R. Fürth. 122pp. 5⅜ × 8½.
60304-0 Pa. $4.95

NUMERICAL METHODS FOR SCIENTISTS AND ENGINEERS, Richard Hamming. Classic text stresses frequency approach in coverage of algorithms, polynomial approximation, Fourier approximation, exponential approximation, other topics. Revised and enlarged 2nd edition. 721pp. 5⅜ × 8½. 65241-6 Pa. $14.95

AN INTRODUCTION TO STATISTICAL THERMODYNAMICS, Terrell L. Hill. Excellent basic text offers wide-ranging coverage of quantum statistical mechanics, systems of interacting molecules, quantum statistics, more. 523pp. 5⅜ × 8½. 65242-4 Pa. $11.95

ELEMENTARY DIFFERENTIAL EQUATIONS, William Ted Martin and Eric Reissner. Exceptionally clear, comprehensive introduction at undergraduate level. Nature and origin of differential equations, differential equations of first, second and higher orders. Picard's Theorem, much more. Problems with solutions. 331pp. 5⅜ × 8½. 65024-3 Pa. $8.95

STATISTICAL PHYSICS, Gregory H. Wannier. Classic text combines thermodynamics, statistical mechanics and kinetic theory in one unified presentation of thermal physics. Problems with solutions. Bibliography. 532pp. 5⅜ × 8½.
65401-X Pa. $11.95

ROTARY-WING AERODYNAMICS, W.Z. Stepniewski. Clear, concise text covers aerodynamic phenomena of the rotor and offers guidelines for helicopter performance evaluation. Originally prepared for NASA. 537 figures. 640pp. 6⅛ × 9¼.
64647-5 Pa. $14.95

DIFFERENTIAL GEOMETRY, Heinrich W. Guggenheimer. Local differential geometry as an application of advanced calculus and linear algebra. Curvature, transformation groups, surfaces, more. Exercises. 62 figures. 378pp. 5⅜ × 8½.
63433-7 Pa. $7.95

INTRODUCTION TO SPACE DYNAMICS, William Tyrrell Thomson. Comprehensive, classic introduction to space-flight engineering for advanced undergraduate and graduate students. Includes vector algebra, kinematics, transformation of coordinates. Bibliography. Index. 352pp. 5⅜ × 8½.   65113-4 Pa. $8.95

A SURVEY OF MINIMAL SURFACES, Robert Osserman. Up-to-date, in-depth discussion of the field for advanced students. Corrected and enlarged edition covers new developments. Includes numerous problems. 192pp. 5⅜ × 8½.
64998-9 Pa. $8.95

ANALYTICAL MECHANICS OF GEARS, Earle Buckingham. Indispensable reference for modern gear manufacture covers conjugate gear-tooth action, gear-tooth profiles of various gears, many other topics. 263 figures. 102 tables. 546pp. 5⅜ × 8½.   65712-4 Pa. $11.95

SET THEORY AND LOGIC, Robert R. Stoll. Lucid introduction to unified theory of mathematical concepts. Set theory and logic seen as tools for conceptual understanding of real number system. 496pp. 5⅜ × 8¼.   63829-4 Pa. $10.95

A HISTORY OF MECHANICS, René Dugas. Monumental study of mechanical principles from antiquity to quantum mechanics. Contributions of ancient Greeks, Galileo, Leonardo, Kepler, Lagrange, many others. 671pp. 5⅜ × 8½.
65632-2 Pa. $14.95

FAMOUS PROBLEMS OF GEOMETRY AND HOW TO SOLVE THEM, Benjamin Bold. Squaring the circle, trisecting the angle, duplicating the cube: learn their history, why they are impossible to solve, then solve them yourself. 128pp. 5⅜ × 8½.   24297-8 Pa. $3.95

MECHANICAL VIBRATIONS, J.P. Den Hartog. Classic textbook offers lucid explanations and illustrative models, applying theories of vibrations to a variety of practical industrial engineering problems. Numerous figures. 233 problems, solutions. Appendix. Index. Preface. 436pp. 5⅜ × 8½.   64785-4 Pa. $9.95

CURVATURE AND HOMOLOGY, Samuel I. Goldberg. Thorough treatment of specialized branch of differential geometry. Covers Riemannian manifolds, topology of differentiable manifolds, compact Lie groups, other topics. Exercises. 315pp. 5⅜ × 8½.   64314-X Pa. $8.95

HISTORY OF STRENGTH OF MATERIALS, Stephen P. Timoshenko. Excellent historical survey of the strength of materials with many references to the theories of elasticity and structure. 245 figures. 452pp. 5⅜ × 8½. 61187-6 Pa. $10.95

GEOMETRY OF COMPLEX NUMBERS, Hans Schwerdtfeger. Illuminating, widely praised book on analytic geometry of circles, the Moebius transformation, and two-dimensional non-Euclidean geometries. 200pp. 5⅜ × 8¼.
63830-8 Pa. $6.95

MECHANICS, J.P. Den Hartog. A classic introductory text or refresher. Hundreds of applications and design problems illuminate fundamentals of trusses, loaded beams and cables, etc. 334 answered problems. 462pp. 5⅜ × 8½. 60754-2 Pa. $8.95

TOPOLOGY, John G. Hocking and Gail S. Young. Superb one-year course in classical topology. Topological spaces and functions, point-set topology, much more. Examples and problems. Bibliography. Index. 384pp. 5⅜ × 8¼.
65676-4 Pa. $8.95

STRENGTH OF MATERIALS, J.P. Den Hartog. Full, clear treatment of basic material (tension, torsion, bending, etc.) plus advanced material on engineering methods, applications. 350 answered problems. 323pp. 5⅜ × 8½. 60755-0 Pa. $7.50

ELEMENTARY CONCEPTS OF TOPOLOGY, Paul Alexandroff. Elegant, intuitive approach to topology from set-theoretic topology to Betti groups; how concepts of topology are useful in math and physics. 25 figures. 57pp. 5⅜ × 8½.
60747-X Pa. $2.95

ADVANCED STRENGTH OF MATERIALS, J.P. Den Hartog. Superbly written advanced text covers torsion, rotating disks, membrane stresses in shells, much more. Many problems and answers. 388pp. 5⅜ × 8½. 65407-9 Pa. $9.95

COMPUTABILITY AND UNSOLVABILITY, Martin Davis. Classic graduate-level introduction to theory of computability, usually referred to as theory of recurrent functions. New preface and appendix. 288pp. 5⅜ × 8½. 61471-9 Pa. $6.95

GENERAL CHEMISTRY, Linus Pauling. Revised 3rd edition of classic first-year text by Nobel laureate. Atomic and molecular structure, quantum mechanics, statistical mechanics, thermodynamics correlated with descriptive chemistry. Problems. 992pp. 5⅜ × 8½. 65622-5 Pa. $19.95

AN INTRODUCTION TO MATRICES, SETS AND GROUPS FOR SCIENCE STUDENTS, G. Stephenson. Concise, readable text introduces sets, groups, and most importantly, matrices to undergraduate students of physics, chemistry, and engineering. Problems. 164pp. 5⅜ × 8½. 65077-4 Pa. $6.95

THE HISTORICAL BACKGROUND OF CHEMISTRY, Henry M. Leicester. Evolution of ideas, not individual biography. Concentrates on formulation of a coherent set of chemical laws. 260pp. 5⅜ × 8½. 61053-5 Pa. $6.95

THE PHILOSOPHY OF MATHEMATICS: An Introductory Essay, Stephan Körner. Surveys the views of Plato, Aristotle, Leibniz & Kant concerning propositions and theories of applied and pure mathematics. Introduction. Two appendices. Index. 198pp. 5⅜ × 8½. 25048-2 Pa. $6.95

THE DEVELOPMENT OF MODERN CHEMISTRY, Aaron J. Ihde. Authoritative history of chemistry from ancient Greek theory to 20th-century innovation. Covers major chemists and their discoveries. 209 illustrations. 14 tables. Bibliographies. Indices. Appendices. 851pp. 5⅜ × 8½. 64235-6 Pa. $17.95

DE RE METALLICA, Georgius Agricola. The famous Hoover translation of greatest treatise on technological chemistry, engineering, geology, mining of early modern times (1556). All 289 original woodcuts. 638pp. 6¾ × 11.
60006-8 Pa. $17.95

SOME THEORY OF SAMPLING, William Edwards Deming. Analysis of the problems, theory and design of sampling techniques for social scientists, industrial managers and others who find statistics increasingly important in their work. 61 tables. 90 figures. xvii + 602pp. 5⅜ × 8½.
64684-X Pa. $15.95

THE VARIOUS AND INGENIOUS MACHINES OF AGOSTINO RAMELLI: A Classic Sixteenth-Century Illustrated Treatise on Technology, Agostino Ramelli. One of the most widely known and copied works on machinery in the 16th century. 194 detailed plates of water pumps, grain mills, cranes, more. 608pp. 9 × 12. (EBE)
25497-6 Clothbd. $34.95

LINEAR PROGRAMMING AND ECONOMIC ANALYSIS, Robert Dorfman, Paul A. Samuelson and Robert M. Solow. First comprehensive treatment of linear programming in standard economic analysis. Game theory, modern welfare economics, Leontief input-output, more. 525pp. 5⅜ × 8½.
65491-5 Pa. $13.95

ELEMENTARY DECISION THEORY, Herman Chernoff and Lincoln E. Moses. Clear introduction to statistics and statistical theory covers data processing, probability and random variables, testing hypotheses, much more. Exercises. 364pp. 5⅜ × 8½.
65218-1 Pa. $9.95

THE COMPLEAT STRATEGYST: Being a Primer on the Theory of Games of Strategy, J.D. Williams. Highly entertaining classic describes, with many illustrated examples, how to select best strategies in conflict situations. Prefaces. Appendices. 268pp. 5⅜ × 8½.
25101-2 Pa. $6.95

MATHEMATICAL METHODS OF OPERATIONS RESEARCH, Thomas L. Saaty. Classic graduate-level text covers historical background, classical methods of forming models, optimization, game theory, probability, queueing theory, much more. Exercises. Bibliography. 448pp. 5⅜ × 8¼.
65703-5 Pa. $12.95

CONSTRUCTIONS AND COMBINATORIAL PROBLEMS IN DESIGN OF EXPERIMENTS, Damaraju Raghavarao. In-depth reference work examines orthogonal Latin squares, incomplete block designs, tactical configuration, partial geometry, much more. Abundant explanations, examples. 416pp. 5⅜ × 8¼.
65685-3 Pa. $10.95

THE ABSOLUTE DIFFERENTIAL CALCULUS (CALCULUS OF TENSORS), Tullio Levi-Civita. Great 20th-century mathematician's classic work on material necessary for mathematical grasp of theory of relativity. 452pp. 5⅜ × 8½.
63401-9 Pa. $9.95

VECTOR AND TENSOR ANALYSIS WITH APPLICATIONS, A.I. Borisenko and I.E. Tarapov. Concise introduction. Worked-out problems, solutions, exercises. 257pp. 5⅜ × 8¼.
63833-2 Pa. $6.95

THE FOUR-COLOR PROBLEM: Assaults and Conquest, Thomas L. Saaty and Paul G. Kainen. Engrossing, comprehensive account of the century-old combinatorial topological problem, its history and solution. Bibliographies. Index. 110 figures. 228pp. 5⅜ × 8½. 65092-8 Pa. $6.95

CATALYSIS IN CHEMISTRY AND ENZYMOLOGY, William P. Jencks. Exceptionally clear coverage of mechanisms for catalysis, forces in aqueous solution, carbonyl- and acyl-group reactions, practical kinetics, more. 864pp. 5⅜ × 8½. 65460-5 Pa. $19.95

PROBABILITY: An Introduction, Samuel Goldberg. Excellent basic text covers set theory, probability theory for finite sample spaces, binomial theorem, much more. 360 problems. Bibliographies. 322pp. 5⅜ × 8½. 65252-1 Pa. $8.95

LIGHTNING, Martin A. Uman. Revised, updated edition of classic work on the physics of lightning. Phenomena, terminology, measurement, photography, spectroscopy, thunder, more. Reviews recent research. Bibliography. Indices. 320pp. 5⅜ × 8¼. 64575-4 Pa. $8.95

PROBABILITY THEORY: A Concise Course, Y.A. Rozanov. Highly readable, self-contained introduction covers combination of events, dependent events, Bernoulli trials, etc. Translation by Richard Silverman. 148pp. 5⅜ × 8¼. 63544-9 Pa. $5.95

THE CEASELESS WIND: An Introduction to the Theory of Atmospheric Motion, John A. Dutton. Acclaimed text integrates disciplines of mathematics and physics for full understanding of dynamics of atmospheric motion. Over 400 problems. Index. 97 illustrations. 640pp. 6 × 9. 65096-0 Pa. $17.95

STATISTICS MANUAL, Edwin L. Crow, et al. Comprehensive, practical collection of classical and modern methods prepared by U.S. Naval Ordnance Test Station. Stress on use. Basics of statistics assumed. 288pp. 5⅜ × 8½. 60599-X Pa. $6.95

DICTIONARY/OUTLINE OF BASIC STATISTICS, John E. Freund and Frank J. Williams. A clear concise dictionary of over 1,000 statistical terms and an outline of statistical formulas covering probability, nonparametric tests, much more. 208pp. 5⅜ × 8½. 66796-0 Pa. $6.95

STATISTICAL METHOD FROM THE VIEWPOINT OF QUALITY CONTROL, Walter A. Shewhart. Important text explains regulation of variables, uses of statistical control to achieve quality control in industry, agriculture, other areas. 192pp. 5⅜ × 8½. 65232-7 Pa. $6.95

THE INTERPRETATION OF GEOLOGICAL PHASE DIAGRAMS, Ernest G. Ehlers. Clear, concise text emphasizes diagrams of systems under fluid or containing pressure; also coverage of complex binary systems, hydrothermal melting, more. 288pp. 6½ × 9¼. 65389-7 Pa. $10.95

STATISTICAL ADJUSTMENT OF DATA, W. Edwards Deming. Introduction to basic concepts of statistics, curve fitting, least squares solution, conditions without parameter, conditions containing parameters. 26 exercises worked out. 271pp. 5⅜ × 8½. 64685-8 Pa. $7.95

TENSOR CALCULUS, J.L. Synge and A. Schild. Widely used introductory text covers spaces and tensors, basic operations in Riemannian space, non-Riemannian spaces, etc. 324pp. 5⅜ × 8¼.   63612-7 Pa. $7.95

A CONCISE HISTORY OF MATHEMATICS, Dirk J. Struik. The best brief history of mathematics. Stresses origins and covers every major figure from ancient Near East to 19th century. 41 illustrations. 195pp. 5⅜ × 8½.   60255-9 Pa. $7.95

A SHORT ACCOUNT OF THE HISTORY OF MATHEMATICS, W.W. Rouse Ball. One of clearest, most authoritative surveys from the Egyptians and Phoenicians through 19th-century figures such as Grassman, Galois, Riemann. Fourth edition. 522pp. 5⅜ × 8½.   20630-0 Pa. $10.95

HISTORY OF MATHEMATICS, David E. Smith. Nontechnical survey from ancient Greece and Orient to late 19th century; evolution of arithmetic, geometry, trigonometry, calculating devices, algebra, the calculus. 362 illustrations. 1,355pp. 5⅜ × 8½.   20429-4, 20430-8 Pa., Two-vol. set $23.90

THE GEOMETRY OF RENÉ DESCARTES, René Descartes. The great work founded analytical geometry. Original French text, Descartes' own diagrams, together with definitive Smith-Latham translation. 244pp. 5⅜ × 8½.
                                                                60068-8 Pa. $6.95

THE ORIGINS OF THE INFINITESIMAL CALCULUS, Margaret E. Baron. Only fully detailed and documented account of crucial discipline: origins; development by Galileo, Kepler, Cavalieri; contributions of Newton, Leibniz, more. 304pp. 5⅜ × 8½. (Available in U.S. and Canada only)   65371-4 Pa. $9.95

THE HISTORY OF THE CALCULUS AND ITS CONCEPTUAL DEVELOP-MENT, Carl B. Boyer. Origins in antiquity, medieval contributions, work of Newton, Leibniz, rigorous formulation. Treatment is verbal. 346pp. 5⅜ × 8½.
                                                                60509-4 Pa. $7.95

THE THIRTEEN BOOKS OF EUCLID'S ELEMENTS, translated with introduction and commentary by Sir Thomas L. Heath. Definitive edition. Textual and linguistic notes, mathematical analysis. 2,500 years of critical commentary. Not abridged. 1,414pp. 5⅜ × 8½.   60088-2, 60089-0, 60090-4 Pa., Three-vol. set $29.85

GAMES AND DECISIONS: Introduction and Critical Survey, R. Duncan Luce and Howard Raiffa. Superb nontechnical introduction to game theory, primarily applied to social sciences. Utility theory, zero-sum games, n-person games, decision-making, much more. Bibliography. 509pp. 5⅜ × 8½.   65943-7 Pa. $11.95

THE HISTORICAL ROOTS OF ELEMENTARY MATHEMATICS, Lucas N.H. Bunt, Phillip S. Jones, and Jack D. Bedient. Fundamental underpinnings of modern arithmetic, algebra, geometry and number systems derived from ancient civilizations. 320pp. 5⅜ × 8½.   25563-8 Pa. $8.95

CALCULUS REFRESHER FOR TECHNICAL PEOPLE, A. Albert Klaf. Covers important aspects of integral and differential calculus via 756 questions. 566 problems, most answered. 431pp. 5⅜ × 8½.   20370-0 Pa. $8.95

CATALOG OF DOVER BOOKS

CHALLENGING MATHEMATICAL PROBLEMS WITH ELEMENTARY SOLUTIONS, A.M. Yaglom and I.M. Yaglom. Over 170 challenging problems on probability theory, combinatorial analysis, points and lines, topology, convex polygons, many other topics. Solutions. Total of 445pp. 5% × 8½. Two-vol. set.

Vol. I 65536-9 Pa. $6.95
Vol. II 65537-7 Pa. $6.95

FIFTY CHALLENGING PROBLEMS IN PROBABILITY WITH SOLU-TIONS, Frederick Mosteller. Remarkable puzzlers, graded in difficulty, illustrate elementary and advanced aspects of probability. Detailed solutions. 88pp. 5% × 8½.
65355-2 Pa. $3.95

EXPERIMENTS IN TOPOLOGY, Stepnen Barr. Classic, lively explanation of one of the byways of mathematics. Klein bottles, Moebius strips, projective planes, map coloring, problem of the Koenigsberg bridges, much more, described with clarity and wit. 43 figures. 210pp. 5% × 8½.
25933-1 Pa. $5.95

RELATIVITY IN ILLUSTRATIONS, Jacob T. Schwartz. Clear nontechnical treatment makes relativity more accessible than ever before. Over 60 drawings illustrate concepts more clearly than text alone. Only high school geometry needed. Bibliography. 128pp. 6⅛ × 9¼.
25965-X Pa. $5.95

AN INTRODUCTION TO ORDINARY DIFFERENTIAL EQUATIONS, Earl A. Coddington. A thorough and systematic first course in elementary differential equations for undergraduates in mathematics and science, with many exercises and problems (with answers). Index. 304pp. 5% × 8½.
65942-9 Pa. $7.95

FOURIER SERIES AND ORTHOGONAL FUNCTIONS, Harry F. Davis. An incisive text combining theory and practical example to introduce Fourier series, orthogonal functions and applications of the Fourier method to boundary-value problems. 570 exercises. Answers and notes. 416pp. 5% × 8½.
65973-9 Pa. $9.95

THE THEORY OF BRANCHING PROCESSES, Theodore E. Harris. First systematic, comprehensive treatment of branching (i.e. multiplicative) processes and their applications. Galton-Watson model, Markov branching processes, electron-photon cascade, many other topics. Rigorous proofs. Bibliography. 240pp. 5% × 8½.
65952-6 Pa. $6.95

AN INTRODUCTION TO ALGEBRAIC STRUCTURES, Joseph Landin. Superb self-contained text covers "abstract algebra": sets and numbers, theory of groups, theory of rings, much more. Numerous well-chosen examples, exercises. 247pp. 5% × 8½.
65940-2 Pa. $6.95